Selected Titles in This Series

14 **Mourad E. H. Ismail, David R. Masson, and Mizan Rahman, Editors,** Special functions, q-series and related topics, 1997

13 **Peter A. Fillmore and James A. Mingo, Editors,** Operator algebras and their applications, 1997

12 **Dan-Virgil Voiculescu, Editor,** Free probability theory, 1997

11 **Colleen D. Cutler and Daniel T. Kaplan, Editors,** Nonlinear dynamics and time series: Building a bridge between the natural and statistical sciences, 1997

10 **Jerrold E. Marsden, George W. Patrick, and William F. Shadwick, Editors,** Integration algorithms and classical mechanics, 1996

9 **W. H. Kliemann, W. F. Langford, and N. S. Namachchivaya, Editors,** Nonlinear dynamics and stochastic mechanics, 1996

8 **Larry M. Bates and David L. Rod, Editors,** Conservative systems and quantum chaos, 1996

7 **William F. Shadwick, Perinkulam Sambamurthy Krishnaprasad, and Tudor Stefan Ratiu, Editors,** Mechanics day, 1996

6 **Anna T. Lawniczak and Raymond Kapral, Editors,** Pattern formation and lattice gas automata, 1996

5 **John Chadam, Martin Golubitsky, William Langford, and Brian Wetton, Editors,** Pattern formation: Symmetry methods and applications, 1996

4 **William F. Langford and Wayne Nagata, Editors,** Normal forms and homoclinic chaos, 1995

3 **Anthony Bloch, Editor,** Hamiltonian and gradient flows, algorithms and control, 1994

2 **K. A. Morris, Editor,** Control of flexible structures, 1993

1 **Michael J. Enos, Editor,** Dynamics and control of mechanical systems: The falling cat and related problems, 1993

Fields Institute Communications

The Fields Institute for Research in Mathematical Sciences

Special Functions, q-Series and Related Topics

Mourad E. H. Ismail
David R. Masson
Mizan Rahman
Editors

American Mathematical Society
Providence, Rhode Island

The Fields Institute for Research in Mathematical Sciences

The Fields Institute is named in honour of the Canadian mathematician John Charles Fields (1863–1932). Fields was a visionary who received many honours for his scientific work, including election to the Royal Society of Canada in 1909 and to the Royal Society of London in 1913. Among other accomplishments in the service of the international mathematics community, Fields was responsible for establishing the world's most prestigious prize for mathematics research—the Fields Medal.

The Fields Institute for Research in Mathematical Sciences is supported by grants from the Ontario Ministry of Education and Training and the Natural Sciences and Engineering Research Council of Canada. The Institute is sponsored by McMaster University, the University of Toronto, the University of Waterloo, and York University and has affiliated universities in Ontario and across Canada.

1991 *Mathematics Subject Classification.* Primary 33-06, 05A15, 05A17, 11P83, 42C05, 68Q40; Secondary 05A19, 05A30, 11R04, 17B37, 40A15, 41A10, 41A63, 68R05.

Library of Congress Cataloging-in-Publication Data

Special functions, q-series, and related topics / Mourad E. H. Ismail, David R. Masson, Mizan Rahman, editors.

p. cm. — (Fields Institute communications, ISSN 1069-5265 ; v. 14)

Includes bibliographical references.

ISBN 0-8218-0524-X (alk. paper)

1. Functions, Special. 2. q-series. I. Ismail, Mourad, 1944– . II. Masson, David R., 1937– . III. Rahman, Mizan. IV. Series.

QA351.S697 1997

515′.5—dc21

96-37968

CIP

This publication was prepared by the Fields Institute.

10 9 8 7 6 5 4 3 2 1 02 01 00 99 98 97

Contents

Preface

The subject of Special Functions has the unique feature of interacting with many branches of Mathematics and its Applications. Although it has had periods of stagnation, the subject is once again undergoing rapid growth and development. Over the past decade, the subtopic of q-Series has been a particularly active area. This is in a large part due to the publications of George Andrews which were inspired by the work of Ramanujan on the one hand and the discovery of Askey-Wilson polynomials on the other. It is appropiate that both the work of Ramanujan and Askey-Wilson polynomials provide the background to several of the contributions to this volume. More recently we have seen some surprising applications with the work of Macdonald on root systems and a host of activity in connection with quantum groups and the multivariate aspects of the theory.

The idea of having a Fields programme on Special Functions, q-Series and Related Topics goes back to the Summer of 1993. The Editors of this volume originally planned on a more extensive programme but eventually opted for the two week programme that was held at University College on the University of Toronto Campus, June 12–23, 1995.

The first week provided an introduction to the subject through five mini-courses. These were:

Topic	**Lecturer**
q-Series	George Gasper
Orthogonal Polynomials	Walter Van Assche
Asymptotics of Orthogonal Polynomials	Paul Nevai
Computer Algebra and Special Functions	Doron Zeilberger
Quantum Groups and q-Special Functions	Tom Koornwinder

Three of these are represented in this volume. They are:

"Elementary derivations of summation and transformation formulas for q-Series" by G. Gasper where new elementary derivations on q-series are given which are simpler and shorter than those given in the now classic book "Basic Hypergeometric Series" by G. Gasper and M. Rahman, (pp. 55–70).

"Orthogonal polynomials in the complex plane and on the real line" by W. Van Assche which provides a review of orthogonal polynomials which emphasizes extremal properties, zeros, recurrence relations and quadrature and specializes to the real line and unit circle, (pp. 213–247).

"Special functions and q-commuting variables" by T. H. Koornwinder which surveys and extends special function theory involving variables x and y where $xy = qyx$, (pp. 133–168).

The second week consisted of research presentations in the form of fifteen one hour invited talks and over forty half hour contributed talks. For an alphabetical listing of speakers and their presentations see Appendix I, p. 273. Each day the emphasis was on a different topic. These topics are listed below together with a brief description of the contributions representing them that appear in this volume. In addition there were two evening sessions related to quantum groups and algebras which were spontaneously organized by Luc Vinet and others.

Orthogonal Polynomials and Special Functions

"$\int_{n/4}^{m/6} \ln \Gamma(z) dz$" by R. Wm. Gosper evaluates these integrals through a Fourier expansion of the integrand, (pp. 71–76).

"On a q-analogue of Gauss equation and some q-Riccati equations" by F. A. Grünbaum and L. Haine. Here the Askey-Wilson polynomial recurrence is considered from a new viewpoint where it is a discrete version of the Gauss equation, (pp. 77–81).

"Some generating functions for q-polynomials" by M. E. H. Ismail, D. R. Masson and S. K. Suslov uses the product and connection formulas for some q-polynomials to derive some new bilinear and linear generating functions. The $q \to 1$ limits of these are familiar expansions, (pp. 91–109).

Multivariable Theory

"Determinant evaluations and $U(n)$ extensions of Heine's ${}_2\phi_1$-transformations" by R. A. Gustafson and C. Krattenthaler where the classical transformations of Heine are given a multiparameter extension using certain determinant evaluations, (pp. 83–89).

"On orthogonal polynomials in several variables" by Y. Xu provides an overview of a new approach to orthogonal polynomials in several variables. This attacks a long outstanding problem by using a vector-matrix notation to parallel results in the single variable theory, (pp. 249–272).

Number Theory and Combinatorics

"Refinements of Rogers-Ramanujan type identities" by K. Alladi extends the theory of partitions based on q-series by using an interesting new technique called *the method of weighted words*, (pp. 1–35).

"Ramanujan's class invariants with applications to the values of q-continued fractions and theta functions" by B. C. Berndt, H. H. Chan and L-C. Zhang follows up some work by G. N. Watson of some 65 years ago and describes Ramanujan's determinations of what are called *class invariants* and their connection to the Rogers-Ramanujan continued fraction, modular equations and theta functions, (pp. 37–53).

Computer Algebra

"A Mathematica q-analogue of Zeilberger's algorithm based on an algebraically motivated approach to q-hypergeometric telescoping" by P. Paule and A. Riese. This paper describes an implementation in Mathematica of algorithms for proving q-hypergeometric summation and transformation formulas. Such algorithms are q-analogues of algorithms due to Gosper and Zeilberger. Central to these are the necessity of telescoping a series. Here the authors describe a new approach to q-hypergeometric telescoping using a new concept of *q-greatest factorial factorization*, (pp. 181–212).

Quantum Groups and Representation Theory

"Addition formulas for q-special functions" by Eric Koelink obtains a very general addition formula for two parameter Askey-Wilson polynomials. The derivation uses properties of the quantum $SU(2)$ group. The addition formula for general Askey-Wilson remains as an open problem, (pp. 111–131).

"Multivariable Askey-Wilson polynomials and quantum complex Grassmannians" by M. Noumi, M. S. Dijkhuizen and T. Sugitani deals with a family of quantum homogenous $U_q(n)$-spaces having zonal spherical functions which are expressed in terms of Koornwinder's multivariable Askey-Wilson polynomials, (pp. 169–179).

We hope that the contents of this volume will help the reader appreciate some of the depth and breadth of the current area of Special Functions and inspire future contributions.

The organizing committee which chose the above topics and the invited speakers consisted of the current editors together with G. E. Andrews, R. A. Askey, C. F. Dunkl, T. H. Koornwinder, S. K. Suslov and D. Zeilberger. Each of these made a valuable contribution to the success of the Programme and we thank them for their patience in responding to our many messages and their advice on scientific matters.

The most important contribution to any meeting is that made by the speakers and, of course, the participants who decided to attend what they considered to be a worthwhile programme and contributed to the lively discussions. There was great diversity in the backgrounds of the participants. It varied from the world famous Boris Levitan of Gelfand-Levitan fame to many bright young mathematicians and graduate students to whom, eventually, the current generation will pass the torch. We thank you all, and particularly those of you who contributed to this volume and/or refereed the submissions. For a list of participants see Appendix II, p. 277.

The programme was not without its lighter side with ample opportunity for participants to meet, socialize and exchange ideas on an informal basis. At a barbeque during the first week, Bill Gosper amused everyone with his gadgets that seemed to defy every law of physics known to mankind. At the final banquet we were treated to a most entertaining piano recital by Christian Krattenthaler.

Finally we wish to express our thanks for the financial support given by The Ontario Ministry of Education and Training, the Natural Sciences and Engineering Research Council of Canada, Carleton University, and the Department of Mathematics of the University of Toronto. We also thank University College for the use of their facilities.

Mourad E. H. Ismail
David R. Masson
Mizan Rahman

August 1996

Fields Institute Communications
Volume **14**, 1997

Refinements of Rogers-Ramanujan Type Identities

Krishnaswami Alladi
Department of Mathematics
University of Florida
Gainesville, FL 32611-8000 USA

Abstract. A Rogers-Ramanujan type identity is one which connects partitions into parts satisfying difference conditions to partitions whose parts satisfy congruence conditions. Usually the generating function for the partitions given by difference conditions is an infinite q-series, while the generating function for partitions satisfying congruence conditions is an infinite product. A unified approach to a wide class of such identities is presented using a new technique called *the method of weighted words.* The method yields substantial generalizations and refinements of many Rogers–Ramanujan type identities including Schur's celebrated partition theorem and extensions due to Andrews, a deep partition theorem of Göllnitz, and Capparelli's theorems arising from Lie algebras. Applications include some remarkable 2-adic partition identities, a new interpretation for Jacobi's Triple Product Identity and a bijection for Göllnitz' deep theorem.

1 Introduction

A central portion of the theory of partitions and $q-$series concerns the study of identities of Rogers–Ramanujan type. Such identities are usually expressed analytically in the form of an infinite $q-$series (sometimes a multiple $q-$series) admitting a product representation. The infinite $q-$series is often the generating function for partitions satisfying certain difference conditions while the product is the generating function for partitions whose parts satisfy congruence conditions. Rogers–Ramanujan type identities have arisen in a variety of settings. They have been used to explain certain models in statistical mechanics (see Chapter 8 of Andrews' [1985] monograph) and have arisen in the study of vertex operators in Lie algebras (see Lepowsky and Wilson [1981], [1982], [1984]). Chapter 5 of Andrews' [1976] encyclopedia is devoted to Rogers–Ramanujan type identities and Slater [1952] gives a list of 130 such identities.

The generic name identities of Rogers–Ramanujan type stems from the following two celebrated identities due independently to L. J. Rogers [1894] and

1991 *Mathematics Subject Classification.* Primary: 05A15, 05A17, 11P83; Secondary: 05A19.
Research supported in part by the National Science Foundation Grant DMS 9400191.

S. Ramanujan [1919] which are the prototype:

$$\sum_{\nu=0}^{\infty} \frac{q^{\nu^2}}{(1-q)(1-q^2)\dots(1-q^\nu)} = \prod_{j=1}^{\infty} \frac{1}{(1-q^{5j-1})(1-q^{5j-4})}, \tag{1.1}$$

and

$$\sum_{\nu=0}^{\infty} \frac{q^{\nu^2+\nu}}{(1-q)(1-q^2)\dots(1-q^\nu)} = \prod_{j=1}^{\infty} \frac{1}{(1-q^{5j-2})(1-q^{5j-3})}. \tag{1.2}$$

Indeed, even today (1.1) and (1.2) are unmatched in simplicity, elegance and depth. The Rogers–Ramanujan identities have very pretty combinatorial interpretations as observed by MacMahon and Schur [1973; Gesammelte Abhandlungen]:

Theorem R *For $i = 1, 2$, the number of partitions of an integer n into parts differing by ≥ 2 and each part $\geq i$ equals the number of partitions of n into parts $\equiv \pm i (mod\ 5)$.*

The case $i = 1$ in Theorem R corresponds to (1.1) while $i = 2$ is the interpretation of (1.2).

Our purpose here is to discuss certain Rogers–Ramanujan type identities which can be refined. Here by a *refinement* of an identity we mean the introduction of free parameters $a, b, c, \dots$ in the analytic form whose powers represent various statistics in the partitions being counted. Such refinements have important consequences. In many refined identities, by using the transformations

$$\text{(dilation) } q \mapsto q^M \text{ and (translations) } a \mapsto aq^\mu, \dots, \tag{1.3}$$

new partition theorems can be generated. More importantly, refined partition identities can usually be proved combinatorially because the refinements suggest the bijection to be set up. In some instances a specific dilation and set of translations are particularly suited for combinatorial treatment (see Section 6).

In this paper we describe a new technique called *the method of weighted words* which provides a unified approach to a number of identities of Rogers–Ramanujan type. The method yields substantial generalizations and refinements of many well-known identities as well as some new ones. We shall describe the major progress achieved by this method against the background of classical work.

Unfortunately no refinement of the original Rogers–Ramanujan identities is known! This is partly the reason that there is no simple combinatorial proof of (1.1) and (1.2). The combinatorial proof of (1.1) and (1.2) due to Garcia and Milne [1981] runs to fifty pages! The proof due to Bressoud and Zeilberger [1982], although very short, is quite difficult.

Perhaps the best example of an identity that can be refined is the celebrated 1926 partition theorem of Schur [1973; Gesammelte Abhandlungen]:

Theorem S *Let $S(n)$ denote the number of partitions of n into distinct parts $\equiv \pm 1 (mod\ 3)$. Let $S_1(n)$ denote the number of partitions of n into parts differing by ≥ 3 and such that there are no consecutive multiples of 3. Then*

$$S(n) = S_1(n).$$

Gleissberg [1928] showed that Theorem S could be refined to

$$S(n;k) = S_1(n;k), \tag{1.4}$$

where $S(n;k)$ and $S_1(n;k)$ enumerate partitions of the type counted by $S(n)$ and $S_1(n)$ with the additional restriction that there are precisely k parts and the convention that parts $\equiv 0 \pmod 3$ are counted twice. Combinatorial proofs based on this refinement have been given by Bressoud [1980] and Alladi–Gordon [1993b].

The method of weighted words was introduced by Alladi and Gordon [1993b], [1995], to obtain refinements, generalizations and companions of Schur's theorem (see Section 3 for a description of the method). Subsequently, the method was improved by Alladi, Andrews and Gordon [1995a] yielding substantial refinements and generalizations of a deep and difficult theorem of Göllnitz [1967]; this work is described in Section 4. The improved method also yielded proof and refinements Alladi, Andrews and Gordon [1995b] of partition conjectures of Capparelli [1993] which arose in a study of Lie algebras (see Section 8). From the point of view of the method of weighted words, the Capparelli conjectures are seen as extensions of a classical identity of Lebesgue (see Section 9).

Starting from Schur's theorem, Andrews [1968], [1969a] obtained two hierarchies of partition theorems yielding two results for each modulus $M \geq 2^r - 1$, $r = 2, 3, \ldots$. In this scheme, Schur's theorem corresponds to the case $r = 2$ and this is the only case where the two Andrews theorems are one and the same (see Section 7). Recently, Alladi and Gordon [in preparation] have obtained an approach to the Andrews hierarchies by the method of weighted words (see Section 7 for a description). In this approach there is only one hierarchy of general theorems and the two results that Andrews obtained for each modulus $M = 2^r - 1$, $r \geq 3$, are two special cases under dilations and translations of a certain basic result.

The main idea in the method of weighted words is to consider positive integers which occur in various colors denoted by $a, b, c, \ldots$. We then form words whose letters are symbols $a, b, c, \ldots$ with subscripts, where the subscript denotes the integer or part of the partition and the alphabets $a, b, c, \ldots$, represent the color of that part. Special types of words are considered by imposing order rules on the symbols and gap conditions on the subscripts. The usefulness of the method lies in the fact that the alphabets $a, b, c, \ldots$, play a dual role; on the one hand they represent colors, and on the other they are free parameters when computing generating functions. One consequence is that the partition theorem in question falls out as a special case under certain dilations and translations and the free parameters provide the refinement of the partition identity. Another advantage is that by changing the order rules for the symbols, new companion partition identities are generated (see Section 3 for a discussion of companions to Schur's partition theorem).

The method of weighted words operates at the base level without any dilations. Often the integer M in (1.3) will be the smallest integer such that based on the colors used, we are guaranteed under suitable translations a set of distinct residue classes (mod M) in the partition theorem under discussion. For example, while discussing Schur's theorem we take $M = 3$ and for Göllnitz' (Big) theorem we have $M = 6$. If a smaller value of M is used, then certain residue classes (mod M) will overlap and so the partitions π will have to be counted with weights $w(\pi) \geq 1$. We have recently noticed (see Alladi [1997a], [1997b]) that some very interesting partition theorems can be obtained by using smaller dilations. For example, if we use the dilation $q \mapsto q^3$ in the key identity (4.3) underlying Göllnitz' theorem, we obtain a weighted partition theorem which turns out to be an important reformulation (see Section 5). It had been an open problem for nearly thirty years to find a combinatorial proof or approach to Göllnitz' (Big) theorem. This was all the more frustrating because Göllnitz' theorem when first proved in 1967 was established in a refined form. It turns out that the reformulation obtained by the $q \mapsto q^3$ dilation is one which can be explained by a combinatorial bijection (see Section 5). This $q \mapsto q^3$ dilation for Göllnitz' theorem has several important consequences as shown in Section 6. Besides yielding a new interpretation for the famous Jacobi triple product identity, it also suggests the remarkable result that for each integer $k \geq 1$, the congruence

$$Q(n) \equiv 0 (\mathrm{mod}\ 2^k) \tag{1.5}$$

holds for almost all n. Here $Q(n)$ is the number of partitions of n into distinct parts.

Finally, in Section 10, some problems for further investigation are suggested.

2 Preliminaries

We adopt the standard notation

$$(a)_n = (a;q)_n = \prod_{j=0}^{n-1}(1-aq^j) \tag{2.1}$$

for any complex number a, and

$$(a)_\infty = \lim_{n\to\infty}(a)_n = \prod_{j=0}^{\infty}(1-aq^j), \text{ for } |q| < 1. \tag{2.2}$$

If $|q| < 1$, then (2.2) can be used to define $(a)_n$ for all real n by means of the relation

$$(a)_n = \frac{(a)_\infty}{(aq^n)_\infty}. \tag{2.3}$$

From (2.3) it follows that

$$\frac{1}{(q)_n} = 0, \quad n = -1, -2, \ldots. \tag{2.4}$$

For a positive integer t, the $q-$multinomial coefficient of order t is defined by

$$\begin{bmatrix} n_1+n_2+\cdots+n_t \\ n_1, n_2, \ldots, n_t \end{bmatrix} = \begin{bmatrix} n_1+n_2+\cdots+n_t \\ n_1, n_2, \ldots, n_t \end{bmatrix}_q = \frac{(q)_{n_1+n_2+\cdots+n_t}}{(q)_{n_1}(q)_{n_2}\cdots(q)_{n_t}}. \tag{2.5}$$

If the base of the multinomial coefficient is q, then as on the left in (2.5) we suppress it. If the base is anything other than q, it will be displayed. Multinomial coefficients of order $t = 3$ will appear in the discussion of Schur's theorem (Section 3), while those of order $t = 6$ will occur in our treatment of Göllnitz' theorem (Section 6). While discussing the Andrews hierarchy in Section 7, we will encounter coefficients of all orders t. When $t = 2$, $\begin{bmatrix} n_1+n_2 \\ n_1, n_2 \end{bmatrix}$ is the $q-$binomial coefficient which is commonly denoted by $\begin{bmatrix} n_1+n_2 \\ n_1 \end{bmatrix}$. Binomial coefficients to bases q and q^2 will be used while treating Capparelli's conjectures in Section 8.

Given a partition π, we let $\nu(\pi)$ denote the number of parts of π and $\sigma(\pi)$, the sum of its parts. Quite often we will be counting parts of a specific type and this will be indicated by a subscript. For example, $\nu_a(\pi)$ will denote the number of $a-$parts of π.

If $\pi : b_1 + b_2 + \ldots$ and $\pi' : b_1' + b_2' + \ldots$ are two partitions whose parts as usual are written in decreasing order, then by $\pi'' = \pi + \pi'$ we mean the partition $b_1'' + b_2'' + \ldots$ where $b_j'' = b_j + b_j'$. If $\nu(\pi) \neq \nu(\pi')$, then in place of b_j or b_j', whichever is missing, the number zero is substituted. Thus $\nu(\pi + \pi') = \max(\nu(\pi), \nu(\pi'))$.

A minimal partition is one for which $\sigma(\pi)$ is minimal subject to a given set of conditions. For partitions π which are given by lower bound gap conditions, their generating function can be calculated by first considering the minimal partitions of the type under discussion.

For instance, consider the generating function for all partitions π into n distinct parts. The minimal partition into n distinct parts is $n + (n-1) + \cdots + 1$ of the nth-triangular number $T_n = \frac{n(n+1)}{2}$. Thus the required generating function is

$$\frac{q^{n(n+1)/2}}{(q)_n}.$$

Next consider the generating function of all partitions into n parts differing by ≥ 2. The minimal partition in this case is $(2n-1) + (2n-3) + \cdots + 3 + 1$ of the integer n^2. So the required generating function is

$$\frac{q^{n^2}}{(q)_n}$$

which is the $n^{th}-$term in the Rogers–Ramanujan series (1.1).

More generally, consider the generating function

$$G(q) = \sum_{\pi} q^{\sigma(\pi)} \tag{2.6}$$

of partitions π which are given by fixing $\nu(\pi) = m$ and by imposing certain lower bound gap conditions on π. Let

$$H(q) = \sum_{\pi^*} q^{\sigma(\pi^*)} \tag{2.7}$$

be the generating function for all minimal partitions subject to these conditions. Then $G(q)$ and $H(q)$ are connected by the relation given by

Lemma 1 $G(q) = H(q)/(q)_m$.

Proof Lemma 1 is an immediate consequence of the observation that the partitions π counted in (2.6) can all be realized in the form $\pi = \pi^* + \pi'$, where π' is a partition satisfying $\nu(\pi') \leq \nu(\pi^*)$. The best way to see this is to draw the Ferrers graphs of π, π^* and π'. Since the generating function of π' is $1/(q)_m$, Lemma 1 follows from (2.6) and (2.7).

The lower bound gap conditions defining π can be quite complicated. In such situations Lemma 1 is very useful.

3 Schur's theorem and companions

Schur had originally stated the equality of three partition functions

$$T(n) = S(n) = S_1(n),$$

where $T(n)$ is the number of partitions of n into parts $\equiv \pm 1(\mod 6)$. The equality $T(n) = S(n)$ is trivial because

$$\begin{aligned}\sum_{n=0}^{\infty} S(n)q^n &= \prod_{m=0}^{\infty}(1+q^{3m+1})(1+q^{3m+2}) = \prod_{m=0}^{\infty}\frac{(1-q^{6m+2})(1-q^{6m+4})}{(1-q^{3m+1})(1-q^{3m+2})} \\ &= \prod_{m=0}^{\infty}\frac{1}{(1-q^{6m+1})(1-q^{6m+5})} = \sum_{n=0}^{\infty} T(n)q^n.\end{aligned} \tag{3.1}$$

Thus the hard part of Schur's theorem is the equality $S(n) = S_1(n)$ and it is also this equality which can be refined as in (1.4). Such refinements are not possible with $T(n)$ and so we will not discuss $T(n)$ any further in this paper.

Andrews [1967] gave a generating function proof of Theorem S. This was the starting point of a series of investigations by him. Extending this method, Andrews [1969b] gave a generating function proof of Göllnitz' theorem and also obtained two hierarchies of general partition theorems (see Andrews [1968], [1969a]). Later, in 1971, while using a computer to search for identities of Rogers–Ramanujan type Andrews [1971] noticed the following companion to Schur's theorem.

Theorem A *Let $S_2(n)$ denote the number of partitions of n in the form $b_1 + b_2 + \ldots$, such that $b_i - b_{i+1} \geq 3, 2$ or 5 depending on $b_i \equiv 1, 2$ or $3 (mod\ 3)$. Then $S_1(n) = S_2(n)$.*

Andrews' proof of Theorem A was similar to his proof of Theorem S. But the exact connections between Theorems A and S were not clear. We shall now describe the approach to Theorem S by the method of weighted words. Besides giving generalizations and refinements of Schur's theorem, this approach shows how exactly Theorems A and S are connected. It turns out that $S_1(n)$ and $S_2(n)$ are only two of six partition functions all equal (see Theorem 6 below).

We assume that each integer $j \geq 2$ occurs in three colors, red, blue and purple and that the integer 1 occurs only in two colors, red and blue. We introduce the symbols a_j, b_j and c_j to represent the integer j in colors red, blue and purple. Sometimes we refer to a_j as an $a-$part, b_j as a $b-$part and c_j as a $c-$part. We think of the letters a, b, c as free parameters. Under the transformations

$$\begin{cases} \text{(dilation) } q \mapsto q^3, \\ \text{(translations) } a \mapsto aq^{-1}, b \mapsto bq^{-2}, c \mapsto cq^{-3}, \end{cases} \tag{3.2}$$

the symbols become

$$a_j = 3j-2, b_j = 3j-1, c_j = 3j-3, \tag{3.3}$$

and so the a_j and b_j for $j \geq 1$ and c_j for $j \geq 2$ represent all positive integers in residue classes $1, 2$ and $3 (\mathrm{mod}\ 3)$ respectively. We call the transformations in (3.2) as the *standard transformations* for Schur's theorem. In order to make a transition from $S_1(n)$ to $S(n)$ below we will need to choose $c = ab$. That is why we think of c as representing the secondary color purple which is the sum of the primary colors red and blue.

The weight of a symbol is its subscript. The gap between any two symbols representing colored integers is the absolute value of the difference between their weights. Thus the gap is a non–negative integer without color. For example, the gap between a_4 and b_6 is 2, and between a_5 and c_5 is 0.

A word is an ordered collection of symbols. Thus $a_4c_5a_1a_1b_2c_3$ is a word. In order to define partitions we need an ordering of the symbols. The ordering we start with is

$$\underline{\text{Scheme 1}}: \; a_1 < b_1 < c_2 < a_2 < b_2 < c_3 < a_3 < b_3 < \ldots . \tag{3.4}$$

The motivation for choosing this particular ordering is that under the transformations (3.2) and (3.3), Scheme 1 ordering becomes

$$1 < 2 < 3 < 4 < 5 < 6 \ldots,$$

the standard ordering among the positive integers. Later we will consider other schemes (orderings).

Given an ordering (in this case Scheme 1), by a partition π we mean a word whose symbols are in decreasing order with respect to the ordering. Thus π : $b_4a_4c_4c_2c_2b_1$ is a partition (under Scheme 1). By $\sigma(\pi) = n$ we mean the sum of the weights of the symbols in a partition and we think or n as being partitioned in terms of the colored integers. In the above example, $\sigma(\pi) = n = 17$. Sometimes we write $\pi : b_4 + a_4 + c_4 + c_2 + c_2 + b_1$ as a sum. Given π, we let $\nu_a(\pi), \nu_b(\pi)$ and $\nu_c(\pi)$ denote the number of $a-$parts, $b-$parts and $c-$parts of π respectively. In this example $\nu_a(\pi) = 1$, $\nu_b(\pi) = 2$ and $\nu_c(\pi) = 3$.

Next we consider partitions $\pi : e_1 + e_2 + \cdots + e_\nu$, where the e_i are symbols from Scheme 1 such that the gap between the symbols is ≥ 1 with the added restriction that the gap between e_ℓ and $e_{\ell+1}$ is > 1 if

$$\begin{cases} e_\ell \text{ is an } a\text{−part and } e_{\ell+1} \text{ is a } b\text{−part,} \\ \text{or if } e_\ell \text{ is a } c\text{−part.} \end{cases} \tag{3.5}$$

We call such a partition π as a Type 1 partition. At first glance the definition of Type 1 partitions might seem a bit artificial, but under the standard transformations the gap conditions (3.5) translate to the difference conditions of the partition function $S_1(n)$ in Theorem S. Also, Type 1 partitions are interesting because their generating function has an elegant form as observed in Alladi and Gordon [1993b]:

Lemma 2 *Let*

$$G = G(i, j, k) = \sum_{\pi} q^{\sigma(\pi)}$$

be the generating function of all Type 1 partitions π for which $\nu_a(\pi) = i, \nu_b(\pi) = j$ and $\nu_c(\pi) = k$. Then

$$G = \frac{q^{T_{i+j+k} + T_k}}{(q)_i (q)_j (q)_k}. \tag{3.6}$$

Lemma 2 can be proved using minimal partitions. Since all partitions counted by $G(i, j, k)$ have $\nu(\pi) = i + j + k$ fixed, it follows from Lemma 1 that Lemma 2 is equivalent to the statement that

$$H = H(i, j, k) = G(i, j, k)(q)_{i+j+k} = q^{T_{i+j+k} + T_k} \begin{bmatrix} i + j + k \\ i, j, k \end{bmatrix}. \tag{3.7}$$

Here $H(i,j,k)$ is the generating function of all minimal Type 1 partitions π^* such that $\nu_a(\pi^*) = i$, $\nu_b(\pi^*) = j$ and $\nu_c(\pi^*) = k$. Since multinomial coefficients satisfy recurrence relations, it turns out that (3.7) can be established by induction on $i+j+k$, the length of the word, as shown in Alladi and Gordon [1993b].

For the purpose of the induction it is convenient to consider the decomposition

$$H = H_a + H_b + H_c,$$

where H_a, H_b and H_c are generating functions of partitions enumerated by H with the additional restriction that the smallest part is an a−part, b−part, and c−part respectively. Then it follows by induction on $i+j+k$ that

$$\begin{aligned} H_a &= q^{T_{i+j+k}+T_k} \begin{bmatrix} i+j+k-1 \\ i-1, j, k \end{bmatrix}, \\ H_b &= q^{T_{i+j+k}+T_{k+i}} \begin{bmatrix} i+j+k-1 \\ i, j-1, k \end{bmatrix}, \\ H_c &= q^{T_{i+j+k}+T_{k+i+j}} \begin{bmatrix} i+j+k-1 \\ i, j, k-1 \end{bmatrix}. \end{aligned} \tag{3.8}$$

Once this is done, (3.7) follows from (3.8) because of the recurrence relation

$$\begin{bmatrix} i+j+k \\ i,j,k \end{bmatrix} = \begin{bmatrix} i+j+k-1 \\ i-1,j,k \end{bmatrix} + q^i \begin{bmatrix} i+j+k-1 \\ i,j-1,k \end{bmatrix} + q^{i+j} \begin{bmatrix} i+j+k-1 \\ i,j,k-1 \end{bmatrix} \tag{3.9}$$

satisfied by the multinomial coefficients.

Next, let $B_1(n;i,j,k)$ denote the number of Type 1 partitions π of n such that $\nu_a(\pi) = i$, $\nu_b(\pi) = j$ and $\nu_c(\pi) = k$. Then by Lemma 2

$$\sum_{i,j,k,n} B_1(n;i,j,k) a^i b^j c^k q^n = \sum_{i,j,k} a^i b^j c^k \frac{q^{T_{i+j+k}+T_k}}{(q)_i (q)_j (q)_k}. \tag{3.10}$$

In order to get a generalization and refinement of Schur's theorem out of this, we need to take $c = ab$ in (3.10). To see the effect of this, put

$$k = m,\ i = r - m, j = s - m. \tag{3.11}$$

Then the main result in Alladi and Gordon [1993b] is the following lemma for which two proofs are given, one being combinatorial.

Lemma 3

$$\sum_{0 \le m \le \min(r,s)} \frac{q^{T_{r+s-m}+T_m}}{(q)_{r-m}(q)_{s-m}(q)_m} = \frac{q^{T_r+T_s}}{(q)_r (q)_s}.$$

Notice that

$$\sum_{r,s} \frac{a^r b^s q^{T_r+T_s}}{(q)_r (q)_s} = (-aq)_\infty (-bq)_\infty, \tag{3.12}$$

and so (3.10), (3.12) and Lemma 3 can be combined into

$$\sum_{r,s} a^r b^s \sum_{0 \le m \le \min(r,s)} \frac{q^{T_{r+s-m}+T_m}}{(q)_{r-m}(q)_{s-m}(q)_m} = \sum_{r,s} \frac{a^r b^s q^{T_r+T_s}}{(q)_r (q)_s} = (-aq)_\infty (-bq)_\infty. \tag{3.13}$$

This is the *key identity* for Schur's Theorem as we shall presently see.

The expression in (3.12) is the generating function of $V(n;r,s)$, the number of vector bi–partitions $(\pi_1;\pi_2)$ of n such that π_1 has r distinct a–parts and π_2 has s distinct b–parts. Thus (3.13) has the following combinatorial interpretation.

Theorem 1 $V(n;r,s) = \sum_{i+k=r,j+k=s} B_1(n;i,j,k)$.

Note that Theorem 1 gives a two parameter refinement of Schur's theorem under the standard transformations (3.2) and (3.3). The relations (3.11) imply

$$i + j + 2k = r + s, \tag{3.14}$$

and so the $c = ab$–parts are counted twice as in (1.4). More generally, under the transformations

$$\begin{cases} \text{(dilation) } q \mapsto q^M, \\ \text{(translations) } a \mapsto aq^{\alpha-M}, b \mapsto bq^{\beta-M}, \end{cases} \tag{3.15}$$

applied to (3.13), Theorem 1 yields

Theorem 2 *Let $M \geq 3$ and α, β integers such that $0 < \alpha < \beta < M \leq \alpha + \beta$.*
Let $A(n;r,s)$ denote the number of partitions of n into r distinct parts $\equiv \alpha(\text{mod } M)$ and s distinct parts $\equiv \beta(\text{mod } M)$.
Let $B(n;i,j,k)$ denote the number of partitions of n into i parts $\equiv \alpha(\text{mod } M)$, j parts $\equiv \beta(\text{mod } M)$ and k parts $\equiv \alpha + \beta(\text{mod } M)$ such that
(i) the difference between any two parts is $\geq M$,
(ii) the difference between parts $\equiv \alpha + \beta(\text{mod } M)$ is $> M$.
Then

$$A(n;r,s) = \sum_{\substack{i+k=r \\ j+k=s}} B_1(n;i,j,k).$$

In Theorem 2 and in the rest of the paper we adopt the convention that $n \equiv j(\text{mod } m)$ means that $n = j + km$ with $n > 0$ and $k \geq 0$.

We now discuss the work of Alladi–Gordon [1995] on companions to Schur's theorem obtained by other orderings of the symbols. But first we need to reformulate the definition of Type 1 partitions.

Let $x_m = x_m^{(1)}$ be the symbol occupying position m in Scheme 1. That is, $x_1 = a_1, x_2 = b_1$, $x_3 = c_2, \ldots$. Then Type 1 partitions are those of the form $x_{m_1} + x_{m_2} + \ldots$, such that

$$m_\ell - m_{\ell+1} \geq 3 \text{ with strict inequality if } x_{m_\ell} \text{ is a } c\text{–part.} \tag{3.16}$$

We refer to the inequalities in (3.16) as the *standard gap conditions* (for Schur's Theorem).

Now consider a different ordering of the symbols, namely,

$$\underline{\text{Scheme 2}}: a_1 < c_2 < b_1 < a_2 < c_3 < b_2 < a_3 < \ldots . \tag{3.17}$$

Under the standard transformations (3.2) and (3.3), Scheme 2 becomes

$$1 < 3 < 2 < 4 < 6 < 5 < \ldots$$

which is not the natural ordering of the integers.

Next, let $x_m = x_m^{(2)}$ be the symbol occupying position m in Scheme 2. Define Type 2 partitions to be those of the form $x_{m_1} + x_{m_2} + \ldots$, such that $x_{m_\ell} = x_{m_\ell}^{(2)}$ satisfies the standard gap conditions (3.16).

Now let $B_2(n;i,j,k)$ denote the number of Type 2 partitions π of n such that $\nu_a(\pi)=i$, $\nu_b(\pi)=j$ and $\nu_c(\pi)=k$. We then have

Theorem 3 $B_1(n;i,j,k)=B_2(n;i,j,k)$.

Under the standard transformations (3.2), Theorem 3 implies Theorem A. One way of proving Theorem 3 is to show that the generating functions of $B_1(n;i,j,k)$ and $B_2(n;i,j,k)$ are the same, that is

$$\sum_{i,j,k,n} B_2(n;i,j,k)a^ib^jc^kq^n = \sum_{i,j,k} a^ib^jc^k \frac{q^{T_{i+j+k}+T_k}}{(q)_i(q)_j(q)_k} \tag{3.18}$$

as in (3.10). The proof of (3.18) proceeds as in the case of $B_1(n;i,j,k)$ by the use of minimal partitions, except that here another recurrence for the multinomial coefficients has to be used, namely,

$$\begin{bmatrix} i+j+k \\ i,j,k \end{bmatrix} = \begin{bmatrix} i+j+k-1 \\ i-1,j,k \end{bmatrix} + q^i \begin{bmatrix} i+j+k-1 \\ i,j,k-1 \end{bmatrix} + q^{i+k} \begin{bmatrix} i+j+k-1 \\ i,j-1,k \end{bmatrix}. \tag{3.19}$$

More precisely, let $H^{(2)}=H^{(2)}(i,j,k)$ denote the generating function of all minimal Type 2 partitions π^* with $\nu_a(\pi^*)=i$, $\nu_b(\pi^*)=j$ and $\nu_c(\pi^*)=k$. Then (3.18) is equivalent to showing that

$$H^{(2)} = q^{T_{i+j+k}+T_k} \begin{bmatrix} i+j+k \\ i,j,k \end{bmatrix}. \tag{3.20}$$

To prove (3.20), consider the decomposition as in (3.7), namely

$$H^{(2)} = H_a^{(2)} + H_b^{(2)} + H_c^{(2)}. \tag{3.21}$$

It can be shown by induction on $i+j+k$ (see Alladi and Gordon [1995] for details) that

$$\begin{aligned} H_a^{(2)} &= q^{T_{i+j+k}+T_k} \begin{bmatrix} i+j+k-1 \\ i-1,j,k \end{bmatrix}, \\ H_c^{(2)} &= q^{T_{i+j+k}+T_{k+i}} \begin{bmatrix} i+j+k-1 \\ i,j,k-1 \end{bmatrix}, \\ H_b^{(2)} &= q^{T_{i+j+k}+T_{k+i+k}} \begin{bmatrix} i+j+k-1 \\ i,j-1,k \end{bmatrix}. \end{aligned} \tag{3.22}$$

Now (3.20) follows from (3.21), (3.22) and (3.19) and this in turn proves (3.18).

More generally, one may consider six schemes given by the $3!=6$ permutations of the starting symbols a_1, b_1, c_2. They are

Scheme 1: $a_1 < b_1 < c_2 < a_2 < b_2 < c_3 < \ldots$

Scheme 2: $a_1 < c_2 < b_1 < a_2 < c_3 < b_2 < \ldots$

Scheme 3: $c_2 < a_1 < b_1 < c_3 < a_2 < b_2 < \ldots$

Scheme 4: $b_1 < a_1 < c_2 < b_2 < a_2 < c_3 < \ldots$

Scheme 5: $b_1 < c_2 < a_1 < b_2 < c_3 < a_2 < \ldots$

Scheme 6: $c_2 < b_1 < a_1 < c_3 < b_2 < a_2 < \ldots$

Now define $x_m = x_m^{(\mu)}$ to be the symbol occupying position m in Scheme μ, for $\mu = 1,2,\ldots,6$. Next, let Type μ partitions be those of the form $x_{m_1}+x_{m_2}+\ldots$ where the $x_{m_\ell} = x_{m_\ell}^{(\mu)}$ are symbols of Scheme μ satisfying the standard gap conditions. Also let $B_\mu(n;i,j,k)$ denote the number of Type μ partitions π of n for which $\nu_a(\pi)=i$, $\nu_b(\pi)=j$ and $\nu_c(\pi)=k$. Extending Theorem 3 we have

Theorem 4 $B_1(n;i,j,k) = B_\mu(n;i,j,k)$ *for* $\mu = 2,3,\ldots,6$.

In establishing the generating function of $B_1(n;i,j,k)$, we made use of the recurrence (3.9) for the multinomial coefficient where the indices i,j,k were decreased by 1 in succession. This corresponded to the starting triple for Scheme 1 which is $a_1 < b_1 < c_2$. Next, in showing that the generating function for $B_2(n;i,j,k)$ is the same, we made use of the recurrence (3.19) where the indices i,k and j are reduced by 1 in succession. The order i,k,j corresponds to the starting triple for Scheme 2 which is $a_1 < c_2 < b_1$. Similarly, since the starting triple for Scheme 3 is $c_2 < a_1 < b_1$, the generating function of $B_3(n;i,j,k)$ can be shown to be $G(i,j,k)$ by making use of the following third recurrence for the multinomial coefficients:

$$\begin{bmatrix} i+j+k \\ i,j,k \end{bmatrix} = \begin{bmatrix} i+j+k-1 \\ i,j,k-1 \end{bmatrix} + q^k \begin{bmatrix} i+j+k \\ i-1,j,k \end{bmatrix} + q^{k+i} \begin{bmatrix} i+j+k-1 \\ i,j-1,k \end{bmatrix}. \quad (3.23)$$

Of the $3! = 6$ recurrences for the multinomial coefficients of order $t = 3$, there are three more that we will not write down. We can summarize the above discussion on companions to Schur's theorem in the form of

Theorem 5 *The six schemes and the six partition functions* $B_\mu(n;i,j,k)$, $\mu = 1,2,\ldots,6$, *correspond to the six ways in which the multinomial coefficent* $\begin{bmatrix} i+j+k \\ i,j,k \end{bmatrix}$ *can be expanded in terms of* $\begin{bmatrix} i+j+k-1 \\ i-1,j,k \end{bmatrix}$, $\begin{bmatrix} i+j+k-1 \\ i,j-1,k \end{bmatrix}$ *and* $\begin{bmatrix} i+j+k \\ i,j,k-1 \end{bmatrix}$.

Under the standard transformations the functions $B_\mu(n;i,j,k)$ become partition functions $S_\mu(n;i,j,k)$ enumerating partitions $\pi : e_1 + e_2 + \ldots$ into parts e_ℓ such that $\nu_1(\pi) = i$, $\nu_2(\pi) = j$ and $\nu_3(\pi) = k$ (here $\nu_i(\pi)$ is the number of parts of n that are $\equiv i (\text{mod } k)$, and where the e_ℓ satisfy gap conditions as given below:

Differences for S_1: $e_\ell - e_{\ell+1} \geq 3, 3$ or 4, if $e_\ell \equiv 1, 2$ or $3 (\text{mod } 3)$.

Differences for S_2: $e_\ell - e_{\ell+1} \geq 3, 2,$ or 5, if $e_\ell \equiv 1, 2$ or $3 (\text{mod } 3)$.

Differences for S_3: $e_\ell - e_{\ell+1} \begin{cases} = 1 \text{ or } \geq 3, & \text{if } e_\ell \equiv 1 (\text{mod } 3), \\ \geq 2 \text{ or } 6, & \text{if } e_\ell \equiv 2 \text{ or } 3 (\text{mod } 3). \end{cases}$

Differences for S_4: $e_\ell - e_{\ell+1} \begin{cases} \geq 2 \text{ or } 4, & \text{if } e_\ell \equiv 1 \text{or } 3 (\text{mod } 3), \\ = 3 \text{ or } \geq 5, & \text{if } e_\ell \equiv 2 (\text{mod } 3). \end{cases}$

Differences for S_5: $e_\ell - e_{\ell+1} \begin{cases} \geq 1, & \text{if } e_\ell \equiv 1 (\text{mod } 3), \\ = 3 \text{ or } \geq 5, & \text{if } e_\ell \equiv 2 (\text{mod } 3), \\ = 4 \text{ or } \geq 6, & \text{if } e_\ell \equiv 3 (\text{mod } 3). \end{cases}$

Differences for S_6: $e_\ell - e_{\ell+1} \begin{cases} \geq 1 \text{ or } 6, & \text{if } e_\ell \equiv 1 \text{ or } 3 (\text{mod } 3), \\ = 2, 3 \text{ or } \geq 5, & \text{if } e_\ell \equiv 2 (\text{mod } 3). \end{cases}$

Thus under the standard transformations, Theorem 4 becomes

Theorem 6 $S_1(n;i,j,k) = S_\mu(n;i,j,k), \mu = 2,3,\ldots,6$.

The gap conditions for S_1 are precisely those for Schur's theorem, while those for S_2 are as in Andrews' Theorem A. The difference conditions defining S_μ for $\mu \geq 3$ are more complicated and so did not show up in the computer search of Andrews [1971]. Actually, there is a gap *gap conservation law* that is common to all functions S_μ. For example, observe that for the differences defining S_1 and S_2

above, we have the relation

$$3 + 3 + 4 = 3 + 2 + 5 = 10.$$

Similarly, for S_3, interpreting the average of 1 and 3 as 2, we have

$$2 + 2 + 6 = 10.$$

The gap conservation law is more elegant at the base level without dilations because the standard gap conditions are uniform for all Type μ partitions. For a complete discussion of this gap conservation law, see Alladi and Gordon [1995].

We conclude this section by giving a combinatorial proof of Theorems 1, 3 and 4. The method is a variation of a technique due to Bressoud [1979].

Let r and s be given. We start with a bi–partition $(\pi_1; \pi_2)$ counted by $V(n; r, s)$. In what follows several steps still be illustrated with

$$\pi_1 : a_7 + a_6 + a_3 + a_2 + a_1 \text{ and } \pi_2 : b_{13} + b_{12} + b_8 + b_7 + b_5 + b_4 + b_2.$$

Step 1: Decompose π_2 into $\pi_4 \cup \pi_5$, where π_4 has the parts of π_2 whose weights are $\leq r$ and π_5 has remaining parts. Let $\nu(\pi_4) = m$. So $\nu(\pi_5) = s - m$.

$$\pi_4 : b_5 + b_4 + b_2 \text{ and } \pi_5 = b_{13} + b_{12} + b_8 + b_7.$$

Step 2: Consider the conjugate π_4^* of the Ferrers graph of π_4 and circle the bottom node of each column. Construct the partition $\pi_6 = \pi_1 + \pi_4^*$, where the number of nodes in each row of the graph of π_6 in the sum of the number of nodes in the corresponding rows of of π_1 and π_4^*. The parts of π_6 ending in the circled nodes are colored purple and the remaining parts of π_6 are red. Thus π_6 has $r - m$ red parts and m purple parts.

$$\pi_6 = \pi_1 + \pi_4^* : a_{10} + c_9 + a_5 + c_4 + c_2.$$

Step 3: Write the parts of π_5 in a column in descending order and below them write the parts of π_6 in descending order.

Step 4: Subtract 0 from the bottom element, 1 from the element above that, 2 from the next one above and so on, and display the new values and the subtracted amounts in two adjacent columns $C_1|C_2$. The elements of C_2 have no color while those of C_1 retain the color of the parts from which they were derived.

Step 5: Rearrange the elements of C_1 in descending order according to Scheme μ to get a new column C_1^R.

Step 6: Add the elements in the corresponding positions in columns C_1^R and C_2 to get a partition counted by $B_\mu(n; r - m, s - m, m)$. As usual, $i = r - m, j = s - m, k = m$. Every one of these steps is reversible, and so this gives a combinatorial proof of Theorems 1, 3 and 4.

We illustrate Steps 3–6 below for Schemes 1 and 2.

Step 3
π_5/π_6
b_{13}
b_{12}
b_8
b_7
a_{10}
c_9
a_5
c_4
c_2

Step 4	
C_1	C_2
b_5	8
b_5	7
b_2	6
b_2	5
a_6	4
c_6	3
a_3	2
c_3	1
c_2	0

Step 5 – Scheme 1	
C_1^R	C_2
a_6	8
c_6	7
b_5	6
b_5	5
a_3	4
c_3	3
b_2	2
b_2	1
c_2	0

Step 6
π_3
a_{14}
c_{13}
b_{11}
b_{10}
a_7
c_6
b_4
b_3
c_2

Step 5 – Scheme 2	
C_1^R	C_2
a_6	8
b_5	7
b_5	6
c_6	5
a_3	4
b_2	3
b_2	2
c_3	1
c_2	0

Step 6
π_3
a_{14}
b_{12}
b_{11}
c_{11}
a_7
b_5
b_4
c_4
c_2

To simply get a bijection between Type μ partitions counted by $B_\mu(n;i,j,k)$ and Type ω partitions counted by $B_\omega(n;i,j,k)$, proceed from Step 6 to Step 5, and in Step 5 replace Scheme μ ordering in C_1^R by Scheme ω ordering. Then revert back to Step 6.

The combinatorial proof given above leads to further improvements of Theorems 1 and 4 based on the size of the largest part. For a full discussion see Alladi and Gordon [1995].

4 Göllnitz' (Big) Theorem

One of the deepest results in the theory of partitions was proved by Göllnitz [1967]:

Theorem G *Let $C(n)$ denote the number of partitions of n into distinct parts $\equiv 2, 4$ or $5 (mod\ 6)$.*

Let $D(n)$ denote the number of partitions of n in the form $h_1 + h_2 + \cdots + h_\nu$, such that $h_\nu \neq 1$ or 3, $h_i - h_{i+1} \geq 6$ with strict inequality if $h_i \equiv 6, 7$ or $9 (mod\ 6)$. Then

$$C(n) = D(n).$$

Like Schur, Göllnitz had stated that the equality of three partition functions

$$F(n) = C(n) = D(n),$$

where $F(n)$ is the number of partitions of n into parts $\equiv 2, 5$ or $11 \pmod{12}$. Here the equality $F(n) = C(n)$ is trivial because

$$\sum_{n=0}^{\infty} C(n)q^n = (-q^2; q^6)_\infty(-q^4; q^6)_\infty(-q^5; q^6)_\infty = \frac{(q^4; q^{12})_\infty(q^8; q^{12})_\infty(q^{10}; q^{12})_\infty}{(q^2; q^6)_\infty(q^4; q^6)_\infty(q^5; q^6)_\infty}$$
$$= \frac{1}{(q^2; q^{12})_\infty(q^5; q^{12})_\infty(q^{11}; q^{12})_\infty} = \sum_{n=0}^{\infty} F(n)q^n. \tag{4.1}$$

Compare (4.1) with (3.1). Thus it is the equality $C(n) = D(n)$ which is the real challenge. The proof of (4.1) by Göllnitz [1967] was extremely complicated but he managed to establish it in the refined form

$$C(n; k) = D(n; k), \tag{4.2}$$

where $C(n; k)$ and $D(n; k)$ enumerate partitons of the type counted by $C(n)$ and $D(n)$ with the added restriction that there are precisely k parts and the convention that parts $\equiv 6, 7$ or $9 (\text{mod } 6)$ are counted twice. Compare (4.2) and (1.4). Such refinements are not possible with $F(n)$ and so we will not discuss $F(n)$ any further in this paper.

Göllnitz also obtained two generalizations of Theorem G (Göllnitz [1967], Satze 4.8 and 4.10). In his Satz 4.8 the modulus 6 is replaced by $M + 4$ with $M \geq 2$, and the residues $2, 4, 5, 6, 7$ and $9 (\text{mod } 6)$ are replaced by $M, M+2, M+3, 2M+2, 2M+3$ and $2M + 5 (\text{mod } M + 4)$. The difference conditions in Satz 4.8 are that $h_i - h_{i+1} \geq M + 4$ with strict inequality if $h_i \equiv 2M + 2$, $2M + 3$ or $2M + 5 (\text{mod } M + 4)$. In his Satz 4.10 also the modulus 6 is replaced by $M + 4$, but the residues $2, 4, 5, 6, 7$ and $9 (\text{mod } 6)$ are replaced by 2, $M + 2$, $M + 3$, $M + 4$, $M + 5$ and $2M + 5$ (mod $M + 4$). Both results are established in the refined form (4.2).

Andrews [1969b] gave a simpler proof of Theorem G using generating functions. Subsequently (see Andrews [1985]), he gave another proof where he used computer algebra to simplify certain calculations and then asked for a proof which would lend more insight into the structure underlying (4.2). We believe that the approach to Theorem G via the method of weighted words due to Alladi–Andrews–Gordon [1995a] does provide the required insight.

From the point of view of the method of weighted words, a three parameter refinement of Theorem G is viewed as emerging out of the incredible *key identity*

$$\begin{aligned}\sum_{i,j,k} a^i b^j c^k &\sum_{\substack{s=\alpha+\beta+\gamma+\delta+\epsilon+\phi\\ i=\alpha+\delta+\epsilon, j=\beta+\delta+\phi, k=\gamma+\epsilon+\phi}} \frac{q^{T_s+T_\delta+T_\epsilon+T_{\phi-1}}(1-q^\alpha(1-q^\phi))}{(q)_\alpha(q)_\beta(q)_\gamma(q)_\delta(q)_\epsilon(q)_\phi} \\ &= \sum_{i,j,k} \frac{a^i b^j c^k q^{T_i+T_j+T_k}}{(q)_i(q)_j(q)_k} = (-aq)_\infty(-bq)_\infty(-cq)_\infty\end{aligned} \tag{4.3}$$

under the transformations

$$\begin{cases} \text{(dilation) } q \mapsto q^6, \\ \text{(translations) } a \mapsto aq^{-4}, b \mapsto bq^{-2}, c \mapsto cq^{-1}. \end{cases} \tag{4.4}$$

We refer to (4.4) as the *standard tranformations* (for Göllnitz' theorem). While it is clear that the product on the right in (4.3) under the transformations (4.4) is the generating function for the three parameter refinement of $C(n)$ in Theorem G, the combinatorial interpretation of the multiple sum on the left is not at all obvious and we discuss this now.

We consider integers $j \geq 2$ as occuring in six colors, three of which, a, b and c are primary, and three colors ab, ac and bc are secondary. The integer 1 occurs only in the primary colors a, b and c. We use the alphabets d, e and f to mean ab, ac and bc. It is to be noted that a, b, c, d, e and f play a dual role; on the other hand they represent colors, and on the other they are free parameters. In order to convert the series on the left in (4.3) into the product on the right we need to take

$$d = ab, e = ac, f = bc.$$

As before, a_j will mean the integer j in color a, with similar interpretations for $b_j, c_j, \ldots, f_j$. The terms weight, gap, word and partition have the same meaning as before. Here also, to define a partition, we need an ordering of the symbols, and the one we start with is

$$\underline{\text{Scheme 1}}: \underline{e}_1 < a_1 < \underline{f}_1 < b_1 < c_1 < d_2 < e_2 < a_2 < f_2 < b_2 < c_2 < \ldots. \tag{4.5}$$

The reason for choosing this ordering is that under the standard transformations (4.4), the symbols become

$$\begin{cases} a_j \mapsto 6j-4, b_j \mapsto 6j-2, c_j \mapsto 6j-1, \text{ for } j \geq 1, \\ d_j \mapsto 6j-6, e_j \mapsto 6j-5, f_j \mapsto 6j-3, \text{ for } j \geq 2, \end{cases} \tag{4.6}$$

and so under (4.6), Scheme 1 becomes

$$\underline{1} < 2 < \underline{3} < 4 < 5 < 6 < \ldots,$$

the standard ordering of the positive integers. In (4.5), the symbols $\underline{e}_1, \underline{f}_1$ are underlined to indicate that they will never occur in the partitions under discussion. This explains why 1 and 3 never occur as parts enumerated by $D(n)$ in Theorem G.

Next, let $x_m = x_m^{(1)}$ be the symbol occupying position m in Scheme 1. That is, $x_1 = \underline{e}_1, x_2 = a_1, \ldots$. Then Type 1 partitions are those of the form $x_{m_1} + x_{m_2} + \ldots$, where the x_{m_ℓ} are symbols from the non–underlined list in Scheme 1 such that

$$m_\ell - m_{\ell+1} \geq 6 \text{ with strict inequality if } x_{m_\ell} \text{ is of secondary color.} \tag{4.7}$$

We call (4.7) as the *standard gap conditions* for Göllnitz' theorem. The main partition result proved in Alladi, Andrews and Gordon [1995a] is

Theorem 7 *Let $D(n;\alpha,\beta,\gamma,\delta,\epsilon,\phi)$ be the number of Type 1 partitions π of n such that $\nu_a(\pi)=\alpha$, $\nu_b(\pi)=\beta,\ldots,\nu_f(\pi)=\phi$.*

Let $C(n;i,j,k)$ denote the number of vector partitions $(\pi_1;\pi_2;\pi_3)$ of n such that π_1 has i distinct $a-$parts, π_2 has j distinct $b-$parts and π_3 has k distinct $c-$parts. Then

$$C(n;i,j,k)=\sum_{i=\alpha+\delta+\epsilon,j=\beta+\delta+\phi,k=\gamma+\epsilon+\phi} D(n;\alpha,\beta,\gamma,\delta,\epsilon,\phi).$$

Clearly, the generating function of $C(n;i,j,k)$ is

$$\sum_n C(n;i,j,k)q^n=\frac{q^{T_i+T_j+T_k}}{(q)_i(q)_j(q)_k}. \tag{4.8}$$

The proof of Theorem 7 given in Alladi, Andrews and Gordon [1995a] is in two stages. The first stage is show that the generating function of $D(n;\alpha,\beta,\gamma,\delta,\epsilon,\phi)$ is

$$G(\alpha,\beta,\gamma,\delta,\epsilon,\phi)=\sum_n D(n;\alpha,\beta,\gamma,\delta,\epsilon,\phi)q^n=\frac{q^{T_s+T_\delta+T_\epsilon+T_{\phi-1}}(1-q^\alpha(1-q^\phi))}{(q)_\alpha(q)_\beta(q)_\gamma(q)_\delta(q)_\epsilon(q)_\phi}, \tag{4.9}$$

where

$$s=\alpha+\beta+\gamma+\delta+\epsilon+\phi.$$

The proof of (4.9) is quite difficult and utilizes minimal partitions. More precisely, let $H(\alpha,\beta,\gamma,\delta,\epsilon,\phi)$ be the generating function of all minimal Type 1 partitions π^* such that $\nu_a(\pi^*)=\alpha$, $\nu_b(\pi^*)=\beta,\ldots,\nu_f(\pi^*)=\phi$. Since all partitions enumerated have exactly s parts (fixed) it follows from Lemma 1 that (4.9) is equivalent to

$$H(\alpha,\beta,\gamma,\delta,\epsilon,\phi)=q^{T_s+T_\delta+T_\epsilon+T_{\phi-1}}(1-q^\alpha(1-q^\phi))\begin{bmatrix} s \\ \alpha,\beta,\gamma,\delta,\epsilon,\phi\end{bmatrix}. \tag{4.10}$$

Equation (4.10) can be established by induction on s, the length of the word, by considering $H_a(\alpha,\beta,\gamma,\delta,\epsilon,\phi)$, $H_b(\alpha,\beta,\gamma,\delta,\epsilon,\phi),\ldots$, which are the generating functions of partitions of the type enumerated by H with the added restriction that the smallest part is an $a-$part, $b-$part, $\ldots$, and using recurrences satisfied by the multinomial coefficients of order $t=6$. The proof of this is quite involved; for details see Alladi, Andrews and Gordon [1995a].

Having established (4.10), the second stage is to prove the key identity (4.3) which is very deep. Its proof utilizes not only Watson's $q-$analog of Whipple's theorem but the ${}_6\psi_6$ summation of Bailey as well. This is substantially more than what is required to prove the Rogers–Ramanujan identities.

Under the standard transformations, Theorem 7 (and (4.3)) yield a three parameter refinement of Göllnitz' Theorem G where the residues $2,4,5 \pmod 6$ are to be treated as primary, and $6,7,9 \pmod 6$ are secondary. More generally, under the transformations

$$\begin{cases} \text{(dilation) } q\mapsto q^M, \\ \text{(translations) } a\mapsto aq^{r_1-M}, b\mapsto bq^{r_2-M}, c\mapsto cq^{r_3-M}, \end{cases} \tag{4.11}$$

we get the following result from which Theorem G and Göllnitz' Satze (4.8) and (4.10) fall out as special cases.

Theorem 8 *Let $M \geq 6$ and r_1, r_2, r_3 integers satisfying*

$$0 < r_1 < r_2 < r_3 < M \leq r_1 + r_2.$$

Let $C(n; i, j, k)$ denote the number of partitions of n into i distinct parts $\equiv r_1 \,(\text{mod } M)$, j distinct parts $\equiv r_2 \,(\text{mod } M)$ and k distinct parts $\equiv r_3 \,(\text{mod } M)$.

Let $D(n; \alpha, \beta, \gamma, \delta, \epsilon, \phi)$ denote the number of partitions π of n of the form $h_1 + h_2 + \cdots + h_\nu$ such that each $h_i \equiv r_1, r_2, r_3, r_1 + r_2, r_1 + r_3$ or $r_2 + r_3$ (mod M) and satisfying

(i) $h_i - h_{i+1} \geq M$ with strict inequality if $h_i \equiv r_1 + r_2, r_1 + r_3$ or $r_2 + r_3$ (mod M),

(ii) the number of parts of π which are $\equiv r_1, r_2, r_3, r_1 + r_2, r_1 + r_3$ and $r_2 + r_3$ (mod M) is $\alpha, \beta, \gamma, \delta, \epsilon$ and ϕ respectively. Then

$$C(n; i, j, k) = \sum_{i=\alpha+\delta+\epsilon, j=\beta+\delta+\phi, k=\gamma+\epsilon+\phi} D(n; \alpha, \beta, \gamma, \delta, \epsilon, \phi).$$

Note that in Theorem 8,

$$i + j + k = \alpha + \beta + \gamma + 2(\delta + \epsilon + \phi),$$

and in this form the parts of secondary color are counted twice as in (4.2). Observe also that unlike Göllnitz' Satze (4.8) and (4.10) where the residue classes are prescribed, we have almost total freedom in choosing the residues in Theorem 8.

Companions to Göllnitz theorem can be obtained by considering other orderings. Since we have six colors here, we have a total of $6! = 720$ schemes yielding 720 companions! Among these schemes we discuss only one here, namely,

$$\underline{\text{Scheme 2}}: \underline{e}_1 < a_1 < \underline{f}_1 < b_1 < d_2 < c_1 < e_2 < a_2 < f_2 < b_2 < d_3 < c_2 < \ldots. \tag{4.12}$$

Under the standard transformations (4.6), Scheme 2 yields

$$\underline{1} < 2 < \underline{3} < 4 < 6 < 5 < 7 < 8 < 9 < 10 < 12 < 11 < \ldots.$$

Next, let $x_m = x_m^{(2)}$ be the symbol occupying position m in Scheme 2. By a Type 2 partition we mean one of the form $x_{m_1} + x_{m_2} + \ldots$, where the x_{m_ℓ} are the non–underlined symbols of Scheme 2 and satisfying the standard gap conditions (4.7). Also, let $D_2(n; \alpha, \beta, \gamma, \delta, \epsilon, \phi)$ denote the number of Type 2 partitions π of n such that $\nu_a(\pi) = \alpha$, $\nu_b(\pi) = \beta, \ldots, \nu_f(\pi) = \phi$. We then have

Theorem 9 $D(n, \alpha, \beta, \gamma, \delta, \epsilon, \phi) = D_2(n; \alpha, \gamma, \delta, \epsilon, \phi)$.

Under the standard transformations, Theorem 9 yeilds the following result which is stated for simplicity in the unrefined form.

Theorem 10 *Let $D_2(n)$ denote the number of partitions of n in the form $h_1 + h_2 + \cdots + h_\nu$ such that $h_\nu \neq 1$ or 3, and $h_i - h_{i+1} \geq 7, 6, 7, 6, 5$ or 8 depending on $h_i \equiv 7, 2, 9, 4, 5$ or 6 (mod 6). Then*

$$D(n) = D_2(n).$$

Theorem 10 is a companion to Theorem G in the same way as Theorem A is a companion to Theorem S. To see the link between the difference conditions of $D(n)$ and $D_2(n)$ more clearly, rewrite the difference conditions of $D(n)$ in the form $h_i - h_{i+1} \geq 7, 6, 7, 6, 6$, or 7 depending on $h_i \equiv 7, 2, 9, 4, 5$ or $6 (\bmod 6)$. Thus, here too we have a *conservation law* because

$$7 + 6 + 7 + 6 + 6 + 7 = 7 + 6 + 7 + 6 + 5 + 8.$$

For a discussion of the conservation law for all schemes, see Alladi, Andrews and Gordon [1995a].

A combinatorial proof of Theorem 9 may be given as follows.

Step 1: Given a partition counted by $D(n; \alpha, \beta, \gamma, \delta, \epsilon, \phi)$, write the parts in a column in descending order.

Step 2: Subtract 0 from the bottom element, 1 from the element that, 2 from the next one above and so on, and display the new values and the ones subtracted in two adjacent columns $C_1 | C_2$. The elements of C_2 have no color while those of C_1 retain the color of the parts from which they were derived. The standard gap conditions ensure that the elements of C_1 from the bottom upwards are in increasing order according to Scheme 1. The symbols of primary color alone may repeat in C_1.

Step 3: Rearrange the elements of C_1 such that from the bottom upwards they are increasing order according to Scheme 2. This gives a rearranged column C_1^R.

Step 4: Add the corresponding elements of C_1^R and C_2 to get a partition counted by $D_2(n; \alpha, \beta, \gamma, \delta, \epsilon, \phi)$, whose parts will retain the color of the elements of C_1^R from which they are derived.

While it was possible to give a simple combinatorial proof of Theorem 9, it is a totally different story to give a bijective proof of Theorem 7. In fact, it had remained an open problem to give a combinatorial proof/approach for Göllnitz' (Big) theorem. This is very surprising because, when Theorem G was first proved in 1967, it was established in the refined form (4.2). The multiparameter refinement we have obtained by the method of weighted words is a further compelling reason that there ought to be a combinatorial explanation for Göllnitz' theorem. I noticed recently (see Alladi [1997b]) that for a combinatorial approach it is better to consider the effect of a different set of transformations on (4.3) instead of the standard transformations (4.4). This is described in the next section.

5 A reformulation of Göllnitz' (Big) theorem

Consider the effect of the *cubic transformation*

$$\begin{cases} \text{(dilation) } q \mapsto q^3, \\ \text{(translations) } a \mapsto aq^{-2}, b \mapsto bq^{-1}, c \mapsto c, \end{cases} \tag{5.1}$$

on the key identity (4.3). Firstly, the product on the right in (4.3) becomes

$$(-aq; q^3)_\infty (-bq^2; q^3)_\infty (-cq^3; q^3)_\infty \tag{5.2}$$

which is the refined generating function for partitions into distinct parts. The effect of the cubic transformation on the symbols is

$$\begin{cases} a_j \mapsto 3j-2, b_j \mapsto 3j-1, c_j \mapsto 3j, \text{ for } j \geq 1, \\ d_j \mapsto 3j-3, e_j \mapsto 3j-2, f_j \mapsto 3j-1, \text{ for } j \geq 2. \end{cases} \tag{5.3}$$

Notice from (5.3) that each integer $N \geq 3$ occurs in two colors, one primary and one secondary. More precisely, the integers $\equiv 4 (\text{mod } 3)$ occur in colors a and ac, those $\equiv 5 (\text{mod } 3)$ occur in colors b and bc, and finally, the ones $\equiv 3 (\text{mod } 3)$ occur in colors c and ab. The integers 1 and 2 occur only in colors a and b respectively. It turns out that under the cubic transformation (5.1), the series on the left of (4.3) is the generating function of partitions with difference ≥ 3 between parts, with each such partition being counted by the weight given in (5.4) below. For a derivation of this weight formula see Alladi [1997b].

Let $\tilde{\pi} : h_1 + h_2 + \cdots + h_\nu$ be a partition with difference ≥ 3 between parts. For the purpose of computing gaps between parts in the table below, put $h_{\nu+1} = -1$. The weight $w(h_i)$ of each part h_i is defined as follows:

Congruence	Gap	Weight $w(h_i)$
$h_i \equiv 1(\text{mod } 3)$	$h_i - h_{i+1} \leq 3$	a
$h_i \equiv 2(\text{mod } 3)$	$h_i - h_{i+1} \leq 3$	b
$h_i \equiv 3(\text{mod } 3)$	$h_i - h_{i+1} \leq 3$	c
$h_i \equiv 1(\text{mod } 3)$	$h_i - h_{i+1} \geq 4$	$a + ac$
$h_i \equiv 2(\text{mod } 3)$	$h_i - h_{i+1} \geq 4$	$b + bc$
$H_i \equiv 3(\text{mod } 3)$	$h_i - h_{i+1} \geq 4$	$c + ab$

Finally, the weight $w(\tilde{\pi})$ of the partition $\tilde{\pi}$ is defined multiplicatively as

$$w(\tilde{\pi}) = \prod_{i=1}^{\nu} w(h_i). \tag{5.4}$$

Then under the cubic transformation, identity (4.3) is equivalent to

Theorem 11 *Let $\mathcal{D}$ denote the set of all partitions π' into distinct parts. For $\pi' \in \mathcal{D}$, let $\nu_i(\pi')$ denote the number of parts of π' which are $\equiv i (\text{mod } 3)$, $i = 1, 2, 3$.*

Let D_3 denote the set of all partitions $\tilde{\pi}$ into parts differing by ≥ 3. For $\tilde{\pi} \in \mathcal{D}_3$ let $w(\tilde{\pi})$ be given by (5.4). Then

$$\sum_{\substack{\pi' \in \mathcal{D} \\ \sigma(\pi')=n}} a^{\nu_1(\pi')} b^{\nu_2(\pi')} c^{\nu_3(\pi')} = \sum_{\substack{\tilde{\pi} \in \mathcal{D}_3 \\ \sigma(\tilde{\pi})=n}} w(\tilde{\pi}).$$

For a proof that Theorem 11 is equivalent to (4.3) under (5.1), see Alladi [1997b]. If we put $a = b = c = 1$ in Theorem 11, we get the following striking result.

Theorem 12 *: Let $Q(n)$ denote the number of partitions of n into distinct parts.*

Let $g_3(n; k)$ denote the number of partitions of n in the form $h_1 + h_2 + \cdots + h_\nu$ with difference ≥ 3 between parts and such that there are precisely k gaps ≥ 4 between the integers $h_1, h_2, \ldots, h_\nu$ and $h_{\nu+1} = -1$. Then

$$Q(n) = \sum_k g_3(n; k) 2^k.$$

I had actually proved Theorem 12 in 1990 but it is only recently that I realized that it had such a nice three parameter refinement like Theorem 11. It is to be noted that Theorem 11 is a reformulation of and equivalent to Theorem 7 (and therefore equivalent to a three parameter refinement of Theorem G), but Theorem

12 is obviously not equivalent to Theorem G. The way I first arrived at Theorem 12 was through a combinatorial analysis of the following well known identity of Sylvester [1882]:

$$(-zq)_\infty = 1 + \sum_{\nu=1}^{\infty} \frac{z^\nu q^{(3\nu^2-\nu)/2}(1+zq^{2\nu})(-zq)_{\nu-1}}{(q)_\nu}. \tag{5.5}$$

The precise combinatorial interpretation of (5.5) making use of the free parameter z is as follows.

Theorem 13 *Let $Q(n;k)$ denote the number of partitions of n into k distinct parts.*

Let $g_3(n;\nu,k)$ denote the number of partitions of the type enumerated by $g_3(n;k)$ above with the added restriction that there exactly ν parts. Then

$$\sum_k Q(n;k)z^k = \sum_\nu \sum_k g_3(n;\nu,k)z^\nu(1+z)^k.$$

Theorem 13 is shown in Alladi [1997a] to be equivalent to (5.5). Clearly Theorem 12 is the special case $z = 1$ in Theorem 13. Andrews [1976] has explained the significance of (5.5) in the theory of partitions. But the combinatorial interpretation given in Theorem 13 is new.

Theorem 12 is an example of what one may call as a weighted partition identity. Recently (see Alladi [1997a]) I have undertaken the study of the following general problem: Consider two sets of partitions S and T with $S \subseteq T$. Let $p_S(n)$ (resp. $p_T(n)$) denote the number of partitions π of n with $\pi \in S$ (resp. $\pi \in T$). Clearly, $p_S(n) \le p_T(n)$. The problem is to determine integral weights $w(\pi) \ge 1$ in a natural way such that

$$p_T(n) = \sum_{\pi \in S, \sigma(\pi)=n} w(\pi). \tag{5.6}$$

In Alladi [1997a] several interesting examples of (5.6) involving classical partition functions are given. In all examples discussed in Alladi [1997a], the weights $w(\pi)$ are defined multiplicatively as in (5.4). Theorem 12 is the case $S = \mathcal{D}_3$, $T = \mathcal{D}$ in (5.6).

We now describe briefly how to set up a bijection to explain Theorem 11. For a complete discussion see Alladi [1997b].

A direct proof of Theorem 12 without recourse to Sylvester's identity can be given using what we can call as the *hook operation* ρ on partitions into distinct parts. That is, given a partition π, form a new partition $\tilde{\pi} = \rho(\pi)$ whose parts are obtained by counting nodes along hooks (rectangular bends) of the Ferrers graph of π. It turns out that ρ provides a surjective map from $\mathcal{D}$ to $\mathcal{D}_3$ that is

$$\begin{aligned} \rho :& \mathcal{D} \to \mathcal{D}_3 \\ & \pi' \to \tilde{\pi} = \rho(\pi'). \end{aligned} \tag{5.7}$$

It is shown in Alladi [1997b] that under this map, given $\tilde{\pi} \in \mathcal{D}_3$ having exactly k gaps ≥ 4 in the sense described in Theorem 12, there corresponds exactly 2^k partitions $\pi' \in \mathcal{D}$ for which $\rho(\pi') = \tilde{\pi}$. Theorem 12 follows from this immediately.

Next, given a partition π' into distinct parts, first assign to it the weight $a^{\nu_1(\pi')}b^{\nu_2(\pi')}c^{\nu_3(\pi')}$ as in Theorem 11. This amounts to drawing the Ferrers graph of π', interpreting the rows of the graph as parts of π', and assigning weight a, b or

c to each row depending on the residue class of the part modulo 3. We will now assign a different weight $\tilde{w}(\pi')$ to π' as follows.

Consider the Durfee square of the Ferrers graph of π' and circle the nodes at the end of each row *below* the square. Next consider the partition $\tilde{\pi} = \rho(\pi')$ whose parts $h_1, h_2, \ldots, h_\nu$ are obtained by counting nodes along hooks of π'. If the hook corresponding to h_i *does not* start from a circled node, define

$$\tilde{w}(h_i) = a, b \text{ or } c \text{ depending on } h_i \equiv 1, 2 \text{ or } 3 \pmod 3. \tag{5.8}$$

If the hook corresponding to h_i *starts* from a circled node, then define

$$\tilde{w}(h_i) = ac, bc \text{ or } ab \text{ depending on } h_i \equiv 1, 2 \text{ or } 3 \pmod 3. \tag{5.9}$$

Finally, define $\tilde{w}(\pi')$ to be

$$\tilde{w}(\pi') = \prod_{i=1}^{\nu} \tilde{w}(h_i), \tag{5.10}$$

where the product is taken over the parts h_i of $\tilde{\pi}$, not of π'.

Using what we have called as the *sliding operation* on Ferrers graphs, it is shown in Alladi [1997b], that given $\tilde{\pi} \in \mathcal{D}_3$

$$w(\tilde{\pi}) = \sum_{\rho(\pi')=\tilde{\pi}} \tilde{w}(\pi'). \tag{5.11}$$

That is, the sum of $\tilde{w}(\pi')$ taken over all partitions $\pi' \in \mathcal{D}$ for which $\rho(\pi) = \tilde{\pi}$, is equal to the weight $w(\tilde{\pi})$ as in Theorem 11. Thus Theorem 11 is equivalent to the statement

Theorem 14 $$\sum_{\pi' \in \mathcal{D}, \sigma(\pi')=n} a^{\nu_1(\pi')} b^{\nu_2(\pi')} c^{\nu_3(\pi')} = \sum_{\pi' \in \mathcal{D}, \sigma(\pi')=n} \tilde{w}(\pi').$$

The hook operation has produced the natural bijection as described above which explains how the weights on the left in Theorem 14 are converted to $\tilde{w}(\pi')$. In general these weights are not equal, that is

$$a^{\nu_1(\pi')} b^{\nu_2(\pi')} c^{\nu_3(\pi')} \neq \tilde{w}(\pi'),$$

but the weights have the same degree as monomials in a, b, c. That is

$$\deg(\tilde{w}(\pi')) = \nu_1(\pi') + \nu_2(\pi') + \nu_3(\pi') = \nu(\pi'). \tag{5.12}$$

Thus the proof of the equality of the two sums in Theorem 14 is deep and goes via (4.3) under the transformations (5.1). The bijection described here is very subtle and only works for the cubic transformation and not for the standard transformations. This bijection underlying Göllnitz' theorem had remained elusive for nearly thirty years.

Next we discuss applications of Theorem 11.

6 Applications

Refined identities as well as weighted identities have several fundamental applications and from this point of view, Theorem 11 is extremely interesting.

Theorem 12, which is a special case of Theorem 11, suggests the remarkable result that for any integer $k \geq 1$, the congruence

$$Q(n) \equiv 0 (\bmod\ 2^k) \tag{6.1}$$

is valid for almost all n. I had used Theorem 12 in 1990 to study (6.1) for small values of k. Congruences for partition functions are usually approached via the theory of modular forms. Recently, using modular forms, Gordon and Ono [1997; to appear] have proved that for each $k \geq 1$, (6.1) is valid for almost all n. But Theorem 12 offers an elementary approach to (6.1) and we describe this briefly now.

First, in order to decide when $Q(n)$ is odd, it suffices to check when $g_3(n;0)$ is odd. Now, $g_3(n;0)$ counts partitions $\tilde{\pi} : h_1 + h_2 + \cdots + h_\nu$ into parts differing by ≥ 3, and with no gaps ≥ 4 between $h_1, h_2, \ldots, h_\nu$ and $h_{\nu+1} = -1$. So there are only two cases to consider.

Case 1: $h_\nu = 1$, and $h_i - h_{i+1} = 3$, for $1 \leq i < \nu$.

In this case $\tilde{\pi} : 1 + 4 + 7 + \cdots + 3\nu - 2 = (3\nu^2 - \nu)/2$.

Case 2: $h_\nu = 2$, and $h_i - h_{i+1} = 3$, for $1 \leq i < \nu$.

In this case $\tilde{\pi} : 2 + 5 + 8 + \cdots + 3\nu - 1 = (3\nu^2 + \nu)/2$.

Thus we quickly conclude that

$$Q(n) \equiv 0(\text{mod } 2) \Leftrightarrow n \neq \frac{3\nu^2 \pm \nu}{2}(\text{Pentagonal}). \tag{6.2}$$

Next, to decide when $Q(n) \not\equiv 0(\text{mod } 4)$ we need to see when $g_3(n;1)$ makes a contribution. By analyzing $g_3(n;1)$ it can be shown that

$$Q(n) \equiv 0(\text{mod } 4) \text{ unless } n = \frac{3\nu^2 \pm \nu}{2} + \mu^2, \tag{6.3}$$

equals a pentagonal number plus a square. Thus (6.1) holds for almost all n when $k = 2$. The exceptional set is actually thinner than indicated in (6.3) and can be determined more precisely. As k gets larger, the combinatorial analysis of (6.1) using Theorem 12 gets more complicated and so it is best to use modular forms as in Gordon and Ono [1997; to appear]. But it is nice to have an elementary approach and in Alladi [in preparation], Theorem 12 will be used to study (6.1) for $k \leq 4$.

Next we show how Jacobi's Triple Product Identity can be deduced from Theorem 11.

Consider the choices $c = -1$, $ab = 1$ in Theorem 11. Thus $a + ac = b + bc = c + ab = 0$. So from the table of weights we see that $w(\tilde{\pi}) = 0$ for any partition $\tilde{\pi}$ having at least one gap ≥ 4 between its parts $h_1 + h_2, \ldots, h_\nu$ and $h_{\nu+1} = -1$. Thus the only partitions $\tilde{\pi}$ that can make a contribution are those in Cases 1 and 2. So the right hand side of Theorem 11 yields

$$\sum_n \left(\sum_{\tilde{\pi} \in \mathcal{D}_3, \sigma(\tilde{\pi}) = n} w(\tilde{\pi}) \right) q^n = \sum_{\nu=-\infty}^{\infty} a^\nu q^{(3\nu^2 - \nu)/2}. \tag{6.4}$$

With the choices $c = -1$ and $ab = -1$, the left hand side of Theorem 11 yields

$$\sum_n \left(\sum_{\pi' \in \mathcal{D}, \sigma(\pi') = n} a^{\nu_1(\pi')} b^{\nu_2(\pi')} c^{\nu_3(\pi')} \right) q^n = (-aq; q^3)_\infty (-a^{-1}q^2; q^3)_\infty (q^3; q^3)_\infty. \tag{6.5}$$

Thus by combining (6.4) and (6.5) we get

$$(-aq; q^3)_\infty (-a^{-1}q^2; q^3)_\infty (q^3; q^3)_\infty = \sum_{\nu=-\infty}^{\infty} a^\nu q^{(3\nu^2 - \nu)/2} \tag{6.6}$$

which is equivalent to Jacobi's Triple Product Identity (see Gasper and Rahman [1990]).

Next, take $c = 0$ in Theorem 11. This means that $a + ac = a$, $b + bc = b$ and $c + ab = ab$. So, regardless of what $h_i - h_{i+1}$ is, $w(h_i) = a$ if $h_i \equiv 1 (\text{mod } 3)$ and $w(h_i) = b$ if $h_i = 2 (\text{mod } 3)$. Also, $w(h_i) = ab$ for $h_i \equiv 3 (\text{mod } 3)$ implies that $h_i - h_{i+1} > 3$ in this case. This means we cannot have consecutive multiples of 3. Thus the choice $c = 0$ in Theorem 11 yields the two parameter refinement of Schur's Theorem established in Section 3.

Finally, Theorem 11 is also useful in obtaining a simpler key identity for Göllnitz theorem in (4.3). Recently, by a further combinatorial study I showed that Theorem 1 is equivalent to the *new key identity*

$$\sum_{i,j,k} \frac{a^i b^j c^k q^{T_{i+j+k}} (-c)_i (-c)_j \left(\frac{-abq}{c}\right)_k (-cq)_{i+j}}{(q)_i (q)_j (q)_k (-c)_{i+j}} = (-aq)_\infty (-bq)_\infty (-cq)_\infty \quad (6.7)$$

under the cubic transformations (5.1). George Andrews has subsequently given a purely q−series proof of (6.7) making use of Jackson's q−analog of Dougall's summation. Although simpler than (4.3), Andrews' proof shows that (6.7) is still quite deep. In Alladi and Andrews [preprint], the q−series proof of (6.7) as well as the combinatorial equivalence of (6.7) and Theorem 11 are presented. We will not discuss (6.7) any further here except to point out that choosing $c = 0$ in (6.7) yields the key identity (3.13) for Schur's Theorem, and after applying the cubic transformations, the choices $c = -1$, $ab = 1$ in (6.7) yields Jacobi's Triple Product Identity in the form (6.6).

We have presented here only applications of Theorem 11, a result which is obtained from (4.3) by using the partial dilation $q \mapsto q^3$ instead of the full dilation $q \mapsto q^6$. Similarly, the partial dilation $q \mapsto q^2$ on the key identity (3.13) underlying Schur's theorem has interesting consequences. In Alladi [1997a] a weighted identity connecting $Q(n)$ with partitions of n with difference ≥ 2 between parts is derived by using transformations $q \mapsto q^2$, $a \mapsto aq^{-1}$, $b \mapsto b$ in (3.13). We have recently noticed that the transformations $q \mapsto q^2$, $a \mapsto aq^{-1}$, $b \mapsto bq^{-1}$ in (3.13) leads to some new links between the partitions of the Schur type and those of Göllnitz–Gordon (namely, partitions with difference ≥ 2 between parts and without consecutive even numbers).

7 The Andrews hierarchies

Why is Göllnitz' (Big) theorem so difficult to prove? The principal reason is that while the product on the right in (4.3) deals with three primary colors a, b and c, the series on the left does not involve the complete alphabet of colors that can be generated using a, b and c. In Göllnitz' theorem, the colors a, b, c and ab, ac, bc are accepted but the ternary color abc is dropped. In contrast, the product is Schur's theorem utilizes two primary colors a and b, while the series there involves the complete alphabet of colors a, b and ab. This explains the substantial increase in depth while passing from Schur's theorem to Göllnitz' theorem.

Extending the ideas in his proof of Schur's theorem, Andrews [1967], [1968], [1969a], obtained two hierarchies of partition theorems making use of the completeness of alphabets. To describe these results, let $a_1, a_2, \ldots, a_r$ be r distinct positive integers such that

$$\sum_{i=1}^{k-1} a_i < a_k, \quad 1 \leq k \leq r. \tag{7.1}$$

Condition (7.1) ensures that $2^r - 1$ sums $\sum \varepsilon_i a_i$, where $\varepsilon_i = 0$ or 1, not all $\varepsilon_i = 0$, are all distinct. Let these sums in increasing order be denoted by $\alpha_1, \alpha_2, \ldots, \alpha_{2^r-1}$.

Next let $N \geq \sum_{i=1}^{r} a_i \geq 2^r - 1$ be a modulus, and A_N denote the set of all positive integers congruent to some $a_i(\text{mod } N)$. Similarly, letA'_N denote the set of all positive integers congruent to some $\alpha_i(\text{mod } N)$ Also let $\beta_N(m)$ denote the least positive residue of $m(\text{mod } N)$. Finally, if $m = \alpha_j$ for some j, let $\phi(m)$ denote the number of terms appearing in the defining sum of m and $\psi(m)$ the smallest a_i appearing in this sum. Then the first general theorem of Andrews [1968] is:

Theorem A_1 *Let* $C(A_N; n)$ *denote the number of partitions of* n *into distinct parts taken from* A_N.

Let $D(A'_N; n)$ *denote the number of partitions of* n *into parts* $b_1, b_2, \ldots, b_\nu$ *from* A'_N *such that*

$$b_i - b_{i+1} \geq N\phi(\beta_N(b_{i+1})) + \psi(\beta_N(b_{i+1})) - \beta_N(b_{i+1}). \tag{7.2}$$

Then

$$C(A_N; n) = D(A'_N; n).$$

To describe the second general theorem of Andrews [1969a], let a_i, α_i and N be as above. Now let $-A_N$ denote the set of all positive integers congruent to some $-a_i(\text{mod } N)$, and $-A'_N$ the set of all positive integers congruent to some $-\alpha_i$(mod N). The quantities $\beta_N(m), \phi(m), \psi(m)$ are also as above. We then have

Theorem A_2 *Let* $C^*(-A_N; n)$ *denote the number of partitions of* n *into distinct parts taken from* $-A_N$.

Let $D^*(-A'_N; n)$ *denote the number of partitions of* n *into parts* $b_1, b_2, \ldots, b_\nu$, *taken from* $-A'_N$ *such that*

$$b_i - b_{i+1} \geq N\phi(\beta_N(-b_i)) + \psi(\beta_N(-b_i)) - \beta_N(-b_i) \tag{7.3}$$

and also

$$b_\nu \geq N(\phi(\beta_N(-b_s) - 1)). \tag{7.4}$$

Then

$$C^*(-A_N; n) = D^*(-A'_N; n).$$

The main differences between Theorems A_1 and A_2 are that the gap conditions for $b_i - b_{i+1}$ in (7.2) are given in terms of b_{i+1}, whereas in (7.3) they are given in terms of b_i. Also, Theorem A_2 has condition (7.4) on the smallest part b_ν, while there is no such condition in Theorem A_1.

When $r = 2$, $a_1 = 1$, $a_2 = 2$, $N = 3 = 2^r - 1$, Theorems A_1 and A_2 both become Theorem S. Conditions (7.2) and (7.3) may appear complicated but can be understood better by classifying b_{i+1} (in Theorem A_1) and b_i (in Theorem A_2) in terms of their residue classes (mod N). In particular, with $r = 3$, $a_1 = 1$, a_2, $a_3 = 4$ and $N = 7 = 2^3 - 1$, Theorems A_1 and A_2 yield the following corollaries.

Corollary 1 *Let* $C_1(n)$ *denote the number of partitions of* n *into distinct parts* $\equiv 1, 2$ *or* 4 *(mod* 7*).*

Let $D_1(n)$ *denote the number of partitions of* n *in the form* $b_1 + b_2 + \ldots, b_\nu$ *such that* $b_i - b_{i+1} \geq 7, 7, 12, 7, 10, 10$ *or* 15 *if* $b_{i+1} \equiv 1, 2, 3, 4, 5, 6$ *or* 7 *(mod* 7*). Then*

$$C_1(n) = D_1(n).$$

Corollary 2 *Let* $C_2(n)$ *denote the number of partitions of* n *in distinct parts* $\equiv 3, 5$ *or* 6 *(mod* 7*).*

Let $D_2(n)$ *denote the number of partitions of* n *in the form* $b_1 + b_2 + \cdots + b_\nu$ *such that* $b_i - b_{i+1} \geq 10, 10, 7, 12, 7, 7$ *or* 15 *if* $b_i \equiv 8, 9, 3, 11, 5, 6$ *or* 14 *(mod* 7*) and* $b_\nu \neq 1, 2, 4$ *or* 7*. Then*

$$C_2(n) = D_2(n).$$

Andrews' proofs of Theorems A_1 and A_2 are extensions of his proof of Theorem S and not as difficult as the proof of Göllnitz' theorem.

By considering the conjugates of the Ferrers graphs of partitions counted by $D_1(A_N; n)$, Andrews and Olsson [1991] obtained an important reformulation and simplification of Theorem A_1 and in that process proved the following more general result.

Theorem C *Let* $A = \{a_1, a_2, \ldots, a_r\}$ *be a set of* r *distinct positive integers arranged in increasing order and let* $N > a_r$*. Let* $P_1(A; N; n)$ *denote the number of partitions of* n *into distinct parts each congruent to some* a_i *(mod* N*). Let* $P_2(A; N; n)$ *denote the number of partitions of* n *into parts* $\equiv 0$ *or some* a_i *(mod* N*) such that only parts divisible by* N *may repeat, the smallest part is* $< N$*, and the difference between parts is* $\leq N$*, and* $< N$ *is either part is divisible by* N*. Then*

$$P_1(A; N; n) = P_2(A; N; n).$$

It is to be noted that condition (7.1) does not appear in Theorem C. Under condition (7.1) it turns out by using $N-$modular diagrams that the partitions counted by $P_2(A; N; n)$ are the conjugates of the partitions counted by $D_1(A_N; n)$. Owing to conjugation the lower bound gap condition (7.2) in Theorem A_1 is replaced by upper bound gap conditions in Theorem C. Bessenrodt [1991] has given a combinatorial proof of Theorem C.

Gordon and I have recently obtained an approach to the Andrews hierarchies via the method of weighted words, and for this purpose we found Theorem C to be very useful (see a forthcoming paper of Alladi and Gordon [in preparation]). Consider the expansion of the product involving r free parameters $z_1, z_2, \ldots, z_r$,

$$(-z_1 q)_\infty (-z_2 q)_\infty \ldots (-z_r q)_\infty = \sum_{i_1, i_2, \ldots, i_r, n} V(n; i_1, i_2, \ldots, i_r) z_1^{i_1} z_2^{i_2} \ldots z_r^{i_r} q^n. \quad (7.5)$$

In (7.5), the z_t represent free parameters as well as colors. Thus $V(n; i_1, i_2, \ldots, i_r)$ is the number of vector partitions $(\pi_1, \pi_2, \ldots, \pi_r)$ of n, where π_t has t distinct z_t-parts. Under the transformations

$$\begin{cases} \text{(dilation) } q \mapsto q^N, \\ \text{(translations)} z_t \mapsto z_t q^{a_t - N}, \end{cases} \quad (7.6)$$

the product in (7.5) is the refined generating function of $P_1(A; N, n)$ in Theorem C.

The Andrews-Olsson gap conditions for $P_2(A;N,n)$ generalize as follows: Let $D(n;i_1,i_2,\ldots,i_r)$ denote the number of partitions of n into i_1 z_1-parts, i_2 z_2-parts, $\ldots$, i_r z_r-parts, and parts of no color (neutral parts) such that

(i) the smallest part is not a neutral part,

(ii) the gap between consecutive parts is ≤ 1 with strict inequality if the smaller part is of lower order (for a given weight, the order of the colors is *neutral* $< z_1 < z_2 < \ldots < z_r$),

(iii) only neutral parts can repeat.

Then the main result in Alladi and Gordon [in preparation] is

Theorem 15 $V(n;i_1,i_2,\ldots,i_r) = D(n;i_1,i_2,\ldots,i_r)$.

The parts of neutral color in Theorem 15 correspond to the parts $\equiv 0 (\text{mod } N)$ in Theorem C.

By means of what we call an *amalgamation process*, the generating function of $D(n;i_1,i_2,\ldots,i_r)$ can be shown to be

$$\sum_n D(n;i_2,i_2,\ldots,i_r)q^n = \frac{q^{T_{i_1}+T_{i_2}+\cdots+T_{i_r}}}{(q)_{i_1}(q)_{i_2}\cdots(q)_{i_r}}. \tag{7.7}$$

In establishing (7.7), minimal partitions are considered and their generating function is

$$q^{T_{i_1}+T_{i_2}+\cdots+T_{i_r}}\begin{bmatrix} i_1+i_2+\cdots+i_r \\ i_1,i_2,\ldots,i_r \end{bmatrix}. \tag{7.8}$$

In order to show that the generating function of the minimal partitions is the expression in (7.8), recurrences for the multinomial coefficients of order r are employed, as well as properties of inversions of permutations. Theorem 15 follows from (7.7) because the expression on the right in (7.7) is clearly the generating function of $V(n;i_1,i_2,\ldots,i_r)$.

The advantages in this approach are many. First, Theorem 15 is a refinement and generalization of Theorem C. Second, in this approach there is only one hierarchy of general results, and Theorems A_1 and A_2 emerge by using two different sets of transformations, having the same dilation but differing in translations. Also, the condition $b_\nu \neq 1,2,4$ or 7 in Corollary 2 (equivalently (7.4) in Theorem A_2) is easily explained. In Corollary 2, the residue classes 3.5, and 6(mod 7) correspond to primary colors. Since secondary colors start only with symbols of weight ≥ 2, the first integer in the residue class $3+5(\text{mod } 7)$ will be 8 and so $b_\nu \neq 1$. Similarly, $b_\nu \neq 2$ or 4 or 7. Finally, the difference in the gap conditions of (7.2) and (7.3) being expressed in terms of the lower and upper part is just a matter of convenience. In the case of Schur's Theorem, the gap conditions refining $S_1(n)$ could also be expressed in the form

$$b_i - b_{i+1} \geq 3, 3 \text{ or } 4 \text{ if } b_{i+1} \equiv 1, 2 \text{ or } 3(\text{mod } 3).$$

This is because the gap condition here is symmetric. If we try to express the difference conditions for $S_2(n)$ in Theorem A in terms of the lower part, we get

$$b_i - b_{i+1}\begin{cases} \geq 3, \text{ if } b_{i+1} \equiv 1(\text{mod } 3), \\ = 3 \text{ or } \geq 5, \text{ if } b_{i+1} \equiv 2(\text{mod } 3), \\ = 2 \text{ or } \geq 4, \text{ if } b_{i+1} \equiv 3(\text{mod } 3). \end{cases} \tag{7.9}$$

So, in this case it is more convenient to express the difference conditions for $S_2(n)$ in terms of the upper part b_i as in Theorem A. Notice that the conservation law holds in (7.9). Here we interpret the average of 3 and 5 to be 4, and the average of 2 and 4 to be 3. Therefore $3+4+3=10$ as in the conservation law. Actually, the best way to describe gap conditons is to embody them in a matrix form as done in the next section. One may read this matrix row–wise, in which case the gap conditions are given in terms of b_i, or one may read the matrix column–wise, in which case the difference conditions are given in terms of b_{i+1}. Thus one of the advantages in approaching the Andrews hierarchies by the method of weighted words is that it shows that there is essentially no difference between Theorems A_1 and A_2.

8 The Capparelli Conjectures

In recent years, a number of Rogers–Ramanujan type identities have been found to lie at the heart of the interaction between the vertex operators and affine Lie Algebras (see Lepowsky and Wilson [1981], [1982], [1984]). In the course of a study of standard modules of level 3 for $A_2^{(2)}$, Capparelli [1993] was led to the following two conjectures.

Conjecture 1 *Let $C^*(n)$ denote the number of partitions of n into parts $\equiv \pm 2, \pm 3 \pmod{12}$.*

Let $D(n)$ denote the number of partitions of n into parts >1 with difference ≥ 2, where the difference is ≥ 4 unless consecutive parts add up to a multiple of 3. Then

$$C^*(n) = D(n).$$

Conjecture 2 *Let $E^*(n)$ denote the coefficient of q^n in the expansion of*

$$\prod_{\substack{j>0 \\ j\equiv \pm 1, \pm 3, \pm 5 \pmod{12}}} (1-q^j)^{-1} \prod_{\substack{j>0 \\ j\equiv \pm 2 \pmod{12}}} (1-q^j).$$

Let $F(n)$ denote the number of partitions of n into parts $\neq 2$, and differing by ≥ 2, where the difference is ≥ 4 unless consecutive parts add up to a multiple of 3. Then

$$E^*(n) = F(n).$$

Conjecture 1 was proved by Andrews [1994] using generating functions. The proof of Conjecture 2 is implicit in the method in Andrews [1994]. We now describe briefly how to obtain refinements and generalizations of the Capparelli conjectures by the method of weighted words (see Alladi–Andrews–Gordon [1995b] for a complete treatment).

From the point of view of refinements, it is preferable to replace $C^*(n)$ by $C(n)$ which denotes the number of partitions of n into distinct parts $\equiv 2, 3, 4$ or $6 \pmod 6$. Clearly $C(n) = C^*(n)$ because,

$$\sum_{n=0}^{\infty} C(n) q^n = (-q^2;q^6)_\infty (-q^4;q^6)_\infty (-q^3;q^3)_\infty = \frac{(q^4;q^{12})_\infty (q^8;q^{12})_\infty (q^6;q^6)_\infty}{(q^2;q^6)_\infty (q^4;q^6)_\infty (q^3;q^3)_\infty}$$
$$= \frac{1}{(q^2;q^{12})_\infty (q^3,q^{12})_\infty (q^9;q^{12})_\infty (q^{10}q^{12})_\infty} = \sum_{n=0}^{\infty} C^*(n) q^n. \tag{8.1}$$

Using $C(n)$, the following three parameter refinement of Conjecture 1 was established in Alladi, Andrews and Gordon [1995b].

Theorem 16 *Let $C(n;i,j,k)$ denote the number of partitions of the type counted by $C(n)$ with the additional restriction that there are exactly i parts $\equiv 4 (\text{mod } 6)$, j parts $\equiv 2 (\text{mod } 6)$, and of those $\equiv 0 (\text{mod } 3)$, exactly k are $> 3(i+j)$.*

Let $D(n;i,j,k)$ denote the number partitions of n of the type counted by $D(n)$ with the additional restriction that there are exactly i parts $\equiv 1 (\text{mod } 3)$, j parts $\equiv 2 (\text{mod } 3)$ and k parts $\equiv 0 (\text{mod } 3)$. Then

$$C(n;i,j,k) = D(n;i,j,k).$$

Similiarly, to refine Conjecture 2, replace $E^*(n)$ by $E(n)$ which is the number of partitions of n into distinct parts $\equiv 1, 3, 5$ or $6 (\text{mod } 6)$. Clearly $E^*(n) = E(n)$ because

$$\begin{aligned} \sum_{n=0}^{\infty} E^*(n) q^n &= \frac{(q^2;q^{12})_\infty (q^{10};q^{12})_\infty}{(q;q^6)_\infty (q^3;q^6)_\infty (q^5;q^6)_\infty} = \frac{(-q;q^6)_\infty (-q^5;q^6)_\infty}{(q^3;q^6)_\infty} \\ &= (-q,q^6)_\infty (-q^5;q^6)_\infty (-q^3;q^3)_\infty = \sum_{n=0}^{\infty} E(n) q^n. \end{aligned} \tag{8.2}$$

There is then a perfectly analogous result to Theorem 16, namely,

Theorem 17 $E(n;i,j,k) = F(n;i,j,k)$.

Theorems 16 and 17 are special cases of the more general Theorem 18 stated below under suitable dilations and translations.

For the Capparelli problem we assume that the integer 1 occurs in colors a and c and that integers ≥ 2 occur in three colors a, b and c. As before a_j, b_j and c_j represent the integer j in colors a, b and c respectively. The ordering of the symbols that we choose is

$$a_1 < b_2 < c_1 < a_2 < b_3 < c_2 < a_3 < b_4 < c_3 \ldots . \tag{8.3}$$

Let $C(n;i,j,k)$ denote the number of vector partitions $(\pi_1;\pi_2;\pi_3)$ of n such that π_1 has distinct even $a-$parts, π_2 has distinct even $b-$parts and π_3 has distinct $c-$parts such that $\nu(\pi_1) = i$, $\nu(\pi_2) = j$ and the number of parts of π_3 which are $> (i+j)$ is k.

Next, let $D(n;i,j,k)$ denote the number of partitions into symbols a_j, b_j and c_j such that each part is $> a_1$ and the gap between the symbols is given by the following matrix:

	a	b	c
a	2	2	1
b	0	2	0
c	2	3	1.

We then have

Theorem 18 $C(n;i,j,k) = D(n;i,j,k)$.

The matrix above may be read row-wise or column-wise. If read row–wise it has the following interpretation. The entries in the first row mean the following. If an $a-$part has weight j, then the next larger part if it is an $a-$part has weight $\geq j+2$, if it is a $b-$part has weight $\geq j+2$ and if it is a $c-$part has weight $\geq j+1$. The interpretation for the other two rows is similar.

From the point of view of generating functions, Theorem 18 is a consequence of

Lemma 4
a) $\sum_{i,j,k,n} C(n;i,j,k)a^i b^j c^k q^n = \sum_{i,j} \frac{a^i b^j q^{2T_i+2T_j}(-q)_{i+j}(-cq^{i+j+1})_\infty}{(q^2;q^2)_i(q^2;q^2)_j}$,

b) $\sum_{i,j,k,n} D(n;i,j,k)a^i b^j c^k q^n = \sum_{i,j,k} \frac{a^i b^j c^k q^{2T_i+2T_j+T_k+(i+j)k}}{(q)_{i+j+k}} \begin{bmatrix} i+j+k \\ i+j,k \end{bmatrix}_q \begin{bmatrix} i+j \\ i,j \end{bmatrix}_{q^2}$.

Theorem 18 is equivalent to the statement that the expressions on the right in parts (a) and (b) of Lemma 4 are equal. To see this equality, sum the expression on the right in Lemma 4(b) over the variable k to get the expression on the right in Lemma 4(a).

While it is clear that the expression on the right in Lemma 4(a) is the generating function of $C(n;i,j,k)$, it is not obvious that the expression on the right in Lemma 4(b) is the generating function of $D(n;i,j,k)$. In order to prove Lemma 4(b), consider for a given i,j and k, the generating function $H(i,j,k)$ of all minimal partitions counted by $D(n;i,j,k)$. Since $i+j+k$ is fixed, it follows from Lemma 1 that Lemma 4(b) is equivalent to the statement

$$H(i,j,k) = q^{2T_i+2T_j+T_k+(i+j)k} \begin{bmatrix} i+j+k \\ i+j,k \end{bmatrix}_q \begin{bmatrix} i+j \\ i,j \end{bmatrix}_{q^2}. \tag{8.4}$$

Equation (8.4) can be established by induction on $i+j+k$, the length of the word, as shown in Alladi, Andrews and Gordon [1995b]. But this is not easy since (8.3) involves a product of Gaussian coefficients to bases q and q^2.

If $c=1$, the generating function in Lemma 4(a) becomes the product

$$(-q)_\infty(-aq^2;q^2)_\infty(-bq^2;q^2)_\infty. \tag{8.5}$$

Under the transformations

$$\begin{cases} \text{(dilation) } q \mapsto q^3, \\ \text{(translations) } a \mapsto aq^{-2}, b \mapsto bq^{-4}, c \mapsto c, \end{cases} \tag{8.6}$$

Theorem 18 implies Theorem 16. Similarly, under the transformations

$$\begin{cases} \text{(dilation) } q \mapsto q^3, \\ \text{(translations) } a \mapsto aq^{-1}, b \mapsto bq^{-5}, c \mapsto c, \end{cases} \tag{8.7}$$

Theorem 18 yields Theorem 17.

Under the transformations (8.6) the symbols become

$$a_j \mapsto 3j-2, b_j \mapsto 3j-4, c_j \mapsto 3j,$$

and so the ordering (8.3) becomes

$$1 < 2 < 3 < 4 < 5 < 6 \ldots,$$

the natural ordering of the integers. So Capparelli's first conjecture corresponds to this ordering. Under transformations (8.7), the symbols become

$$a_j \mapsto 3j-1, b_j \mapsto 3j-5, c_j \mapsto 3j,$$

and so the ordering (8.3) yields

$$2<1<3<5<4<6<8<7<9\ldots. \tag{8.8}$$

So Capparelli's second conjecture corresponds to the ordering (8.8) among the integers.

In discussing Schur's theorem we made use of multinomial coefficients of order $t=3$. These coefficients satisfy the relation

$$\begin{bmatrix} i+j+k \\ i,j,k \end{bmatrix}_q = \begin{bmatrix} i+j+k \\ i+j,k \end{bmatrix}_q \begin{bmatrix} i+j \\ i,j \end{bmatrix}_q, \tag{8.9}$$

both to base q. The main difference now is that for the Capparelli conjectures we have made use of the product

$$\begin{bmatrix} i+j+k \\ i,j,k \end{bmatrix}_q \begin{bmatrix} i+j \\ i,j \end{bmatrix}_{q^2}. \tag{8.10}$$

It is possible to give a combinatorial proof of Theorem 18 by utilizing a variation in the combinatorial proof of Schur's theorem given in Section 3 (see Alladi, Andrews and Gordon [1995b]).

Recently two Lie theoretic proofs of the Capparelli conjectures have been announced Capparelli [preprint], Tamba and Xie [preprint]. But the refinements we have obtained are yet to be realized via Lie algebras.

9 Lebesgue's identity

Although (8.9) and (8.10) show the connection between the Capparelli and Schur theorems, the Capparelli theorems should really be viewed as extensions of the well known Lebesgue identity (see Andrews [1976], Ch. 2)

$$\sum_{n=0}^{\infty} \frac{q^{n(n+1)/2}(-bq)_n}{(q)_n} = (-bq^2;q^2)_\infty(-q)_\infty. \tag{9.1}$$

Indeed taking $a=0$ and $c=1$ in Lemma 4(a) leads to (9.1). Compare (9.1) with (8.5). It is more elegant to write (9.1) in the form

$$\sum_{n=0}^{\infty} \frac{q^{n(n+1)/2}(-bq)_n}{(q)_n} = \frac{(-bq^2;q^2)_\infty}{(q;q^2)_\infty} \tag{9.2}$$

because (9.2) with $b=0$ reduces to Euler's famous theorem on the equality of partitions into odd parts and distinct parts.

In Lemma 4 the best dilation was $q \mapsto q^3$ and associated to this were two translations as given in (8.6) and (8.7). Similarly, $q \mapsto q^2$ is the ideal dilation in (9.1) and there are two interesting translations associated with this. More precisely, with

$$(\text{dilation })q \mapsto q^2 \text{ and (translation) } b \mapsto bq^{-3} \tag{9.3}$$

in (9.1) we get

$$\sum_{n=0}^{\infty} \frac{q^{n^2+n}(-bq^{-1};q^2)_n}{(q^2;q^2)_n} = (-bq;q^4)_\infty(-q^2;q^4)_\infty(-q^4;q^4)_\infty. \tag{9.4}$$

Similarly, with

$$\text{(dilation) } q \mapsto q^2 \text{ and (translation) } b \mapsto bq^{-1} \tag{9.5}$$

in (9.1), we obtain

$$\sum_{n=0}^{\infty} \frac{q^{n^2+n}(-bq;q^2)_n}{(q^2;q^2)_n} = (-bq^3;q^4)_\infty(-q^2;q^4)_\infty(-q^4;q^4)_\infty. \tag{9.6}$$

Identities (9.4) and (9.6) have nice combinatorial interpretations as was noticed by Göllnitz [1967].

Theorem G_2 *For $i = 1, 3$, let $A_i(n;k)$ denote the number of partitions of n into distinct parts $\equiv i, 2$ or $4 \pmod 4$ such that there are k parts $\equiv i \pmod 4$.*

For $i = 1, 3$, let $B_i(n;k)$ denote the number of partitions of n into parts differing by ≥ 2 and with no consecutive odd integers as parts such that each part is $\geq i$ and the number of the odd parts is k. Then

$$A_i(n;k) = B_i(n;k).$$

Bressoud [1979] has given a combinatorial proof of Theorem G_2 and the proof of Theorem 1 in Section 3 utilizes an extension of this method of Bressoud. Theorem G_2 of Göllnitz is well known and not at all as complicated to prove as Theorem G. In order to make a distinction between the two results we have referred to Theorem G as Göllnitz' (Big) theorem.

If we take $b = 1$ in (9.4) and (9.6), then the products mod 4 get converted to products mod 8, and these have obvious partition interpretations:

$$\sum_{n=0}^{\infty} \frac{q^{n^2+n}(-q^{-1};q^2)_n}{(q^2;q^2)_n} = \frac{1}{(q;q^8)_\infty(q^5;q^8)_\infty(q^6;q^8)_\infty}, \tag{9.7}$$

and

$$\sum_{n=0}^{\infty} \frac{q^{n^2+n}(-q;q^2)_n}{(q^2;q^2)_n} = \frac{1}{(q^2;q^8)_\infty(q^3;q^8)_\infty(q^7;q^8)_\infty}. \tag{9.8}$$

Identities (9.7) and (9.8) are valid only in the unrefined form.

Very closely related to (9.7) and (9.8) are the two identities

$$\sum_{n=0}^{\infty} \frac{q^{n^2}(-q;q^2)_n}{(q^2;q^2)_n} = \frac{1}{(q;q^8)_\infty(q^4;q^8)_\infty(q^7;q^8)_\infty}, \tag{9.9}$$

and

$$\sum_{n=0}^{\infty} \frac{q^{n^2+2n}(-q;q^2)_n}{(q^2;q^2)_n} = \frac{1}{(q^3;q^8)_\infty(q^4;q^8)_\infty(q^5;q^8)_\infty}. \tag{9.10}$$

Identities (9.9) and (9.10) are (36) and (34) in Slater's [1952] list but it was Göllnitz [1967] and Gordon [1965] who independently realized that they had elegant partition interpretations.

Theorem $G-G$ *For $i = 1, 3$, the number of partitions of n into parts differing by ≥ 2 and no consecutive even numbers as parts and each part $\geq i$ equals the number of partitions of n into parts $\equiv \pm i$ or 4 (mod 8).*

Identities (9.9) and (9.10) are known as the Göllnitz–Gordon identities and they are the perfect analogues to the modulus 8 to what the Rogers–Ramanujan identities are to the modulus 5. No refinement of the Göllnitz–Gordon identities is known and so at this point we do not have a bijective proof of (9.9) and (9.10).

Since (9.9) and (9.10) are so similar to (9.7) and (9.8) which are descended from Lebesgue's identity, it is natural to ask whether it is possible to derive the Göllnitz–Gordon identities from (9.1). Recently (see Alladi [1995]) we found a new and simple proof of (9.9) and (9.10) by taking $b = -1$ in (9.1) and splitting the resulting series into its odd and even parts.

In Alladi and Gordon [1993a] a combinatorial proof of the Lebesgue identity is given. As a consequence it is shown that (9.2) is equivalent to the following partition result:

Theorem 19 *Let $g(n; k)$ denote the number of partitions of n in which even parts do not repeat such that there are exactly k even parts.*

Let $g_1(n; k)$ denote the number of partitions of n into distinct parts $b_1, b_2, \ldots, b_\nu$ such that there are exactly k gaps ≥ 2 between $b_1, b_2, \ldots, b_\nu$ and 0. Then

$$\sum_k g(n; k) b^k = \sum_k g_1(n; k)(1 + b)^k.$$

In particular taking $b = 1$ in Theorem 19 we get

Theorem 20 *Let $g(n)$ denote the number of partitions of n in which even parts do not repeat. Then*

$$g(n) = \sum_k g_1(n; k) 2^k.$$

Note the striking similarity between Theorems 20 and 12. Theorem 20 suggests the result that for any integer $k \geq 1$, the congruence

$$g(n) \equiv 0 (\text{mod } 2^k) \tag{9.11}$$

is valid for almost all n and offers an elementary approach to (9.11).

In order to decide when $g(n)$ is odd, it suffices to determine when $g_1(n; 0)$ is odd. Now $g_1(n; 0)$ enumerates partitions $b_1 + b_2 + \ldots b_\nu$ of n into distinct parts such that there are no gaps ≥ 2 among $b_1, b_2, \ldots, b_\nu$ and 0. The only such partition is $1 + 2 + \cdots + \nu$ of the integer $n = \nu(\nu + 1)/2$. So we conclude that

$$g(n) \neq 0 (\text{mod } 2) \Leftrightarrow n = \frac{\nu(\nu + 1)}{2}$$

is a triangular number. Next, to determine when $g(n) \not\equiv 0$ (mod 4) we need to investigate $g_1(n; 1)$. By analyzing this it can be shown that

$$g(n) \equiv 0 (\text{mod } 4) \text{ unless } n = \frac{\nu(\nu + 1)}{2} + \mu^2, \tag{9.12}$$

equals a triangular number plus a perfect square. The exceptional set in (9.12) is even thinner and can be determined precisely. The triangular numbers play a role in the study of the congruence (9.11) analogous to the role of the pentagonal numbers in the study of (6.1). In a forthcoming paper (see Alladi [in preparation]), Theorem 20 will be utilized to study (9.11) for small values of k.

10 Further problems

What lies beyond Göllnitz' theorem? In this direction we wish to investigate the situation involving r primary colors for $r > 3$ on the one hand, and a *proper subset* of the $2^r - 1$ colors that can be formed using these primary colors. Even the case $r = 4$ is difficult. For example, in considering expansions of

$$(-aq)_\infty(-bq)_\infty(-cq)_\infty(-dq)_\infty$$

involving four primary colors a, b, c and d, it is unclear whether in addition to the quaternary color $abcd$, any of the ternary or secondary colors should be dropped. Taking clue from the work of Andrews and Olsson [1991], Gordon and I are approaching this difficult problem through conjugation of Ferrers graphs.

Section 6 dealt with applications of weighted identities obtained by partial dilation on the key identity underlying Göllnitz' theorem. Now that the Andrews hierarchies of theorems are amenable by the method or weighted words, it might be worthwhile to see whether interesting weighted identities can be constructed by using the partial dilation

$$q \mapsto q^M, \text{ with } M < 2^r - 1$$

on the basic identities underlying Theorem 15.

Acknowledgements

I wish to thank George Andrews for suggesting my name to the organizers of The Fields Institute Conference on Special Functions, q-Series and Related Topics. I would also like to thank the Director of The Fields Institute and the organizers of the conference for the kind invitation and warm hospitality.

References

Alladi, K. [1995] *Some new observations on the Göllnitz–Gordon and Rogers–Ramanujan identities* Trans. A.M.S., **347**, 897-914.

Alladi, K. [1997a] *Partition identities involving gaps and weights* Trans. A.M.S, **347** (to appear).

Alladi, K. [1997b] *A combinatorial correspondence related to Göllnitz' (Big) partition theorem and applications* Trans. A.M.S, **347** (to appear).

Alladi, K. [in preparation] *On the divisibility of certain partition functions by small powers of 2.*

Alladi, K. and Andrews, G. E. [preprint] *A new key identity for Göllnitz' (Big) partition theorem.*

Alladi, K., Andrews, G. E. and Gordon, B. [1995a] *Generalizations and refinements of a partition theorem of Göllnitz* J. Reine Angew. Math., **460**, 165-188.

Alladi, K., Andrews, G. E. and Gordon, B. [1995b] *Refinements and generalizations of Capparelli's conjecture on partitions* J. Algebra, **174**, 636-658.

Alladi, K. and Gordon, B. [1993a] *Partition identities and a continued fraction of Ramanujan* J. Comb. Th. Ser. A, **63**, 275-300.

Alladi, K. and Gordon, B. [1993b] *Generalizations of Schur's partition theorem* Manus. Math., **79**, 113-126.

Alladi, K. and Gordon, B. [1995] *Schur's partition theorem, companions, refinements and generalizations* Trans. A.M.S., **347**, 1591-1608.

Alladi, K. and Gordon, B. [in preparation] *A hierarchy of general partition theorems via the method of weighted words.*

Andrews, G. E. [1967] *On Schur's second partition theorem* Glasgow Math. J., **9**, 127-132.

Andrews, G. E. [1968] *A new generalization of Schur's second partition theorem* Acta Arith., **14**, 429-434.

Andrews, G. E. [1969a] *A general partition theorem with difference conditions* Amer. J. Math., **91**, 18-24.

Andrews, G. E. [1969b] *On a partition theorem of Göllnitz and related formulae* J. Reine Angew. Math., **236**, 37-42.

Andrews, G. E. [1971] *The use of computers in search of identities of the Rogers–Ramanujan type*, Computers in Number Theory (A. O. L. Atkin and B. J. Birch Eds.), Academic Press, New York, pp. 377-387.

Andrews, G. E. [1976] *The theory of partitions* Encyclopedia of Math., **2**, Addison–Wesley, Reading.

Andrews, G. E. [1985] *q−series: Their development and application in Analysis, Number Theory, Combinatorics, Physics and Computer Algebra*, NSF-CBMS Lectures, vol. 66, Amer. Math. Soc., Providence, RI.

Andrews, G. E. [1994] *Schur's theorem, Capparelli's conjecture and the q−trinomial coefficients*, Proc. Rademacher Centenary Conf. 1992, Contemp. Math., vol. 66, pp. 141-154.

Andrews, G. E. and Olsson, J. B. [1991] *Partition identities with an application to group representation theory* J. Reine Angew. Math., **413**, 198-212.

Bessenrodt, C. [1991] *A combinatorial proof of the Andrews–Olsson partition identity* Europ. J. Combinatorics, **12**, 271-276.

Bressoud, D. M. [1980] *Combinatorial proof of Schur's 1926 partition theorem* Proc. A.M.S., **19**, 338-340.

Bressoud, D. M. [1979] *On a partition theorem of Göllnitz* J. Reine Angew. Math., **305**, 215-217.

Bressoud, D. M. and Zeilberger, D. [1982] *A short Rogers-Ramanujan bijection* Discrete Math., **38**, 313-315.

Capparelli, S. [1993] *On some representations of twisted affine Lie algebras and combinatorial identities* J. Algebra, **154**, 335-355.

Capparelli, S. [preprint] *A construction of level 3 modules for the affine algebra* $A_2^{(2)}$ *and a new combinatorial identity of the Rogers–Ramanujan type.*

Garcia, A. M. and Milne, S. C. [1981] *A Rogers–Ramanujan bijection* J. Comb. Th. Ser A, **31**, 289-339.

Gasper, G. and Rahman, M. [1990] *Basic hyper-geometric series*, Encyclopedia of Math, vol. 35, Cambridge University Press, New York.

Gleissberg, W. [1928] *Über einen Satz von Herrn. I. Schur* Math. Zeitschrift, **28**, 372-382.

Göllnitz, H. [1967] *Partitionen mit Differenzenbedingungen,* J. Reine Angew. Math., **225**, 154-190.

Gordon, B. [1965] *Some continued fractions of the Rogers–Ramanujan type* Duke Math. J., **32**, 741-748.

Gordon, B. and Ono, K. [1997] *On the divisibility of certain partition functions by powers of primes* The Ramanujan Journal, **1**, 25-34.

Lepowsky, J. and Wilson, R. L. [1981] *A new family of algebras underlying the Rogers–Ramanujan identities and generalizations* Proc. Nat'l. Acad. Sci. U.S.A., **78**, 7254-7258.

Lepowsky, J. and Wilson, R. L. [1982] *A Lie-theoretic interpretation and proof of the Rogers–Ramanujan identities* Adv. Math., **45**, 21-72.

Lepowsky, J. and Wilson, R. L. [1984] *The Structure of Standard modules I, Universal algebras and the Rogers–Ramanujan identities* Invent. Math., **77**, 199-290.

Ramanujan, S. [1919] *Proof of certain identities in combinatory analysis* Proc. Cambridge Phil. Soc., **19**, 214-216.

Rogers, L. J. [1894] *Second memoir on the expansion of certain infinite products* Proc. London Math. Soc., **25**, 318-343.

Slater, L. J. [1952] *Further identities of the Rogers–Ramanujan type* Proc. London Math. Soc (2), **54**, 147-167.

Schur, I. J. [1973] *Zur Additiven Zahentheorie: Gesammelte Abhandlungen*, vol. 2, Springer, Berlin, pp. 43-50.

Sylvester, I. J. [1882] *A constructive theory of partitions arranged in three Acts, an Interact and an Exodion* Amer. J. Math., **5**, 251-330.

Tamba, M. and Xie, C. [preprint] *Level three standard modules for* $A_2^{(2)}$ *and combinatorial identities.*

Fields Institute Communications
Volume **14**, 1997

Ramanujan's Class Invariants With Applications to the Values of q-Continued Fractions and Theta Functions

Bruce C. Berndt
Department of Mathematics
University of Illinois
1409 West Green Street
Urbana, Illinois 61801 USA

Heng Huat Chan
School of Mathematics
Institute for Advanced Study
Princeton, NJ 08540 USA

Liang-Cheng Zhang
Department of Mathematics
Southwest Missouri State University
Springfield, MO 65804 USA

Abstract. A survey of Ramanujan's unpublished work on class invariants, singular moduli, values of theta functions, the Rogers-Ramanujan continued fraction, and Ramanujan's cubic continued fraction is given.

1 Introduction

As usual, set

$$(a;q)_\infty = \prod_{n=0}^{\infty}(1-aq^n), \qquad |q|<1,$$

and, following Ramanujan, let

$$\chi(q) = (-q;q^2)_\infty.$$

If n is any postitive rational number and $q = \exp(-\pi\sqrt{n})$, the two *class invariants* G_n and g_n are defined by

$$G_n := 2^{-1/4}q^{-1/24}\chi(q) \qquad \text{and} \qquad g_n := 2^{-1/4}q^{-1/24}\chi(-q). \tag{1.1}$$

1991 *Mathematics Subject Classification.* Primary: 33D10; Secondary: 33E05, 40A15, 11R04.

In the notation of H. Weber [43], $G_n =: 2^{-1/4}f(\sqrt{-n})$ and $g_n =: 2^{-1/4}f_1(\sqrt{-n})$. The term "invariant" is due to Weber. If $\mathbb{Q}(\omega)$ is the algebraic number field generated by the complex quadratic integer ω, such that $\{1,\omega\}$ is a basis for the algebraic integral domain, which is also called the maximal order of $\mathbb{Q}(\omega)$, then the absolute class field of $\mathbb{Q}(\omega)$ is generated by the modular invariant $j(\omega)$, which Weber calls a class invariant. More generally, if $\mathfrak{O}_f := [1, f\omega]$, $f \in \mathbb{N}$, is an order of $\mathbb{Q}(\omega)$, then the ring class field of $\mathfrak{O}_f$ is generated by $j(\mathfrak{O}_f)$. Often some power of $f(\tau)$, or $f_1(\tau)$, with $\tau = f\omega$, is in $\mathbb{Q}(\omega, j(\mathfrak{O}_f))$, and, in such cases, Weber calls $f(\tau)$, or $f_1(\tau)$, an invariant as well. As G. N. Watson [37] remarked, "For reasons which had commended themselves to Weber and Ramanujan independently, it is customary to determine G_n for odd values of n, and g_n for even values of n."

If, as usual in the theory of elliptic functions, $k = k(q)$ denotes the modulus, then the singular modulus k_n is defined by $k_n = k(e^{-\pi\sqrt{n}})$, where n is a positive rational integer. Following Ramanujan, set $\alpha = k^2$ and $\alpha_n = k_n^2$. Since [3, p. 124]

$$\chi(q) = 2^{1/6}\{\alpha(1-\alpha)/q\}^{-1/24} \qquad \text{and} \qquad \chi(-q) = 2^{1/6}(1-\alpha)^{1/12}(\alpha/q)^{-1/24},$$

it follows from (1.1) that

$$G_n = \{4\alpha_n(1-\alpha_n)\}^{-1/24} \qquad \text{and} \qquad g_n = 2^{-1/12}(1-\alpha_n)^{1/12}\alpha_n^{-1/24}. \tag{1.2}$$

It is well–known that G_n and g_n are algebraic; for example, see Cox's book [18, p. 214, Theorem 10.23; p. 257, Theorem 12.17]. It follows trivially from (1.2) that α_n is algebraic. However, much more is known [17].

Theorem 1.1
(a) If $n \equiv 1 \pmod 4$, *then* G_n *and* $2\alpha_n$ *are units.*
(b) If $n \equiv 3 \pmod 8$, *then* $2^{-1/12}G_n$ *and* $2^2\alpha_n$ *are units.*
(c) If $n \equiv 7 \pmod 8$, *then* $2^{-1/4}G_n$ *and* $2^4\alpha_n$ *are units.*
(d) If $n \equiv 2 \pmod 4$, *then* g_n *and* α_n *are units.*

Section 2 will be devoted to describing Ramanujan's determinations of G_n and g_n and recent efforts to establish Ramanujan's asserted values. We also offer a few remarks about the determination of α_n.

The celebrated Rogers–Ramanujan continued fractions $R(q)$ and $S(q)$ are defined by

$$R(q) = \frac{q^{1/5}}{1} \, \frac{q}{+1} \, \frac{q^2}{+1} \, \frac{q^3}{+1} + \cdots, \qquad |q| < 1, \tag{1.3}$$

and

$$S(q) = -R(-q).$$

As intimated, (1.3) was first introduced by L. J. Rogers [30] in 1894. However, most of our knowledge about $R(q)$ arises from Ramanujan who communicated several of its properties in his first and second letters to Hardy [28, pp. xxvii, xxviii]. Several of Ramanujan's theorems on $R(q)$ and $S(q)$ can also be found in his notebooks [27] and lost notebook [29].

Ramanujan studied several other q–continued fractions. In particular, in his second letter to Hardy [28, p. xxviii], Ramanujan offered two theorems on the "cubic" continued fraction

$$G(q) := \frac{q^{1/3}}{1} + \frac{q+q^2}{1} + \frac{q^2+q^4}{1} + \frac{q^3+q^6}{1} + \cdots. \tag{1.4}$$

In Section 3, we focus on explicit values for $R(q)$ and $G(q)$. Class invariants and modular equations are central to our determinations.

At the heart of the theorems on class invariants and q–continued fractions are theta functions. In Ramanujan's notation, set, for $|q| < 1$,

$$\varphi(q) := \sum_{n=-\infty}^{\infty} q^{n^2}, \qquad \psi(q) := \sum_{n=0}^{\infty} q^{n(n+1)/2}, \tag{1.5}$$

and

$$f(-q) = \sum_{n=-\infty}^{\infty} (-1)^n q^{n(3n-1)/2} = (q;q)_\infty. \tag{1.6}$$

In more standard notation, $\varphi(q) = \vartheta_3(0,z), 2q^{1/8}\psi(q) = \vartheta_2(0,z/2)$, and $q^{1/24}f(-q) = \eta(z/2)$, where $q = \exp(\pi i z)$, ϑ_2 and ϑ_3 are classical theta functions [44, p. 464], and $\eta(z)$ denotes the Dedekind eta–function.

Section 4 is devoted to explicit determinations of $\varphi(q)$ and certain products of the functions φ and ψ. The determination of $\varphi(q)$ is equivalent to evaluating the complete elliptic integral of the first kind $K(k)$ and the ordinary hypergeometric series ${}_2F_1(\frac{1}{2}, \frac{1}{2}; 1; k^2)$, as seen in the next paragraph. The theta products of Ramanujan which we examine have remarkable properties.

Because modular equations will be frequently mentioned in the sequel, we conclude this introduction by giving a precise definition of a modular equation, as understood by Ramanujan. Let K, K', L, and L' denote complete elliptic integrals of the first kind associated with the moduli k, $k' := \sqrt{1-k^2}$, ℓ, and $\ell' := \sqrt{1-\ell^2}$, respectively, where $0 < k, \ell < 1$. Suppose that

$$n\frac{K'}{K} = \frac{L'}{L} \tag{1.7}$$

for some positive rational integer n. A relation between k and ℓ induced by (1.7) is called a *modular equation of degree n*. Following Ramanujan, set

$$\alpha = k^2 \qquad \text{and} \qquad \beta = \ell^2.$$

We often say that β has degree n over α. If

$$q = \exp(-\pi K'/K), \tag{1.8}$$

one of the most fundamental relations in the theory of elliptic functions is given by the formula [3, pp. 101, 102]

$$\varphi^2(q) = {}_2F_1(\tfrac{1}{2}, \tfrac{1}{2}; 1; k^2) = \frac{2}{\pi}\int_0^{\pi/2} \frac{d\theta}{\sqrt{1-k^2\sin^2\theta}} =: \frac{2}{\pi}K(k). \tag{1.9}$$

The first equality in (1.9) and elementary theta function identities make it possible to write each modular equation as a theta function identity [3, pp. 122–124]. (The second equality in (1.9) arises from expanding the integrand in a binomial series and integrating termwise.) Lastly, the multiplier m of degree n is defined by

$$m = \frac{\varphi^2(q)}{\varphi^2(q^n)}. \tag{1.10}$$

2 Class Invariants

Weber primarily was motivated to calculate class invariants so that he could construct Hilbert class fields. The table at the end of Weber's book [43, pp. 721–726] contains the values of 105 class invariants, although in some cases he only gives the irreducible polynomials satisfied by G_n or g_n.

Ramanujan apparently had no knowledge of class field theory and independently calculated class invariants for different reasons. Scattered throughout his first notebook [27] are the values of 106 class invariants. On pages 294–299 in his second notebook [27], Ramanujan gives a table of 77 class invariants, three of which are not found in the first notebook. Since the second notebook is an enlarged edition of the first, it is unclear why Ramanujan failed to record 32 class invariants that he offered in the first notebook. While in England, Ramanujan learned of Weber's work, and so his table of 46 class invariants in his paper [26], [28, pp. 23–39] does not contain any that are found in Weber's book [43]. Except for G_{325} and G_{363}, all of the remaining values are found in Ramanujan's notebooks.

The value of G_{1353} was communicated by Ramanujan [28, p. xxix, eq. (23)] in his second letter to Hardy and was first proved by Watson [36]. In a letter of October 1, 1930 written to B. M. Wilson [13, pp. 237, 238], Watson confided, "... but 23 which deals with the singular modulus associated with 1353 is included; I was pleased at getting this out, because the bulk of the singular moduli in the Notebooks can be obtained in the same way; I have not yet written out any of the latter properly, but have worked at many of them enough to make the formal writing out an easy matter. You will be interested to hear how Ramanujan got no. 23, particularly when you look at the length of the answer. I am absolutely convinced that he guessed it; I get out the others mentioned above by the same process of guessing."

In the first of two papers [37] devoted to proving Ramanujan's class invariants, Watson employed this "empirical process" to calculate fourteen of Ramanujan's invariants and opined, "I believe that fourteen were obtained by Ramanujan by means of the empirical process which I described in the discussion of G_{1353}." We emphasize that Watson's process is not rigorous. Watson numerically calculated the values of G_n and certain related invariants. He then numerically calculated certain polynomials of products and quotients of these invariants. Next, he found integers in appropriate real quadratic fields that numerically agreed with these polynomials. Then Watson *assumed* that, in fact, they are equal and so solved for G_n. It seems to us that determining these polynomials, the appropriate real quadratic field, and the algebraic integers that approximate them is extremely difficult without knowing beforehand, as Watson did, the requisite number fields and the appropriate algebraic integers. Thus, we are convinced, as much as Watson was to the contrary, that his "empirical process" was definitely not the method employed by Ramanujan in determining the values of class invariants.

In his second paper [38] devoted to calculating Ramanujan's class invariants, Watson utilized modular equations. Here Watson's methods are likely to be close to those of Ramanujan. However, the method is applicable only when n is a square, a small prime times a square, or twice a certain prime, and so can be applied in only a limited number of cases.

Watson wrote four further papers [39]–[42] on the calculation of class invariants. Among the dozens of invariants calculated by Watson in these papers were three previously unproved invariants found in Ramanujan's paper [26], namely, for $n =$ 81, 147, 289, as well as eleven invariants of Ramanujan that had been previously verified.

Thus, after Watson's papers, a total of eighteen invariants of Ramanujan remained to be verified. In the introduction to [37], Watson remarked, "It is intended to publish the calculations involved in the construction of the set N + Q (the invariants appearing in both Ramanujan's paper [26] and the second notebook) as part of the commentary on the note–books by Dr. B. M. Wilson and myself." Although Watson and Wilson's efforts to edit Ramanujan's notebooks have been preserved in the library at Trinity College, Cambridge, Watson's work on these eighteen invariants cannot be found there. Thus, if Watson actually calculated these invariants, it appears that his work has been lost.

Proofs of Ramanujan's remaining eighteen class invariants recently have been given for the first time by the authors [8], [9]. This work will be briefly described.

Five of the eighteen values of n are divisible by 9, namely, $n =$ 117, 153, 441, 90, and 198. Our starting point is a relation connecting g_n and g_{9n} found on page 318 of Ramanujan's first notebook [27], but not in his second notebook. K. G. Ramanathan [23], [24] noticed this relation, but apparently he never gave a proof. We [8] have proved this formula as well as a companion formula relating G_n and G_{9n}. Variants of these two formulas were previously found by J. M. and P. B. Borwein [14, pp. 145, 149]. They also derived formulas connecting G_{81n} with G_n and G_{9n}, and g_{81n} with g_n and g_{9n}.

We state below our theorem relating G_n and G_{9n}.

Theorem 2.1 *Let*

$$p = G_n^4 + G_n^{-4}. \tag{2.1}$$

Then

$$G_{9n} = G_n \left(p + \sqrt{p^2 - 1}\right)^{1/6}$$

$$\times \left\{\sqrt{\frac{p^2 - 2 + \sqrt{(p^2-1)(p^2-4)}}{2}} + \sqrt{\frac{p^2 - 4 + \sqrt{(p^2-1)(p^2-4)}}{2}}\right\}^{1/3}. \tag{2.2}$$

Our proof relies on a modular equation of degree 3 found as Entry 5(xii) in Chapter 19 of Ramanujan's second notebook [3, p. 231].

Lemma 2.2 (Modular Equation of degree 3) *Let*

$$P = \{16\alpha\beta(1-\alpha)(1-\beta)\}^{1/8} \qquad \textit{and} \qquad Q = \left(\frac{\beta(1-\beta)}{\alpha(1-\alpha)}\right)^{1/4}. \tag{2.3}$$

Then

$$Q + \frac{1}{Q} + 2\sqrt{2}\left(P - \frac{1}{P}\right) = 0. \tag{2.4}$$

Lemma 2.2 is useful in the theory of class invariants, because, by (1.2), the expressions P and Q in (2.3) can be expressed in terms of class invariants.

As an illustration, we briefly indicate how to prove that

$$G_{117} = \frac{1}{2}\left(\frac{3+\sqrt{13}}{2}\right)^{1/4}(2\sqrt{3}+\sqrt{13})^{1/6}\left(3^{1/4}+\sqrt{4+\sqrt{3}}\right).$$

From Weber's treatise [43, p. 721], or Ramanujan's paper [26],

$$G_{13} = \left(\frac{3+\sqrt{13}}{2}\right)^{1/4}.$$

It follows easily from (2.1) that $p = \sqrt{13}$. Thus, by Theorem 2.1,

$$G_{117} = \left(\frac{3+\sqrt{13}}{2}\right)^{1/4}(2\sqrt{3}+\sqrt{13})^{1/6}\left\{\sqrt{\frac{11+6\sqrt{3}}{2}}+\sqrt{\frac{9+6\sqrt{3}}{2}}\right\}^{1/3}.$$

It therefore remains to show that

$$\left\{\sqrt{\frac{11+6\sqrt{3}}{2}}+\sqrt{\frac{9+6\sqrt{3}}{2}}\right\}^{1/3} = \frac{1}{2}\left(3^{1/4}+\sqrt{4+\sqrt{3}}\right),$$

and we refer to [8] for details.

We remark that Watson [37] computed G_{333}, G_{765}, g_{522}, and g_{630} by his "empirical process," but easier proofs can be constructed by using Theorem 2.1 and its analogue for g_n.

The remaining thirteen values of G_n are for $n = 65$, 69, 77, 141, 145, 205, 213, 217, 265, 301, 445, 505, and 553. Note that each value of n has the form pq, where p is a "small" prime (3, 5, or 7), and q is a "large" prime. Quite astonishingly, the class number for each of these thirteen imaginary quadratic fields $\mathbb{Q}(\sqrt{-n})$ equals 8. Moreover, there are precisely two classes per genus in each case. It was considerably more difficult to prove these thirteen values for G_n than it was for the five previously described values.

We devised three approaches [9]. The first employs Kronecker's limit formula and extends a method due to C.L. Siegel and Ramanathan [20], [24]. This method is perhaps the most successful, since it can be utilized to establish all thirteen invariants. Furthermore, Ramanathan had used a simpler theorem based on Kronecker's limit formula to calculate several invariants, and Zhang [45] has developed further theorems of this type to give rigorous proofs of some of the class invariants unrigorously established by Watson [37]. Ramanujan was unfamiliar with Kronecker's limit formula and the additionally needed concepts of ideal classes and their characters, L–series of an algebraic number field, genus theory and characters, class numbers, and units, and so clearly this approach is not close to any of the methods that Ramanujan might have used.

The second method employs class field theory to make Watson's empirical method rigorous in the cases at hand. This procedure can also be used to calculate all thirteen invariants. Besides being rigorous, our version is superior to that of Watson because there is no need to know the value of the invariants in advance. In fact, a couple of new invariants were calculated by using this method. However, Ramanujan was unfamiliar with class field theory, and so this was clearly not the mode employed by Ramanujan.

Our third method employs modular equations. However, Watson's methods [38] using modular equations are inapplicable, because the values of n under consideration here are not of the forms described above, and so a new approach had to be devised. In particular, we used the following general principle. Let p and r denote coprime, positive integers. Set

$$q = \exp(-\pi\sqrt{p/r}) \qquad \text{and} \qquad q' = \exp(-\pi\sqrt{pr}), \tag{2.5}$$

and let β have degree r over α. Then, by (1.2),

$$G_{p/r} = \{4\alpha(1-\alpha)\}^{-1/24} \qquad \text{and} \qquad G_{pr} = \{4\beta(1-\beta)\}^{-1/24}. \tag{2.6}$$

Then, by using (1.7), (1.8), and (2.5), we can conclude that

$$\textit{If } \beta \textit{ has degree } r \textit{ over } \alpha, \textit{ then } \beta \textit{ has degree } p \textit{ over } 1-\alpha. \tag{2.7}$$

(See [9] for more details.) We emphasize that (2.7) is not a statement about modular equations in general; (2.7) is only valid when q and q' have the special values given in (2.5). Unfortunately, this third approach can only be utilized to establish Ramanujan's class invariants for $n = 65$, 69, 77, 141, 145, and 213. To establish the remaining seven invariants by employing modular equations, we would need modular equations of degrees equal to the larger of the two prime factors of n in each case, namely, for degrees 31, 41, 43, 53, 79, 89, and 101. Apparently, only for degree 31 did Ramanujan derive a modular equation, for he recorded no modular equations for the remaining six degrees in his notebooks. An attempt to calculate G_{217} using this modular equation of degree 31 generated horrible equations. Thus, Ramanujan's methods remain elusive, since he somehow must have circumvented the need for modular equations of high degree in at least these seven instances.

To give readers some idea of how modular equations are employed in determining class invariants, we give a brief sketch of a proof of Ramanujan's value

$$G_{69} = \left(\frac{5+\sqrt{23}}{\sqrt{2}}\right)^{1/12} \left(\frac{3\sqrt{3}+\sqrt{23}}{2}\right)^{1/8} \left(\sqrt{\frac{6+3\sqrt{3}}{4}} + \sqrt{\frac{2+3\sqrt{3}}{4}}\right)^{1/2}. \tag{2.8}$$

We need two modular equations of degree 3. The first is given in Lemma 2.2; the second is given in Entry 5(ix) of Chapter 19 in the second notebook [3, p. 231], namely,

$$\{\alpha(1-\beta)\}^{1/2} + \{(1-\alpha)\beta\}^{1/2} = 2\{\alpha\beta(1-\alpha)(1-\beta)\}^{1/8}. \tag{2.9}$$

We also require two modular equations of degree 23 [3, p. 411, Entry 15(i),(ii)]. If β has degree 23 over α, then

$$(\alpha\beta)^{1/8} + \{(1-\alpha)(1-\beta)\}^{1/8} + 2^{2/3}\{\alpha\beta(1-\alpha)(1-\beta)\}^{1/24} = 1 \tag{2.10}$$

and

$$\begin{aligned} 1+&(\alpha\beta)^{1/4} + \{(1-\alpha)(1-\beta)\}^{1/4} + 2^{4/3}\{\alpha\beta(1-\alpha)(1-\beta)\}^{1/12} \\ &= \left\{2\left(1+(\alpha\beta)^{1/2} + \{(1-\alpha)(1-\beta)\}^{1/2}\right)\right\}^{1/2}. \end{aligned} \tag{2.11}$$

We now apply (2.7) with $r = 3$ and $p = 23$. Thus, β has degree 23 over $1-\alpha$, and so we replace α by $1-\alpha$ in (2.10) and (2.11). By judiciously combining these new equations with (2.9), we eventually find that

$$2u^6 - 8\sqrt{2}u^5 + 24u^4 - 22\sqrt{2}u^3 + 24u^2 - 8\sqrt{2}u + 2 = 0, \tag{2.12}$$

where $u = (G_{69}G_{23/3})^{-1}$. The form of (2.12) suggests the substitution $x = u + 1/u$, and so (2.12) takes the shape

$$x^3 - 4\sqrt{2}x^2 + 9x - 3\sqrt{2} = 0.$$

The root $\sqrt{2}$ can be eliminated by numerical considerations, and we find that $x = (3+\sqrt{3})/\sqrt{2}$ and

$$G_{69}G_{23/3} = \frac{1}{u} = \sqrt{\frac{6+3\sqrt{3}}{4}} + \sqrt{\frac{2+3\sqrt{3}}{4}}, \tag{2.13}$$

since $u < 1$. Now observe that, from (2.6) with $p = 23$ and $r = 3$, Lemma 2.2 yields an equation in the two variables u and $v = G_{69}/G_{23/3}$. Using (2.13) in Lemma 2.2, we eventually find that

$$v = \left(\sqrt{748 + 432\sqrt{3}} + \sqrt{747 + 432\sqrt{3}}\right)^{1/6}. \tag{2.14}$$

Combining (2.13) and (2.14), we deduce that

$$G_{69} = \left(\sqrt{748 + 432\sqrt{3}} + \sqrt{747 + 432\sqrt{3}}\right)^{1/12} \left(\sqrt{\frac{6+3\sqrt{3}}{4}} + \sqrt{\frac{2+3\sqrt{3}}{4}}\right)^{1/2}. \tag{2.15}$$

Comparing (2.15) with (2.8), we see that we must establish a certain equality between radicals, which is easily shown. See [9] for complete details of the proof sketched here.

We conclude this section with a few comments about the calculation of singular moduli α_n. If G_n or g_n is known, then α_n is easily found from (1.2) by solving a quadratic equation. However, the representation that is obtained normally does not evince the fact (from Theorem 1.1) that α_n can be expressed in terms of units. In his first notebook, Ramanujan offered 21 values of α_n for even n and twelve values for odd n. For even n, Ramanujan recorded in his first notebook a remarkable formula to facilitate the calculation of α_n. This formula was noticed by Watson [35] who proved it in order to verify Ramanujan's value for α_{210},

$$\begin{aligned}\alpha_{210} =& (\sqrt{2}-1)^4(2-\sqrt{3})^2(\sqrt{7}-\sqrt{6})^4(8-3\sqrt{7})^2 \\ & \times (\sqrt{10}-3)^2(4-\sqrt{15})^4(\sqrt{15}-\sqrt{14})^2(6-\sqrt{35})^2,\end{aligned}$$

communicated by Ramanujan [28, p. xxix] in his second letter to Hardy. For odd n, the calculation of α_n is somewhat more difficult. Proofs for all 33 values of α_n claimed by Ramanujan can be found in our paper [12].

3 q–Continued Fractions

In both his first and second letters to Hardy, Ramanujan [28, pp. xxvii, xxviii] communicated theorems about $R(q)$ and $S(q)$. In particular, in his first letter, he asserted that

$$R(e^{-2\pi}) = \sqrt{\frac{5+\sqrt{5}}{2}} - \frac{\sqrt{5}+1}{2} \tag{3.1}$$

and

$$S(e^{-\pi}) = \sqrt{\frac{5-\sqrt{5}}{2}} - \frac{\sqrt{5}-1}{2}. \tag{3.2}$$

The evaluation (3.1) follows easily from a reciprocity theorem for $R(q)$, which Ramanujan stated in his second letter, and which was first proved by Watson [34]. The evaluation (3.2) follows from a similar reciprocity theorem for $S(q)$, which apparently Ramanujan did not communicate to Hardy, but which is found in his notebooks [27, p. 204], [3, p. 83] and first proved by Ramanathan [21]. However, (3.2) was first proved by Watson [33] in a different manner.

In both his first [27] and lost notebooks [29], Ramanujan recorded several other evaluations of $R(q)$ and $S(q)$. In fact, page 210 in his lost notebook comprises a list of evaluations and (mostly) intended evaluations. The authors [6], [10] have recently proved all the aforementioned evaluations claimed by Ramanujan in his notebooks. In particular, we have proved four general theorems providing explicit formulas for $R(e^{-2\pi\sqrt{n}})$ and $S(e^{-\pi\sqrt{n}})$ in terms of class invariants. The similarity evinced in (3.1) and (3.2) is manifest in these general theorems, which depend upon two formulas for $R(q)$ found as Entry 11(iii) in Chapter 19 in Ramanujan's second notebook [3, pp. 265, 266]. If $f(-q)$ is given by (1.6), then

$$R^{-1}(q) - 1 - R(q) = q^{-1/5}\frac{f(-q^{1/5})}{f(-q^5)} \tag{3.3}$$

and

$$R^{-5}(q) - 11 - R^5(q) = \frac{f^6(-q)}{qf^6(-q^5)}. \tag{3.4}$$

Both (3.3) and (3.4) were first proved by Watson [33]; proofs can also be found in [3, p. 267]. We discuss here only the evaluation of $R(e^{-2\pi\sqrt{n}})$; companion theorems can be established for $S(e^{-\pi\sqrt{n}})$.

First note that, by (3.3) and (3.4), it suffices to evaluate either

$$A := e^{2\pi\sqrt{n}/5}\frac{f(-e^{-2\pi\sqrt{n}/5})}{f(-e^{-10\pi\sqrt{n}})} \qquad \text{or} \qquad A_1 := e^{2\pi\sqrt{n}/6}\frac{f(-e^{-2\pi\sqrt{n}})}{f(-e^{-10\pi\sqrt{n}})}, \tag{3.5}$$

for then $R(e^{-2\pi\sqrt{n}})$ can be found by merely solving a quadratic equation.

Theorem 3.1 *Let A be defined by (3.5), and let*

$$U := \sqrt{\frac{g_{25n}}{g_{n/25}}} \qquad \text{and} \qquad V := \sqrt{\frac{G_{25n}}{G_{n/25}}}.$$

Then

$$\frac{A}{\sqrt{5}U} + \frac{\sqrt{5}U}{A} = (U+U^{-1})^2\left(\frac{U+U^{-1}}{\sqrt{5}} - \frac{\sqrt{5}}{U+U^{-1}}\right)$$

and

$$\frac{A}{\sqrt{5}V} - \frac{\sqrt{5}V}{A} = (V - V^{-1})^2 \left(\frac{V - V^{-1}}{\sqrt{5}} + \frac{\sqrt{5}}{V - V^{-1}} \right).$$

Theorem 3.2 *Let A_1 be defined by (3.5), and let*

$$U_1 := \frac{g_{25n}}{g_n} \qquad \text{and} \qquad V_1 := \frac{G_{25n}}{G_n}.$$

Then

$$\frac{A_1^2}{\sqrt{5}U_1} + \frac{\sqrt{5}U_1}{A_1^2} = \frac{1}{\sqrt{5}} \left(U_1^3 + U_1^{-3} \right)$$

and

$$\frac{A_1^2}{\sqrt{5}V_1} - \frac{\sqrt{5}V_1}{A_1^2} = \frac{1}{\sqrt{5}} \left(V_1^3 - V_1^{-3} \right).$$

The proof of Theorem 3.1 depends upon (3.3) and two modular equations of Ramanujan of degree 25 [3, p. 291, Entry 15 (i), (ii)]. The proof of Theorem 3.2 rests upon (3.4) and two modular equations of degree 5 [3, pp. 281, 282, Entry 13 (xii)].

To illustrate Theorem 3.1, let $n = 9$. Then after determining G_{225} and $G_{9/25}$ and manipulating some radicals, we deduce the following corollary [10], which we formulate in the same manner as Ramanujan [27, p. 311]. Another proof of Corollary 3.3 can be found in [6].

Corollary 3.3 *Let $a = 60^{1/4}$ and $b = 2 - \sqrt{3} + \sqrt{5}$. If $2c = \frac{a+b}{a-b}\sqrt{5} + 1$, then*

$$R(e^{-6\pi}) = \sqrt{c^2 + 1} - c.$$

To illustrate Theorem 3.2, let $n = 1/5$. Then $V_1 = G_5/G_{1/5} = 1$, since $G_n = G_{1/n}$. It follows trivially from Theorem 3.2 that $A_1^2 = \sqrt{5}$. We then conclude from (3.4) that

$$R^5(e^{-2\pi/\sqrt{5}}) = \sqrt{\left(\frac{5\sqrt{5} + 11}{2} \right)^2 + 1} - \frac{5\sqrt{5} + 11}{2}.$$

K. G. Ramanathan devoted several papers, in particular, [21], [22], and [23], to evaluating special cases of the Rogers–Ramanujan continued fraction.

In his first letter to Hardy, Ramanujan [28, p. xxvii] made the very strong assertion,

$$R(e^{-\pi\sqrt{n}}) \text{ can be exactly found if } n \text{ be any positive rational quantity.} \tag{3.6}$$

Watson [33] vaguely discussed (3.6) and claimed that $R(e^{-\pi\sqrt{n}})$ is algebraic. Using a lemma which can be found in Stark's paper [32], we [10] have proved the stronger assertion that $R(e^{-\pi\sqrt{n}})$ is a unit for each positive rational number n.

Ramanujan found many other properties of the Rogers–Ramanujan continued fraction which he recorded at scattered places in the unorganized portion of his second notebook, his third notebook, and his lost notebook. For proofs of some of these theorems, see [2] and [1]. In particular, Ramanujan found beautiful equations relating $R(q)$ with $R(-q)$, $R(q^2)$, $R(q^3)$, $R(q^4)$, or $R(q^5)$. For example [2], if $u = R(q)$ and $v = R(q^2)$, then

$$\frac{v-u^2}{v+u^2} = uv^2. \tag{3.7}$$

These equations can also be used to find certain values of the Rogers–Ramanujan continued fraction, but generally the form of the evaluation that one so obtains is not attractive.

In a fragment published with his lost notebook [29, pp. 363–366], Ramanujan gives a list of fourteen theorems that he established for the Rogers–Ramanujan continued fraction. This is followed by five theorems for his cubic continued fraction $G(q)$, defined in (1.4). In fact, Ramanujan writes, "I have also found empirically the following result ... and many results analogous to the previous continued fraction." Motivated by this claim, Chan [15], [16] has developed much of the theory for $G(q)$ and likely has found several of the "many results" not divulged by Ramanujan.

First, we offer an exact formula for $G(e^{-\pi\sqrt{n}})$ that was proved in [8]. This is analogous to the formulas for $R(e^{-2\pi\sqrt{n}})$ that arise from Theorems 3.1 and 3.2; a similar result can be proved for $G(-e^{-\pi\sqrt{n}})$.

Theorem 3.4 *If $G(q)$ is defined by (1.4) and p is given by (2.1), then, for every positive rational number n,*

$$G(e^{-\pi\sqrt{n}}) = \sqrt{\frac{\sqrt{p^2+1}-p}{\sqrt{p^2+4}+p}} \times \left(\sqrt{\frac{p^2+4+\sqrt{(p^2+1)(p^2+4)}}{2}} - \sqrt{\frac{p^2+2+\sqrt{(p^2+1)(p^2+4)}}{2}}\right).$$

For the first example, let $n = 2$. Then [43, p. 721], [26] $g_2 = 1$. Hence, $p = 0$, and, by Theorem 3.4,

$$G(e^{-\pi\sqrt{2}}) = \frac{1}{\sqrt{2}}\left(\sqrt{\frac{4+2}{2}} - \sqrt{\frac{2+2}{2}}\right) = \frac{\sqrt{6}-2}{2}. \tag{3.8}$$

For the second example, let $n = 10$. Then [43, p. 721] $g_{10} = \sqrt{(\sqrt{5}+1)/2}$. Thus, $p = \sqrt{5}$, and, by Theorem 3.4,

$$G(e^{-\pi\sqrt{10}}) = \frac{\sqrt{9+3\sqrt{6}} - \sqrt{7+3\sqrt{6}}}{(1+\sqrt{5})\sqrt{\sqrt{6}+\sqrt{5}}}. \tag{3.9}$$

The value (3.9) was given by Ramanujan on page 366 in the publication of the lost notebook [29]. Ramanathan [21] also proved (3.9) by another method. But his proof depends upon Kronecker's limit formula, as did his evaluations of the Rogers–Ramanujan continued fraction in [23] and [22].

Further values of $G(q)$ can also be computed from reciprocity theorems satisfied by $G(q)$. For example, using primarily modular equations of degree 3 found in Entry 1 of Chapter 20 of Ramanujan's second notebook [3, pp. 345, 346], Chan [15]; [16, Section 6.3] proved the elegant reciprocity theorems

$$\left(1-2G(-e^{-\pi\alpha})\right)\left(1-2G(-e^{-\pi\beta})\right) = 3$$

and

$$\left(1+G(-e^{-\sqrt{2}\pi\alpha})\right)\left(1+G(-e^{-\sqrt{2}\pi\beta})\right)=\frac{3}{2},$$

where $\alpha, \beta > 0$ and $\alpha\beta = 1$. In particular, if we set $\alpha = \beta = 1$ in the latter equality, we again deduce (3.8).

Like $R(q)$, $G(q)$ satisfies some beautiful modular equations. For example, if $u = G(q)$ and $v = G(q^2)$, then

$$u^2 + 2uv^2 = v,$$

which is an analogue of (3.7).

Ramanujan studied several other q–continued fractions possessing product representations. For example, in Entry 1(ii) of Chapter 19 of his second notebook [3, p. 221], Ramanujan claimed that

$$\begin{aligned} H(q) :=& \frac{\sqrt{q}}{1+q} \,\frac{q^2}{+\,1+q^3} \,\frac{q^4}{+\,1+q^5} \,\frac{q^6}{+\,1+q^7} + \cdots, \qquad |q| < 1, \\ =& \sqrt{q}\frac{(q;q^8)_\infty(q^7;q^8)_\infty}{(q^3;q^8)_\infty(q^5;q^8)_\infty}, \end{aligned} \tag{3.10}$$

which was first proved by B. Gordon [19] in 1965. Another representation for $H(q)$ was given by Ramanujan later in his second notebook [3, p. 290], [2, p. 23, Entry 13]. The continued fraction $H(q)$ also enjoys many beautiful properties that have been discovered by Chan and S.–S. Huang [17].

4 The Values of Theta Functions

Recall that $\varphi(q)$ is defined in (1.5). Observe from (1.9) that an evaluation of any of the functions φ, ${}_2F_1$, or K yields an evaluation of the other two functions. However, such evaluations may not be explicit. For example, if $K(k)$ is known for a certain value of k, it may be difficult (or impossible) to explicitly determine $K' = K(k')$, and so q would then not be explicitly determined. Conversely, it may be possible to evaluate $\varphi(q)$ for a certain value of q, but it may be impossible to determine the corresponding value of k. Recall that k is given by [3, p. 102]

$$k = \sqrt{1 - \frac{\varphi^4(-q)}{\varphi^4(q)}}.$$

In the literature more attention has been devoted to determining ${}_2F_1$ and K, and we refer to our paper [7] for several references. In particular, concentration has focused on the instances when $K'/K = \sqrt{n}$, where n is a positive integer.

At scattered places in his notebooks, particularly in his first notebook, Ramanujan recorded several values of $\varphi(q)$, some of which have not been heretofore proved in print. For example,

$$\frac{\varphi(e^{-3\pi})}{\varphi(e^{-\pi})} = \frac{1}{\sqrt[4]{6\sqrt{3}-9}}, \tag{4.1}$$

$$\frac{\varphi(e^{-5\pi})}{\varphi(e^{-\pi})} = \frac{1}{\sqrt{5\sqrt{5}-10}}, \tag{4.2}$$

and

$$\frac{\varphi^2(e^{-7\pi})}{\varphi^2(e^{-\pi})} = \frac{\sqrt{13+\sqrt{7}}+\sqrt{7+3\sqrt{7}}}{14}(28)^{1/8}. \tag{4.3}$$

The value

$$\varphi(e^{-\pi}) = \frac{\pi^{1/4}}{\Gamma(\frac{3}{4})}$$

is well known [44, p. 525], [3, p. 103], and so (4.1)–(4.3) provide explicit values for $\varphi(e^{-3\pi})$, $\varphi(e^{-5\pi})$, and $\varphi(e^{-7\pi})$, respectively.

As indicated above, as corollaries, values of ${}_2F_1$ can be deduced. For example,

$${}_2F_1\left(\frac{1}{3},\frac{2}{3};1;\frac{3\sqrt{3}-5}{4}\right) = \frac{\sqrt{\pi}}{(12)^{1/8}\sqrt{\sqrt{3}-1}\,\Gamma^2(\frac{3}{4})}.$$

We observe that the values of $\varphi(e^{-n\pi})/\varphi(e^{-\pi})$ given above are algebraic. Indeed, the authors [11] have shown that when n is odd, $\sqrt{2n}\,\varphi(e^{-n\pi})/\varphi(e^{-\pi})$ is an algebraic integer dividing $2\sqrt{n}$. A similar result holds for even n.

J. M. and P. B. Borwein [14, p. 145] first observed that class invariants could be used to calculate certain values of $\varphi(e^{-n\pi})$. Indeed, we have verified all of Ramanujan's values for $\varphi(e^{-n\pi})$ and some new ones as well by combining Ramanujan's class invariants with his modular equations in judicious manners. We illustrate these ideas by giving a proof of (4.2).

Recall that the multiplier m is defined by (1.10). If β has degree 5 and m is the multiplier for degree 5, then from Chapter 19 of Ramanujan's second notebook [3, Entry 13(xii), pp. 281, 282]

$$m = \left(\frac{\beta}{\alpha}\right)^{1/4} + \left(\frac{1-\beta}{1-\alpha}\right)^{1/4} - \left(\frac{\beta(1-\beta)}{\alpha(1-\alpha)}\right)^{1/4} \tag{4.4}$$

and

$$\frac{5}{m} = \left(\frac{\alpha}{\beta}\right)^{1/4} + \left(\frac{1-\alpha}{1-\beta}\right)^{1/4} - \left(\frac{\alpha(1-\alpha)}{\beta(1-\beta)}\right)^{1/4}. \tag{4.5}$$

Set $\alpha = \frac{1}{2}$, so that, by (1.8), $q = e^{-\pi}$. From (4.4) we find that

$$(2\beta)^{1/4} + (2(1-\beta))^{1/4} = m + (4\beta(1-\beta))^{1/4}, \tag{4.6}$$

and, from (4.5) and (4.6), we find that

$$\begin{aligned}\frac{5}{m} &= \frac{(2(1-\beta))^{1/4}+(2\beta)^{1/4}-1}{(4\beta(1-\beta))^{1/4}}\\ &= \frac{m+(4\beta(1-\beta))^{1/4}-1}{(4\beta(1-\beta))^{1/4}}\\ &= \frac{m+G_{25}^{-6}-1}{G_{25}^{-6}},\end{aligned} \tag{4.7}$$

by (1.2), with α_n replaced by β_n. From Ramanujan's paper [26], [28, p. 26] or Weber's tables [43, p. 722],

$$G := G_{25} = \frac{1+\sqrt{5}}{2}. \tag{4.8}$$

Hence, from (4.7) and (4.8), since $G^3 = 2 + \sqrt{5}$,

$$G^3 m - \frac{5}{G^3 m} = G^3 - G^{-3} = 4,$$

from which we deduce that $G^3 m = 5$, or $m = 5(\sqrt{5} - 2)$. The evaluation (4.2) now follows from (1.10).

On page 338 in his first notebook, Ramanujan defines

$$a_{m,n} := n e^{-(\pi/4)(n-1)\sqrt{m/n}} \frac{\psi^2(e^{-\pi\sqrt{mn}})\varphi^2(-e^{-2\pi\sqrt{mn}})}{\psi^2(e^{-\pi\sqrt{m/n}})\varphi^2(-e^{-2\pi\sqrt{m/n}})},$$

where m and n evidently are intended to be positive integers. Then, on pages 338 and 339, he offers a list of 18 particular values, which we present in the following table.

m,n	$a_{m,n}$	m,n	$a_{m,n}$
3,3	$\frac{1}{\sqrt{3}}$	$3,13$	$\left(\sqrt{\frac{5+\sqrt{13}}{8}} - \sqrt{\frac{\sqrt{13}-3}{8}}\right)^8$
3,9	$\frac{1}{(2^{1/3}+1)^2}$	$3,23$	$\left(\sqrt{\frac{7+4\sqrt{3}}{2}} - \sqrt{\frac{5+4\sqrt{3}}{2}}\right)^2$
3,15	$\frac{2-\sqrt{3}}{3}$	$3,71$	$\left(\sqrt{\frac{175+100\sqrt{3}}{2}} - \sqrt{\frac{173+100\sqrt{3}}{2}}\right)^2$
3,5	$\frac{3-\sqrt{5}}{2}$	$5,9$	$(2-\sqrt{3})^2$
3,7	$2-\sqrt{3}$	$5,11$	$\left(\sqrt{\frac{7+\sqrt{5}}{8}} - \sqrt{\frac{\sqrt{5}-1}{8}}\right)^8$
3,11	$2\sqrt{3}-\sqrt{11}$	$5,13$	$\left(\sqrt{\frac{9+\sqrt{65}}{2}} - \sqrt{\frac{7+\sqrt{65}}{2}}\right)^2$
3,19	$2\sqrt{19}-5\sqrt{3}$	$5,17$	$(\sqrt{17}-4)^2$
3,31	$(2-\sqrt{3})^3$	$5,29$	$\left(\sqrt{49+4\sqrt{145}} - \sqrt{48+4\sqrt{145}}\right)^2$
3,59	$102\sqrt{3}-23\sqrt{59}$	$7,9$	$\left(\sqrt{\frac{5+\sqrt{21}}{8}} - \sqrt{\frac{\sqrt{21}-5}{8}}\right)^8$

Upon examining this table, we see that when $(m,n)=1$ each value is a unit in some algebraic number field. In fact, we have proved some general theorems which ensure that $a_{m,n}$ is a unit for large classes of pairs m,n that include the examples above [11]. Several approaches were devised to calculate $a_{m,n}$.

First, we derived formulas for $a_{m,n}$, when $m = 3$, 5, and 7, in terms of class invariants. For example,

$$a_{3,n} = \frac{1}{3}\left(\frac{G_{n/3}^6}{G_{3n}^6} + \frac{2\sqrt{2}}{G_{n/3}^3 G_{3n}^3}\right). \tag{4.9}$$

In each case G_{3n} is known. To find $G_{n/3}$ we use Lemma 2.2 in the form

$$\left(\frac{G_{n/3}}{G_{3n}}\right)^6 + \left(\frac{G_{3n}}{G_{n/3}}\right)^6 + 2\sqrt{2}\left(\frac{1}{G_{n/3}^3 G_{3n}^3} + G_{n/3}^3 G_{3n}^3\right) = 0. \tag{4.10}$$

Analogues of (4.9) and (4.10) exist for $m = 5$ and 7 as well.

Secondly, we were able to find formulas for $a_{m,n}$, in each case $m = 3$, 5, and 7, that are in the spirit of Theorems 3.1 and 3.2. For example, if $V_n = G_{n/3}/G_{3n}$, then

$$a_{3,n} - \frac{1}{a_{3,n}} = \frac{1}{3}\left(V_n^6 - V_n^{-6}\right). \tag{4.11}$$

Equation (4.11) is used in conjunction with (4.10), as in the first approach. Analogues of (4.11) have also been established for $a_{5,n}$ and $a_{7,n}$.

Thirdly, general formulas for $a_{m,n}$ were derived by using Kronecker's limit formula. These formulas involve class numbers and fundamental units in real quadratic fields, and so this approach would have been unknown to Ramanujan.

We give one simple example to illustrate the first approach. Let $n = 5$. Then, from Weber's book [43, p. 721], or from Ramanujan's notebooks [27, vol. 1, p. 289; vol. 2, p. 294],

$$G_{15} = 2^{-1/12}(\sqrt{5}+1)^{1/3}.$$

From (4.9), we easily verify that

$$G_{5/3} = 2^{-1/12}(\sqrt{5}-1)^{1/3}.$$

Using these values in (4.10) with $n = 5$, we readily find that

$$a_{3,5} = \frac{3-\sqrt{5}}{2}.$$

Another interesting product of theta functions is examined by Ramanujan in his lost notebook, and these numbers were studied by Ramanathan [24], [25].

5 Concluding Remarks

Ramanujan recorded several hundred modular equations in his notebooks, and proofs of most of these can be found in [3], [4], and [5]. However, many of our proofs are not those found by Ramanujan, and more enlightening proofs are greatly desired. Perhaps Ramanujan merely enjoyed deriving modular equations. However, since some of them have now been employed in the computation of class invariants and values of theta functions, it is very likely that Ramanujan had applications in mind for many of his modular equations.

It also could be surmised that Ramanujan enjoyed computing class invariants because they are lovely numbers. This is likely true, but we have seen from their applications to resolving values of continued fractions and theta functions that Ramanujan had a broader agenda in mind.

The contemporary stages for much of the work described here are "modern" algebra and algebraic number theory. Ramanujan evidently knew nothing of these subjects. However, he must have discerned the special nature of the arithmetic and number theory in these situations and so had a deep understanding that he would have conveyed in a different language. His calculations of some invariants are particularly enigmatic. Our understanding of these subjects would be greatly enhanced if we could discover some of Ramanujan's thinking.

References

1. Andrews, G. E. [1981] *Ramanujan's "lost" notebook III. The Rogers–Ramanujan continued fraction* Adv. Math. **41**, 186–208.
2. Andrews, G. E., Berndt, B. C., Jacobsen, L. and Lamphere, R. L. [1992] *The Continued Fractions Found in the Unorganized Portions of Ramanujan's Notebooks, Memoir no. 477*, vol. 99, American Mathematical Society, Providence, R.I.
3. Berndt, B. C. [1991] *Ramanujan's Notebooks, Part III*, Springer–Verlag, New York.
4. Berndt, B. C. [1994] *Ramanujan's Notebooks, Part IV*, Springer–Verlag, New York.
5. Berndt, B. C. *Ramanujan's Notebooks, Part V*, Springer–Verlag, New York (to appear).
6. Berndt, B. C. and Chan, H. H. [1995] *Some values for the Rogers–Ramanujan continued fraction* Canad. J. Math. **47**, 897-914.
7. Berndt, B. C. and Chan, H. H. [1995] *Ramanujan's explicit values for the classical theta-function* Mathematika **42**, 278–294.
8. Berndt, B. C., Chan, H. H. and Zhang, L.–C. [1995] *Ramanujan's class invariants and cubic continued fraction* Acta Arith. **73**, 67-85.
9. Berndt, B. C., Chan, H. H. and Zhang, L.–C. *Ramanujan's class invariants, Kronecker's limit formula, and modular equations* Trans. Amer. Math. Soc. (to appear).
10. Berndt, B. C., Chan, H. H. and Zhang, L.–C. *Explicit evaluations of the Rogers–Ramanujan continued fraction* J. Reine Angew. Math. (to appear).
11. Berndt, B. C., Chan, H. H. and Zhang, L.–C. *Ramanujan's remarkable product of theta functions* Proc. Edinburgh Math. Soc. (to appear).
12. Berndt, B. C., Chan, H. H. and Zhang, L.–C. [1997] *Ramanujan's singular moduli* The Ramanujan Journal 1 (to appear).
13. Berndt, B. C. and Rankin, R. A. [1995] *Ramanujan: Letters and Commentary*, American Mathematical Society, Providence; London Mathematical Society, London, 1995.
14. Borwein, J. M. and Borwein, P. B. [1987] *Pi and the AGM*, Wiley, New York.
15. Chan, H. H. [1995] *On Ramanujan's cubic continued fraction* Acta Arith. **73**, 343-355.
16. Chan, H. H. [1995] *Contributions to Ramanujan's continued fractions, class invariants, partition identities and modular equations, Doctoral Thesis*, University of Illinois at Urbana–Champaign.
17. Chan, H. H. and Huang, S.–S. [1997] *On the Ramanujan–Gordon–Göllnitz continued fraction* The Ramanujan Journal **1** (to appear).
18. Cox, D. A. [1989] *Primes of the Form $x^2 + ny^2$*, Wiley, New York.
19. Gordon, B. [1965] *Some continued fractions of the Rogers–Ramanujan type* Duke Math. J. **32**, 741–748.
20. Ramanathan, K. G. [1982] *Remarks on some series considered by Ramanujan* J. Indian Math. Soc. **46**, 107–136.

21. Ramanathan, K. G. [1984] *On Ramanujan's continued fraction* Acta Arith. **43**, 209–226.
22. Ramanathan, K. G. [1984] *On the Rogers–Ramanujan continued fraction* Proc. Indian Acad. Sci. (Math. Sci.) **93**, 67–77.
23. Ramanathan, K. G. [1985] *Ramanujan's continued fraction* Indian J. Pure Appl. Math. **16**, 695–724.
24. Ramanathan, K. G. [1987] *Some applications of Kronecker's limit formula* J. Indian Math. Soc. **52**, 71–89.
25. Ramanathan, K. G. [1989] *On some theorems stated by Ramanujan*, Number Theory and Related Topics, Oxford University Press, Bombay.
26. Ramanujan, S. [1914] *Modular equations and approximations to* π Quart. J. Math. (Oxford) **45**, 350–372.
27. Ramanujan, S. [1957] *Notebooks (2 volumes)*, Tata Institute of Fundamental Research, Bombay.
28. Ramanujan, S. [1962] *Collected Papers*, Chelsea, New York.
29. Ramanujan, S. [1988] *The Lost Notebook and Other Unpublished Papers*, Narosa, New Delhi.
30. Rogers, L. J. [1894] *Second memoir on the expansion of certain infinite products* Proc. London Math. Soc. **25**, 318–343.
31. Siegel, C. L. [1980] *Advanced Analytic Number Theory*, Tata Institute of Fundamental Research, Bombay.
32. Stark, H. M. [1971] *Values of L–functions at* $s = 1$ *I. L–functions for quadratic forms* Adv. Math. **7**, 301–343.
33. Watson, G. N. [1929] *Theorems stated by Ramanujan (VII): theorems on continued fractions* J. London Math. Soc. **4**, 39–48.
34. Watson, G. N. [1929] *Theorems stated by Ramanujan (IX): two continued fractions* J. London Math. Soc. **4**, 231–237.
35. Watson, G. N. [1931] *Theorems stated by Ramanujan (XII): a singular modulus* J. London Math. Soc. **6**, 65–70.
36. Watson, G. N. [1931] *Theorems stated by Ramanujan (XIV): a singular modulus* J. London Math. Soc. **6**, 126–132.
37. Watson, G. N. [1932] *Some singular moduli (I)* Quart. J. Math. **3**, 81–98.
38. Watson, G. N. [1932] *Some singular moduli (II)* Quart. J. Math. **3**, 189–212.
39. Watson, G. N. [1936] *Singular moduli (3)* Proc. London Math. Soc. **40**, 83–142.
40. Watson, G. N. [1936] *Singular moduli (4)* Acta Arith. **140**, 284–323.
41. Watson, G. N. [1937] *Singular moduli (5)* Proc. London Math. Soc. **42**, 377–397.
42. Watson, G. N. [1937] *Singular moduli (6)* Proc. London Math. Soc. **42**, 398–409.
43. Weber, H. [1961] *Lehrbuch der Algebra, dritter Band*, Chelsea, New York.
44. Whittaker, E. T. and Watson, G. N. [1966] *A Course of Modern Analysis, 4th ed.*, University Press, Cambridge.
45. Zhang, L.–C. [1996] *Kronecker's limit formula, class invariants and modular equations (II)* in Analytic Number Theory: Proceedings of a Conference in Honor of Heini Halberstam, vol. 2 (B. C. Berndt, H. G. Diamond, and A. J. Hildebrand, eds.), Birkhäuser, Boston, 817-838.

Fields Institute Communications
Volume **14**, 1997

Elementary Derivations of Summation and Transformation Formulas for q-Series

George Gasper
Department of Mathematics
Northwestern University
Evanston, IL 60208-2730 USA

Abstract. Elementary derivations of some summation and transformation formulas for q-series are presented which provide deeper insights into the theory of q-series. The main emphasis is on methods that can be used to **derive** formulas, rather than to just *verify* previously derived or conjectured formulas. This approach leads to the derivation of a new family of summation formulas for very-well-poised basic hypergeometric series ${}_{6+2k}W_{5+2k}$, $k = 1, 2, \ldots$.

Introduction

We present some elementary derivations of summation and transformation formulas for q-series, which are different from, and in several cases simpler or shorter than, those presented in the Gasper and Rahman [1990] "Basic Hypergeometric Series" book (which we will refer to as BHS), the Bailey [1935] and Slater [1966] books, and in some papers; thus providing deeper insights into the theory of q-series. Our main emphasis is on methods that can be used to **derive** formulas, rather than to just *verify* previously derived or conjectured formulas.

In Section 5 this approach leads to the derivation of a new family of summation formulas for very-well-poised basic hypergeometric series ${}_{6+2k}W_{5+2k}$, $k = 1, 2, \ldots$. Several of the observations in this paper were presented, along with related exercises, in the author's minicourse on "q-Series" at the Fields Institute miniprogram on "Special Functions, q-Series and Related Topics," June 12 –14, 1995. As is customary, we employ the notations used in BHS for the *shifted factorial*

$$(a)_0 = 1,\ (a)_k = a(a+1)\cdots(a+k-1), \qquad k = 1, 2, \ldots ,$$

1991 *Mathematics Subject Classification.* Primary: 33D15, 33D20, 33D65; Secondary: 33C05, 33C20.

This work was supported in part by the National Science Foundation under grant DMS-9401452.

the *q-shifted factorial*

$$(a;q)_0 = 1,\ (a;q)_k = (1-a)(1-aq)\cdots(1-aq^{k-1}), \qquad k = 1,2,\ldots,$$

$$(a;q)_\infty = \lim_{k\to\infty}(a;q)_k = \prod_{k=0}^{\infty}(1-aq^k), \qquad |q|<1,$$

$$(a;q)_\alpha = \frac{(a;q)_\infty}{(aq^\alpha;q)_\infty}, \qquad 0<|q|<1,$$

the ${}_rF_s$ *hypergeometric series*

$${}_rF_s\,(a_1,a_2,\ldots,a_r;\ b_1,\ldots,b_s;z) \equiv {}_rF_s\left[\begin{matrix}a_1,a_2,\ldots,a_r\\ b_1,\ldots,b_s\end{matrix};z\right]$$
$$=\sum_{k=0}^{\infty}\frac{(a_1)_k(a_2)_k\cdots(a_r)_k}{k!(b_1)_k\cdots(b_s)_k}\,z^k,$$

and the ${}_r\phi_s$ *basic hypergeometric series*

$${}_r\phi_s\,(a_1,a_2,\ldots,a_r;b_1,\ldots,b_s;q,z) \equiv {}_r\phi_s\left[\begin{matrix}a_1,a_2,\ldots,a_r\\ b_1,\ldots,b_s\end{matrix};q,z\right]$$

$$=\sum_{k=0}^{\infty}\frac{(a_1,a_2,\ldots,a_r;q)_k}{(q,b_1,\ldots,b_s;q)_k}\left[(-1)^k q^{\binom{k}{2}}\right]^{1+s-r} z^k,$$

where $\binom{k}{2} = k(k-1)/2$, $(a_1,a_2,\ldots,a_r;q)_k = (a_1;q)_k(a_2;q)_k\cdots(a_r;q)_k$ and the principal value of q^α is taken. We also employ the compact notation

$${}_{r+1}W_r\,(a_1;a_4,a_5,\ldots,a_{r+1};q,z)$$

for the *very-well-poised* ${}_{r+1}\phi_r$ series

$${}_{r+1}\phi_r\left[\begin{matrix}a_1,qa_1^{\frac12},-qa_1^{\frac12},a_4,\ldots,a_{r+1}\\ a_1^{\frac12},-a_1^{\frac12},qa_1/a_4,\ldots,qa_1/a_{r+1}\end{matrix};q,z\right]$$

and define the *bilateral basic hypergeometric* ${}_r\psi_s$ series by

$${}_r\psi_s(z) \equiv {}_r\psi_s\left[\begin{matrix}a_1,a_2,\ldots,a_r\\ b_1,b_2,\ldots,b_s\end{matrix};q,z\right]$$

$$=\sum_{k=-\infty}^{\infty}\frac{(a_1,a_2,\ldots,a_r;q)_k}{(b_1,b_2,\ldots,b_s;q)_k}(-1)^{(s-r)k}q^{(s-r)\binom{k}{2}}z^k.$$

For simplicity, unless stated otherwise we shall assume that n is a nonnegative integer, $|q|<1$ in nonterminating q-series, and that the parameters and variables are complex numbers such that the series converge absolutely and any singularities are avoided (which usually leads to isolated conditions on the parameters and variables since the singularities are usually at poles and at limits of sequences of poles). For a discussion of when the above series converge, see Sections 1.2 and 5.1 in BHS (in the third paragraph on p. 5 each of the ratios $|b_1b_2\cdots b_s|/|a_1a_2\cdots a_r|$ should be replaced by $|b_1b_2\cdots b_sq|/|a_1a_2\cdots a_r|$).

1 The q-binomial theorem

The summation formula

$$ {}_1F_0(a;—;z) = \sum_{k=0}^{\infty} \frac{(a)_k}{k!} z^k = (1-z)^{-a}, \qquad |z| < 1, \tag{1.1} $$

is called the *binomial theorem* because, when $-a = n$ is a nonnegative integer and $z = -x/y$, it reduces to the binomial theorem for the n-th power of the binomial $x + y$:

$$ (x+y)^n = \sum_{k=0}^{n} \binom{n}{k} x^k y^{n-k} . \tag{1.2} $$

Since, by l'Hôpital's rule,

$$ \lim_{q \to 1} \frac{1-q^a}{1-q} = a $$

and hence

$$ \lim_{q \to 1} \frac{(q^a;q)_k}{(q;q)_k} = \frac{(a)_k}{k!}, $$

it is natural to consider what happens when the coefficient $(a)_k/k!$ of z^k in the series in (1.1) is replaced by $(q^a;q)_k/(q;q)_k$ or, more generally, by $(a;q)_k/(q;q)_k$. Hence, let us set

$$ f(a,z) = \sum_{k=0}^{\infty} \frac{(a;q)_k}{(q;q)_k} z^k, \qquad |z| < 1, \tag{1.3} $$

with $|q| < 1$. The case when $|q| > 1$ will be considered later. Note that, by the Weierstrass M-test, since $|q| < 1$ the series in (1.3) converges uniformly on compact subsets of the unit disk $\{z : |z| < 1\}$ to a function $f(a,z)$ that is an analytic function of z (and of a) when $|z| < 1$. One way to find a formula for $f(a,z)$ that is a generalization of (1.1) is to first observe that, since $1 - a = 1 - aq^k + aq^k - a = (1-aq^k) - a(1-q^k)$,

$$ \begin{aligned} f(a,z) &= 1 + \sum_{k=1}^{\infty} \frac{(a;q)_k}{(q;q)_k} z^k \\ &= 1 + \sum_{k=1}^{\infty} \frac{(aq;q)_{k-1}}{(q;q)_k} [(1-aq^k) - a(1-q^k)] z^k \\ &= 1 + \sum_{k=1}^{\infty} \frac{(aq;q)_k}{(q;q)_k} z^k - a \sum_{k=1}^{\infty} \frac{(aq;q)_{k-1}}{(q;q)_{k-1}} z^k \\ &= f(aq,z) - azf(aq,z) = (1-az)f(aq,z). \end{aligned} \tag{1.4} $$

By iterating this functional equation $n-1$ times, we find that

$$ f(a,z) = (az;q)_n \ f(aq^n,z), $$

which on letting $n \to \infty$ and using $q^n \to 0$ yields

$$ f(a,z) = (az;q)_\infty \ f(0,z). \tag{1.5} $$

Now set $a = q$ in (1.5) to get

$$f(0,z) = \frac{f(q,z)}{(qz;q)_\infty} = \frac{(1-z)^{-1}}{(qz;q)_\infty} = \frac{1}{(z;q)_\infty},$$

which, combined with (1.3) and (1.5), gives the *q-binomial theorem*

$$ {}_1\phi_0(a;—;q,z) = \sum_{k=0}^{\infty} \frac{(a;q)_k}{(q;q)_k} z^k = \frac{(az;q)_\infty}{(z;q)_\infty}, \qquad |z|<1,\ |q|<1. \tag{1.6}$$

This summation formula was derived by Cauchy [1843], Jacobi [1846], and Heine [1847]. Heine's proof of (1.6), which is reproduced in the books Heine [1878], Bailey [1935, pp. 66], Slater [1966, pp. 92], and in Section 1.3 of BHS along with some motivation from Askey [1980], consists of using series manipulations to derive the functional equation

$$(1-z)f(a,z) = (1-az)f(a,qz), \tag{1.7}$$

iterating (1.7) $n-1$ times, and then letting $n \to \infty$ to get

$$f(a,z) = \frac{(az;q)_n}{(z;q)_n} f(a,q^n z) = \frac{(az;q)_\infty}{(z;q)_\infty} f(a,0) = \frac{(az;q)_\infty}{(z;q)_\infty},$$

which gives (1.6).

Another derivation of the q-binomial theorem can be given by calculating the coefficients $c_k = g_a^{(k)}(0)/k!$, $k = 0,1,2,\dots$, in the Taylor series expansion of the function

$$g_a(z) = \frac{(az;q)_\infty}{(z;q)_\infty} = \sum_{k=0}^{\infty} c_k z^k, \tag{1.8}$$

which is an analytic function of z when $|z|<1$ and $|q|<1$. Clearly $c_0 = g_a(0) = 1$. One may show that $c_1 = g_a'(0) = (1-a)/(1-q)$ by taking the logarithmic derivative of $(az;q)_\infty/(z;q)_\infty$ and then setting $z = 0$. But, unfortunately, the succeeding higher order derivatives of $g_a(z)$ become more and more difficult to calculate for $|z|<1$, and so one is forced to abandon this approach and to search for another way to calculate all of the c_k coefficients. One simple method is to notice that from the definition of $g_a(z)$ as the quotient of two infinite products, it immediately follows that $g_a(z)$ satisfies the functional equation

$$(1-z)\,g_a(z) = (1-az)\,g_a(qz), \tag{1.9}$$

which is of course the same as the functional equation (1.7) satisfied by $f(a,z)$. In a *verification* type proof of the q-binomial theorem, (1.9) provides *substantial* motivation for showing, as in Heine's proof, that the sum of the q-binomial series $f(a,z)$ satisfies the functional equation (1.7).

To calculate the c_k coefficients, we first use (1.9) to obtain

$$\sum_{k=0}^{\infty} c_k\, z^k - \sum_{k=0}^{\infty} c_k\, z^{k+1} = \sum_{k=0}^{\infty} c_k\, q^k z^k - a \sum_{k=0}^{\infty} c_k\, q^k\, z^{k+1},$$

or, equivalently,

$$1+\sum_{k=1}^{\infty}(c_k-c_{k-1})\,z^k=1+\sum_{k=1}^{\infty}(c_kq^k-ac_{k-1}\,q^{k-1})\,z^k,$$

which implies that

$$c_k-c_{k-1}=c_kq^k-ac_{k-1}\,q^{k-1}$$

and hence

$$c_k=\frac{1-aq^{k-1}}{1-q^k}\,c_{k-1},\qquad k=1,2,\dots\,. \tag{1.10}$$

Iterating the recurrence relation (1.10) gives

$$c_k=\frac{(a;q)_k}{(q;q)_k}c_0=\frac{(a;q)_k}{(q;q)_k},\qquad k=0,1,2,\dots,$$

and concludes the derivation of (1.6). For a combinatorial proof of the q-binomial theorem using a bijection between two classes of partitions, see Andrews [1969].

It is of interest to note that if $|q|>1$, then by replacing q in (1.6) by q^{-1}, applying the *inversion identity*

$$(a;q)_k=(a^{-1};q^{-1})_k(-a)^kq^{\binom{k}{2}},\qquad k=0,1,2,\dots, \tag{1.11}$$

replacing a by a^{-1} and then z by az/q, it follows that when $|q|>1$ the q-binomial theorem takes the form

$${}_1\phi_0(a;—;q,z)=\frac{(z/q;q^{-1})_\infty}{(az/q;q^{-1})_\infty},\qquad |az/q|<1,\ |q|>1. \tag{1.12}$$

In the special case when $z=-x/y$ and $a=q^{-n}$, $n=0,1,2,\dots$, both (1.6) and (1.12) give that (1.2) has a q-analogue of the form

$$y^n(-xq^{-n}/y;q)_n=\sum_{k=0}^{n}\frac{(q^{-n};q)_k}{(q;q)_k}(-1)^kx^ky^{n-k}. \tag{1.13}$$

Another q-analogue of (1.2), which we will utilize in the next section, may be easily derived by first observing that from (1.6) we have the product formula

$$\sum_{j=0}^{\infty}\frac{(a;q)_j}{(q;q)_j}\,z^j\sum_{k=0}^{\infty}\frac{(b;q)_k}{(q;q)_k}\,(az)^k=\sum_{n=0}^{\infty}\frac{(ab;q)_n}{(q;q)_n}\,z^n,\qquad |z|<1, \tag{1.14}$$

which is a q-analogue of $(1-z)^{-a}(1-z)^{-b}=(1-z)^{-a-b}$. Then set $j=n-k$ in the product on the left side of (1.14) and compare the coefficients of z^n on both sides of the equation to get

$$\frac{(ab;q)_n}{(q;q)_n}=\sum_{k=0}^{n}\frac{(a;q)_{n-k}(b;q)_k}{(q;q)_{n-k}(q;q)_k}\,a^k, \tag{1.15}$$

which gives a q-analogue of (1.2) in the form

$$(ab;q)_n=\sum_{k=0}^{n}\begin{bmatrix}n\\k\end{bmatrix}_q(a;q)_{n-k}(b;q)_k\,a^k, \tag{1.16}$$

where the *q-binomial coefficient* is defined by

$$\begin{bmatrix} n \\ k \end{bmatrix}_q = \frac{(q;q)_n}{(q;q)_k (q;q)_{n-k}}, \qquad k = 0, 1, \dots, n.$$

Replacing a in (1.16) by q^{1-n}/c and manipulating the q-shifted factorials via the identities (I.8), (I.14), and (I.42) in Appendix I of BHS shows that (1.16) is equivalent to the *q-Chu–Vandermonde summation formula*

$$_2\phi_1(q^{-n}, b; c; q, q) = \frac{(c/b;q)_n}{(c;q)_n} b^n, \tag{1.17}$$

which is a q-analogue of the *Chu–Vandermonde summation formula* (see p. 2 in BHS)

$$F(-n, b; c; 1) = \frac{(c-b)_n}{(c)_n}.$$

By either switching the order of summation or inverting the base q via (1.11), we find that (1.17) is also equivalent to the summation formula

$$_2\phi_1(q^{-n}, b; c; q, cq^n/b) = \frac{(c/b;q)_n}{(c;q)_n}. \tag{1.18}$$

2 Analytic continuations

By manipulating the products on the right side of (1.18) we find that

$$_2\phi_1(z^{-1}, b; c; q, cz/b) = \frac{(c/b, cz;q)_\infty}{(c, cz/b;q)_\infty} = \prod_{k=0}^{\infty} \frac{(1-cq^k/b)(1-czq^k)}{(1-cq^k)(1-czq^k/b)} \tag{2.1}$$

with $z = q^n$, $n = 0, 1, \dots$, and $|q| < 1$. The infinite product on the right side of (2.1) converges uniformly on compact subsets of the disk $\{z : |cz/b| < 1\}$ to an analytic function of z. In view of the identity

$$(z^{-1};q)_k z^k = (z-1)(z-q)\cdots(z-q^{k-1}), \qquad k = 0, 1, \dots,$$

the series on the left side of (2.1) is a sum of analytic functions of z, which also converges uniformly on compact subsets of the disk $\{z : |cz/b| < 1\}$ to an analytic function of z. Since $q^n \to 0$ as $n \to \infty$ when $|q| < 1$, it follows by analytic continuation that (2.1) holds for arbitrary complex values of z when $|cz/b| < 1$ and $|q| < 1$. Hence, by setting $z = a^{-1}$ we have derived the Jacobi [1846] and Heine [1847] *q-Gauss summation formula*

$$_2\phi_1(a, b; c; q, c/ab) = \frac{(c/a, c/b;q)_\infty}{(c, c/ab;q)_\infty}, \qquad |c/ab| < 1. \tag{2.2}$$

Similarly, from (1.17)

$$_2\phi_1(a, b; c; q, q) = \frac{(a/c, b/c;q^{-1})_\infty}{(1/c, ab/c;q^{-1})_\infty}, \qquad |q| > 1, \tag{2.3}$$

with $a = q^{-n}$, $n = 0, 1, \dots$, and, because $q^{-n} \to 0$ as $n \to \infty$ when $|q| > 1$, it follows by analytic continuation that (2.3) holds for arbitrary complex values of a when, for convergence, $|ab/c| < 1$. Formula (2.3) also follows from (2.2) by inverting the base q.

To see that both (2.2) and (2.3) are q-analogues of Gauss' [1813] famous summation formula

$$ {}_2F_1(a,b;c;1) = \frac{\Gamma(c)\Gamma(c-a-b)}{\Gamma(c-a)\Gamma(c-b)}, \qquad \text{Re}\,(c-a-b) > 0, \tag{2.4} $$

it suffices to replace a, b, c by q^a, q^b, q^c, respectively, with $0 < q < 1$ in (2.2) and $q > 1$ in (2.3), and then let $q \to 1$. The q-Gauss formula (2.2) was derived in BHS by using the q-binomial theorem to derive Heine's [1847] ${}_2\phi_1$ transformation formula

$$ {}_2\phi_1(a,b;c;q,z) = \frac{(b,az;q)_\infty}{(c,z;q)_\infty}\, {}_2\phi_1(c/b, z; az; q, b), \qquad |z| < 1,\ |b| < 1, \tag{2.5} $$

setting $z = c/ab$ to reduce the series on the right side of this transformation formula to a ${}_1\phi_0$ series, and then summing that series via the q-binomial theorem.

If we shift the index of summation k in (1.6) by replacing it by $k+n$, we obtain that if $0 < |q| < 1, 0 < |z| < 1$, and $n = 0, 1, 2, \ldots$, then

$$ \begin{aligned} \frac{(az;q)_\infty}{(z;q)_\infty} &= \sum_{k=-\infty}^{\infty} \frac{(a;q)_k}{(q;q)_k} z^k = \sum_{k=-\infty}^{\infty} \frac{(a;q)_{k+n}}{(q;q)_{k+n}} z^{k+n} \\ &= \sum_{k=-\infty}^{\infty} \frac{(a;q)_n (aq^n;q)_k}{(q;q)_n (q^{n+1};q)_k} z^{k+n} = \frac{(a;q)_n}{(q;q)_n} z^n \sum_{k=-\infty}^{\infty} \frac{(aq^n;q)_k}{(q^{n+1};q)_k} z^k, \end{aligned} \tag{2.6} $$

where, as usual, the definition of $(a;q)_k$ is extended to negative integer values of k by defining

$$ (a;q)_{-k} = \frac{1}{(aq^{-k};q)_k} = \frac{(-q/a)^k}{(q/a;q)_k} q^{\binom{k}{2}}, \qquad k = 0, 1, 2, \ldots . $$

After replacing a by aq^{-n} and then setting $q^{n+1} = b$, the left and right sides of (2.6) give

$$ {}_1\psi_1(a;b;q,z) = \sum_{k=-\infty}^{\infty} \frac{(a;q)_k}{(b;q)_k} z^k = \frac{(q, b/a, az, q/az; q)_\infty}{(b, q/a, z, b/az; q)_\infty} \tag{2.7} $$

for $b = q^{n+1}$ when $0 < |q| < 1, 0 < |z| < 1$, and $n = 0, 1, 2, \ldots$. Since $q^{n+1} \to 0$ as $n \to \infty$ when $|q| < 1$, and the infinite series and the infinite product on the left and right sides, respectively, of (2.7) converge to analytic functions of b when $|b| < \min(1, |az|)$ and $|z| < 1$, it follows by analytic continuation that we have derived *Ramanujan's* ${}_1\psi_1$ *summation formula*

$$ {}_1\psi_1(a;b;q,z) = \frac{(q, b/a, az, q/az; q)_\infty}{(b, q/a, z, b/az; q)_\infty}, \qquad |b/a| < |z| < 1, \tag{2.8} $$

which reduces to the q-binomial theorem when $b = q$.

The above derivation of (2.8) is essentially in the reverse order of Ismail's [1977] proof of (2.8), which first *reduces* the proof of (2.8) to the case when $b = q^{n+1}$, where n is a nonnegative integer, and then verifies this case by using a shift in the index of summation to obtain a series that is summable by the q-binomial theorem. For other proofs of (2.8) and historical comments, see Berndt [1993], BHS, and their references. It should be noted that if we set $b = 0$ in (2.8), replace q and z by

q^2 and $-qz/a$, respectively, and then let $a \to \infty$, we obtain Jacobi's [1829] *triple product identity*

$$\sum_{k=-\infty}^{\infty} q^{k^2} z^k = \left(q^2, -qz, -q/z; q^2\right)_\infty . \tag{2.9}$$

The infinite product representations for the *theta functions* $\vartheta_1(x), \vartheta_2(x), \vartheta_3(x)$, and $\vartheta_4(x)$ displayed on p. 13 of BHS are special cases of (2.9).

It is natural to investigate what happens when the shift in index of summation method is applied to the q-Gauss summation formula (2.2). Proceeding as in (2.6) with the parameters a, b, c in (2.2) replaced by $aq^{-n}, bq^{-n}, cq^{-n}$, respectively, we find that if $|q| < 1$ and $d = q^{n+1}$, $n = 0, 1, 2, \ldots$, then

$${}_2\psi_2(a, b; c, d; q, cd/abq) = \frac{d(q, c/a, c/b, d/a, d/b; q)_\infty}{qc^n(q/a, q/b, c, d, cd/abq; q)_\infty}, \tag{2.10}$$

when $0 < |cd/abq| < 1$ and $c \neq q^{n-k}$, $k = 0, 1, 2, \ldots$. Unfortunately, the right side of (2.10) is not an analytic function of d in a neighborhood of the origin because of the c^n factor and the initial $d = q^{n+1}$ condition, and the term with index k in the series on the left side of (2.10) has a pole of order $-k$ at $d = 0$ when k is a negative integer. Therefore we cannot analytically continue (2.10) in d to derive an infinite product representation for the series on the left side of (2.10) that is valid for d in a neighborhood of the origin. This helps to explain why *no one* has been able to extend the q-Gauss summation formula (2.2) to derive an infinite product representation for a ${}_2\psi_2$ series that is a q-analogue of Dougall's [1907] bilateral hypergeometric series summation formula

$$\sum_{k=-\infty}^{\infty} \frac{(a)_k(b)_k}{(c)_k(d)_k} = \frac{\Gamma(c)\Gamma(d)\Gamma(1-a)\Gamma(1-b)\Gamma(c+d-a-b-1)}{\Gamma(c-a)\Gamma(c-b)\Gamma(d-a)\Gamma(d-b)}, \tag{2.11}$$

where Re $(c+d-a-b-1) > 0$ and $(a)_k = (-1)^k/(1-a)_{-k}$, $k = -1, -2, \ldots$, which reduces to Gauss' formula (2.4) when $d = 1$. However, see Ex. 5.20 in BHS for two transformation formulas for ${}_2\psi_2(a, b; c, d; q, z)$ series.

3 The q-Pfaff–Saalschütz summation formula

In order to extend the q-Chu–Vandermonde and q-Gauss summation formulas to ${}_3\phi_2$ series, we use (1.17) in the form

$$\frac{(a;q)_k}{(c;q)_k} = \frac{(a;q)_n}{(c;q)_n}\, {}_2\phi_1(q^{k-n}, c/a; q^{1-n}/a; q, q), \qquad 0 \le k \le n,$$

to obtain

$$\begin{aligned}\sum_{k=0}^{n} \frac{(q^{-n}, a, b; q)_k}{(q, c, d; q)_k} q^k &= \frac{(a;q)_n}{(c;q)_n} \sum_{k=0}^{n} \frac{(q^{-n}, b; q)_k}{(q, d; q)_k} q^k \sum_{j=0}^{n-k} \frac{(q^{k-n}, c/a; q)_j}{(q, q^{1-n}/a; q)_j} q^j \\ &= \frac{(a;q)_n}{(c;q)_n} \sum_{j=0}^{n} \frac{(q^{-n}, c/a; q)_j}{(q, q^{1-n}/a; q)_j}\, {}_2\phi_1(q^{j-n}, b; d; q, q)\end{aligned}$$

by a change in order of summation, which on using (1.17) to sum the ${}_2\phi_1$ series gives the ${}_3\phi_2$ transformation formula

$$\begin{aligned} &{}_3\phi_2(q^{-n}, a, b; c, d; q, q) \\ &\quad = \frac{(a, d/b; q)_n}{(c, d; q)_n} b^n \, {}_3\phi_2(q^{-n}, c/a, q^{1-n}/d; q^{1-n}/a, q^{1-n}b/d; q, q). \end{aligned} \tag{3.1}$$

This formula is a generalization of Heine's ${}_2\phi_1$ transformation formula (2.5), which follows from (3.1) by setting $d = q^{1-n}/z$, letting $n \to \infty$, and then switching a and b. Now notice that if $c/a = q^{1-n}b/d$, i.e. if the *balanced* condition $cd = abq^{1-n}$ holds, then the ${}_3\phi_2$ series on the right side of (3.1) reduces to a terminating ${}_2\phi_1$ series that can be summed by (1.17) to give Jackson's [1910] summation formula for a terminating balanced ${}_3\phi_2$ series

$${}_3\phi_2(a, b, q^{-n}; c, abc^{-1}q^{1-n}; q, q) = \frac{(c/a, c/b; q)_n}{(c, c/ab; q)_n}. \tag{3.2}$$

This formula is usually called the *q-Saalschütz* or the *q-Pfaff–Saalschütz summation formula* because it is a q-analogue of the Pfaff [1797] and Saalschütz [1890] summation formula

$${}_3F_2(a, b, -n; c, 1 + a + b - c - n; 1) = \frac{(c-a)_n(c-b)_n}{(c)_n(c-a-b)_n}.$$

Notice that letting $a \to 0$ in (3.2) gives (1.17), letting $a \to \infty$ gives (1.18), and letting $n \to \infty$ in (3.2) gives the q-Gauss summation formula (2.2).

If, as in our derivation of the q-Gauss formula (2.2), we observe that (3.2) can be rewritten in the form

$${}_3\phi_2(a, b, 1/z; c, abq/cz; q, q) = \frac{(c/a, c/b, cz, cz/ab; q)_\infty}{(c, c/ab, cz/a, cz/b; q)_\infty}, \qquad |q| < 1, \tag{3.3}$$

with $z = q^n$, $n = 0, 1, \ldots$, then the infinite product on the right side of (3.3) clearly converges to an analytic function of z when $|z| < \min(|a/c|, |b/c|)$. However, because $(1/z; q)_k/(abq/cz; q)_k$, $k = 1, 2, \ldots$, has poles at the points $z = abq^j/c$, $1 \le j \le k$, there is no neighborhood of the origin in which all of the terms of the series on the left side of (3.3) are analytic; thus (3.3) cannot be analytically continued in z to a neighborhood of the origin. Nevertheless, see equation (2.10.12) in BHS for a nonterminating extension of (3.2) with the sum of *two* balanced ${}_3\phi_2$ series, and also the bilateral nonterminating extension in Ex. 5.13 of BHS.

The q-Pfaff–Saalschütz formula (3.2) was derived in Bailey [1935] and in Slater [1966] by first using an induction argument to give Jackson's [1921] *verification* type proof of his summation formula for a terminating ${}_8W_7$ series (derived in Section 5)

$${}_8W_7(a; b, c, d, e, q^{-n}; q, q) = \frac{(aq, aq/bc, aq/bd, aq/cd; q)_n}{(aq/b, aq/c, aq/d, aq/bcd; q)_n}, \tag{3.4}$$

where $bcde = a^2q^{n+1}$, and then replacing d by aq/d in (3.4) and letting $a \to \infty$ to get (3.2) as a limit case of (3.4). In BHS, (3.2) was derived by iterating (2.5) twice to derive another of Heine's [1847] transformation formulas

$${}_2\phi_1(a, b; c; q, z) = \frac{(abz/c; q)_\infty}{(z; q)_\infty} \, {}_2\phi_1(c/a, c/b; c; q, abz/c), \tag{3.5}$$

where $|z| < 1$ and $|abz/c| < 1$, using the q-binomial theorem to expand the coefficient of the ${}_2\phi_1$ series on the right side of (3.5) as a power series in powers of z, and then equating the coefficients of z^n on both sides of the resulting formula to get (3.2). Recently, Ismail [1995] used the Askey-Wilson difference operators to derive (3.2). For some combinatorial proofs of (3.2), see Andrews and Bressoud [1984], Goulden [1985], and Zeilberger [1987]. Also see Wilf and Zeilberger [1992] for a computer-constructible WZ proof of (3.2).

4 Summation formulas for some very-well-poised series

Let us start by deriving some transformation formulas for ${}_4W_3(a;b;q,z)$ series. Let $|q| < 1$. Since

$$\begin{aligned}(1-a)(1-b)\,\frac{(qa^{\frac12},-qa^{\frac12};q)_k}{(a^{\frac12},-a^{\frac12};q)_k} &= (1-b)(1-aq^{2k}) = 1+abq^{2k}-(b+aq^{2k})\\ &= (1+abq^{2k}-aq^k-bq^k)-(b+aq^{2k}-aq^k-bq^k)\\ &= (1-aq^k)(1-bq^k)-b(1-aq^k/b)(1-q^k),\end{aligned}$$

we have

$$\begin{aligned}{}_4W_3(a;b;q,z) &= 1+\sum_{k=1}^{\infty}\frac{(aq,bq;q)_{k-1}}{(q,aq/b;q)_k}(1-b)(1-aq^{2k})z^k\\ &= 1+\sum_{k=1}^{\infty}\frac{(aq,bq;q)_k}{(q,aq/b;q)_k}z^k-bz\sum_{k=1}^{\infty}\frac{(aq,bq;q)_{k-1}}{(q,aq/b;q)_{k-1}}z^{k-1}\\ &= (1-bz)\,{}_2\phi_1(aq,bq;aq/b;q,z), \qquad |z|<1.\end{aligned} \tag{4.1}$$

Hence ${}_4W_3(a;b;q,b^{-1}) = 0$ when $|b^{-1}| < 1$, and it follows from the special cases $b = q^{-n}$, $n = 1, 2, \ldots$, and the fact that ${}_4W_3(a;1;q,z) = 0$ that

$${}_4W_3(a;q^{-n};q,q^n) = \delta_{n,0}\,, \tag{4.2}$$

where $\delta_{n,m}$ is the Kronecker delta function. Some recent multidimensional generalizations of (4.2) are given in Bhatnagar and Milne [1996].

This derivation of (4.2) is substantially simpler than that in Section 2.3 of BHS, which used the q-Pfaff–Saalschütz formula (3.2) and the Bailey [1941] and Daum [1942] *q-Kummer summation formula*

$${}_2\phi_1(a,b;aq/b;q,-q/b) = \frac{(-q;q)_\infty(aq,aq^2/b^2;q^2)_\infty}{(aq/b,-q/b;q)_\infty}, \qquad |q/b|<1, \tag{4.3}$$

derived in Section 1.8 of BHS. Formula (4.2) can also be derived by the finite difference method employed in Gasper [1989] and Gasper and Rahman [1990a] to derive bibasic extensions of (4.2), and by the method pointed out in Rahman [1990].

Using (4.2), we obtain the expansion formula

$$
\begin{aligned}
u_0 = \sum_{k\geq 0} u_k\, \delta_{k,0} &= \sum_{k\geq 0} u_k \sum_{j=0}^{k} \frac{(a, qa^{\frac{1}{2}}, -qa^{\frac{1}{2}}, q^{-k}; q)_j}{(q, a^{\frac{1}{2}}, -a^{\frac{1}{2}}, aq^{k+1}; q)_j} q^{jk} \\
&= \sum_{j\geq 0} \frac{(a, qa^{\frac{1}{2}}, -qa^{\frac{1}{2}}; q)_j}{(a^{\frac{1}{2}}, -a^{\frac{1}{2}}, aq^{j+1}; q)_j} (-1)^j q^{\binom{j}{2}} \sum_{k\geq 0} \frac{(q^{j+1}, aq^{j+1}; q)_k}{(q, aq^{2j+1}; q)_k} u_{j+k}
\end{aligned}
\tag{4.4}
$$

by a change in order of summation, where $\{u_k\}$ is a sequence of complex numbers such that the change in order of summation is justified, which is the case when $\{u_k\}$ terminates and when the double series converge absolutely. If $\{u_k\}$ is such that the sum over k on the right side of (4.4) can be evaluated in terms of q-shifted factorials, then (4.4) yields a summation formula for a q-series. In particular, setting

$$
u_k = \frac{(b, c, q^{-n}; q)_k}{(q, aq, bcq^{-n}/a; q)_k} q^k
$$

in (4.4), the sum over k on the right side of (4.4) becomes a multiple of a terminating balanced ${}_3\phi_2$ series that can be summed by means of the q-Pfaff–Saalschütz summation formula (3.2), thus giving the sum of a terminating ${}_6W_5$ series

$$
{}_6W_5(a; b, c, q^{-n}; q, aq^{n+1}/bc) = \frac{(aq, aq/bc; q)_n}{(aq/b, aq/c; q)_n}, \tag{4.5}
$$

which reduces to (4.2) when $c = aq/b$. This formula was derived in BHS by first using (3.2) to derive the expansion formula (2.2.4) in BHS, and then applying (4.2) to a special case of the expansion formula. As in our derivation of (2.2) from (1.18), analytic continuation of (4.5) gives its nonterminating extension

$$
{}_6W_5(a; b, c, d; q, aq/bcd) = \frac{(aq, aq/bc, aq/bd, aq/cd; q)_\infty}{(aq/b, aq/c, aq/d, aq/bcd; q)_\infty}, \qquad |aq/bcd| < 1. \tag{4.6}
$$

This formula was obtained in Section 2.8 of BHS as a limit case of the summation formula (5.6) derived in Section 5.

In order to extend (4.6) to a summation formula for a ${}_6\psi_6$ series, it suffices to proceed as in the derivation of (2.8) from (2.2). Explicitly, replace the index of summation, call it k, in (4.6) by $k + n$, replace the parameters a, b, c, d by $aq^{-2n}, bq^{-n}, cq^{-n}, dq^{-n}$, respectively, and manipulate the infinite products to obtain that

$$
\begin{aligned}
&{}_6\psi_6 \left[\begin{matrix} a/z, qa^{\frac{1}{2}}, -qa^{\frac{1}{2}}, b, c, d \\ zq, a^{\frac{1}{2}}, -a^{\frac{1}{2}}, aq/b, aq/c, aq/d \end{matrix} ; q, \frac{azq}{bcd} \right] \\
&\quad = \frac{(aq, aq/bc, aq/bd, aq/cd, zq/b, zq/c, zq/d, q, q/a; q)_\infty}{(aq/b, aq/c, aq/d, q/b, q/c, q/d, zq, zq/a, azq/bcd; q)_\infty}
\end{aligned}
\tag{4.7}
$$

with $z = q^n$, $n = 0, 1, 2, \ldots$, where $0 < |azq/bcd| < 1$. Since the infinite series and the infinite product on the left and right sides, respectively, of (4.7) converge to analytic functions of z when $|z| < \min(|1/q|, |a/q|, |bcd/aq|)$, by analytic continuation

(4.7) holds when $|z| < \min(|1/q|, |a/q|, |bcd/aq|)$. Thus, setting $z = a/e$, we have derived Bailey's [1936] summation formula for a very-well-poised ${}_6\psi_6$ series

$$
\begin{aligned}
&{}_6\psi_6 \left[\begin{matrix} qa^{\frac{1}{2}}, -qa^{\frac{1}{2}}, b, c, d, e \\ a^{\frac{1}{2}}, -a^{\frac{1}{2}}, aq/b, aq/c, aq/d, aq/e \end{matrix} ; q, \frac{a^2q}{bcde} \right] \\
&\quad = \frac{(aq, aq/bc, aq/bd, aq/be, aq/cd, aq/ce, aq/de, q, q/a; q)_\infty}{(aq/b, aq/c, aq/d, aq/e, q/b, q/c, q/d, q/e, a^2q/bcde; q)_\infty},
\end{aligned}
\tag{4.8}
$$

where $0 < |a^2q/bcde| < 1$. Also see the proofs referred to in Section 5.3 of BHS and, in particular, the extension of Ismail's [1977] proof of (2.8) to a proof of (4.8) presented in Askey and Ismail [1979].

Notice that, when $d = a^{\frac{1}{2}}$ and $e = -a^{\frac{1}{2}}$, (4.8) reduces to the following summation formula for a ${}_2\psi_2$ series that is different from the series considered in (2.10)

$$
\begin{aligned}
&{}_2\psi_2(b, c; aq/b, aq/c; q, -aq/bc) \\
&\quad = \frac{(aq/bc; q)_\infty (aq^2/b^2, aq^2/c^2, q^2, aq, q/a; q^2)_\infty}{(aq/b, aq/c, q/b, q/c, -aq/bc; q)_\infty}, \qquad 0 < |aq/bc| < 1.
\end{aligned}
\tag{4.9}
$$

5 Additional summation and transformation formulas

In view of the important formulas that were derived in Section 4 from the expansion formula (4.4), which followed from the summation formula (4.2), it is of interest to investigate what formulas can be derived by replacing (4.2) by the more general summation formula (4.8). To introduce an integer parameter k, let us start by replacing a, b, c, d, e in (4.8) by $aq^{2k}, bq^k, cq^k, dq^k, eq^{2k}$, respectively, to rewrite (4.8) in the form

$$
\begin{aligned}
&{}_6\psi_6 \left[\begin{matrix} q^{k+1}a^{\frac{1}{2}}, -q^{k+1}a^{\frac{1}{2}}, bq^k, cq^k, dq^k, eq^{2k} \\ q^k a^{\frac{1}{2}}, -q^k a^{\frac{1}{2}}, aq^{k+1}/b, aq^{k+1}/c, aq^{k+1}/d, aq/e \end{matrix} ; q, \frac{a^2q^{1-k}}{bcde} \right] \\
&\quad = A \frac{(aq/b, aq/c, aq/d, be/a, ce/a, de/a; q)_k (a; q)_{2k}}{(b, c, d, bcde/a^2; q)_k (aq, e; q)_{2k}} \left(-\frac{bcd}{a}\right)^k q^{\binom{k}{2}}
\end{aligned}
\tag{5.1}
$$

with

$$
A = \frac{(aq, aq/bc, aq/bd, aq/be, aq/cd, aq/ce, aq/de, q, q/a; q)_\infty}{(aq/b, aq/c, aq/d, aq/e, q/b, q/c, q/d, q/e, a^2q/bcde; q)_\infty}.
\tag{5.2}
$$

Using (5.1) with A written as a (constant) function of k and with j as the index of summation in the ${}_6\psi_6$ series, we find that

$$
\begin{aligned}
A \sum_{k=-\infty}^{\infty} u_k &= \sum_{k=-\infty}^{\infty} A u_k = \sum_{j=-\infty}^{\infty} \frac{(qa^{\frac{1}{2}}, -qa^{\frac{1}{2}}, b, c, d, e; q)_j}{(a^{\frac{1}{2}}, -a^{\frac{1}{2}}, aq/b, aq/c, aq/d, aq/e; q)_j} \left(\frac{a^2q}{bcde}\right)^j \\
&\quad \times \sum_{k=-\infty}^{\infty} \frac{(bcde/a^2, eq^j, eq^{-j}/a; q)_k}{(be/a, ce/a, de/a; q)_k} u_k
\end{aligned}
\tag{5.3}
$$

after replacing j by $j-k$ and changing the order of summation, where it is assumed that $\{u_k\}_{k=-\infty}^{\infty}$ is a bilateral sequence of complex numbers such that the double series on the right side of (5.3) converges absolutely.

If we let $e = a$ in (5.3), which is equivalent to starting with the ${}_6\phi_5$ special case of (5.1), and set

$$u_k = \frac{(b, c, d, aq/ef; q)_k}{(q, aq/e, aq/f, bcd/a; q)_k} q^k, \tag{5.4}$$

where at least one of numerator parameters $b, c, d, aq/ef$ is of the form q^{-n}, $n = 0, 1, 2 \ldots$, so that the sequence $\{u_k\}$ has compact support, then the sum over k on the right side of (5.3) becomes a terminating balance ${}_3\phi_2$ series that is summable via (3.2). This gives the transformation formula

$$\begin{aligned} &{}_8W_7\,(a; b, c, d, e, f; q, a^2q^2/bcdef) \\ &\quad = \frac{(aq, aq/bc, aq/bd, aq/cd; q)_\infty}{(aq/b, aq/c, aq/d, aq/bcd; q)_\infty}\, {}_4\phi_3 \left[\begin{matrix} b, c, d, aq/ef \\ aq/e, aq/f, bcd/a \end{matrix}; q, q\right] \end{aligned} \tag{5.5}$$

provided that the ${}_4\phi_3$ series on the right side terminates and $|a^2q^2/bcdef| < 1$, so that the series on the left side converges. Watson [1929] used induction to prove (5.5) when both of the series terminate; and he commented that it should extend to the case when the series on the left converges and the series on the right terminates, which was subsequently proved by Bailey (see pp.70 in Bailey [1935]). Fifty years after Watson's paper was published, Askey and Ismail [1979] showed that (5.5) follows from Watson's terminating case by analytic continuation in the variable $z = 1/f$. Formula (5.5) can also be derived by observing that the proof of the expansion formula (2.5.2) in BHS extends to the case when q^{-n} is replaced by a complex variable f and aq/bc is a negative integer power of q, so that using (4.6) to sum the resulting nonterminating very-well-poised ${}_6\phi_5$ series on the right side of the expansion formula yields (5.5). For a proof via orthogonal polynomials of (5.5) when both series terminate, see Andrews and Askey [1977].

When $b = q^{-n}$ and $a^2q = bcdef$ the ${}_4\phi_3$ series in (5.5) reduces to a terminating balanced ${}_3\phi_2$ series that can be summed by (3.2) to derive, after replacing f by b, Jackson's [1921] summation formula for a terminating ${}_8W_7$ series

$${}_8W_7(a; b, c, d, e, q^{-n}; q, q) = \frac{(aq, aq/bc, aq/bd, aq/cd; q)_n}{(aq/b, aq/c, aq/d, aq/bcd; q)_n}, \tag{5.6}$$

where $a^2q^{n+1} = bcde$. Letting $n \to \infty$ in (5.6) gives (4.6). Note that the restrictions that the series in (5.6) and the series on the right side of (5.5) terminate cannot be removed by analytic continuation in a neighborhood of the origin of the z-plane by starting with the $z = q^n$, $n = 0, 1, 2, \ldots$, cases because there is no neighborhood of the origin in which all of the terms of the series are analytic. The ${}_4\phi_3$ series in (5.5) also reduces to a terminating balanced ${}_3\phi_2$ series that is summable by (3.2) when $f = aq^{n+1}/e$ and $c = a/b$, which leads to a summation formula for a nonterminating ${}_8W_7$ series

$$\begin{aligned} &{}_8W_7\,(a; b, a/b, d, e, aq^{n+1}/e; q, q^{1-n}/d) \\ &\quad = \frac{(q, aq, aq/bd, bq/d; q)_\infty}{(bq, aq/b, aq/d, q/d; q)_\infty}\, \frac{(aq/be, bq/e; q)_n}{(aq/e, q/e; q)_n}, \qquad |q^{1-n}/d| < 1, \end{aligned} \tag{5.7}$$

where $n = 0, 1, 2, \ldots$.

In order to use (5.7) to derive some new summation formulas, we first introduce a nonnegative integer parameter k by replacing a, b, c, d, e in (5.7) by $aq^{2k}, bq^k, cq^k, dq^k, eq^k$, respectively, to get

$$\begin{aligned} &{}_8W_7\ (aq^{2k}; bq^k, aq^k/b, dq^k, eq^k, aq^{k+n+1}/e; q, q^{1-k-n}/d) \\ &= B_n \frac{(bq, aq/b, aq/d, aq/e, eq^{-n}; q)_k}{(aq;q)_{2k}(d, e, aq^{n+1}/e; q)_k} \quad (-d)^k q^{\binom{k}{2}+kn} \end{aligned} \tag{5.8}$$

with

$$B_n = \frac{(q, aq, aq/bd, bq/d; q)_\infty}{(bq, aq/b, aq/d, q/d; q)_\infty} \frac{(aq/be, bq/e; q)_n}{(aq/e, q/e; q)_n}. \tag{5.9}$$

Then, proceeding as in the derivation of (5.3) with (5.1) replaced by (5.8), we obtain the expansion formula

$$\begin{aligned} B_n \sum_{k=0}^{\infty} u_k = \sum_{j=0}^{\infty} & \frac{(a, qa^{\frac{1}{2}}, -qa^{\frac{1}{2}}, b, a/b, d, e, aq^{n+1}/e; q)_j}{(q, a^{\frac{1}{2}}, -a^{\frac{1}{2}}, aq/b, bq, aq/d, aq/e, eq^{-n}; q)_j} \left(\frac{q^{1-n}}{d}\right)^j \\ & \times \sum_{k=0}^{j} \frac{(aq^j, q^{-j}; q)_k}{(b, a/b; q)_k} u_k, \end{aligned} \tag{5.10}$$

where $\{u_k\}_{k=0}^{\infty}$ is a sequence of complex numbers such that the double series on the right side of (5.10) converges absolutely.

If we let

$$u_k = \frac{(b, a/b, q^{-m}; q)_k}{(q, aq/f, fq^{-m}; q)_k} q^k \tag{5.11}$$

in (5.10), where m is a nonnegative integer, then the sums over k on both the right and left sides of (5.10) are summable via (3.2), yielding the summation formula

$$ {}_{10}W_9(a; b, a/b, d, e, aq^{n+1}/e, f, aq^{m+1}/f; q, q^{1-n-m}/d) = \tag{5.12}$$

$$\frac{1}{(q, aq, aq/bd, bq/d; q)_\infty} (bq, aq/b, aq/d, q/d; q)_\infty \frac{(aq/be, bq/e; q)_n}{(aq/e, q/e; q)_n} \frac{(aq/bf, bq/f; q)_m}{(aq/f, q/f; q)_m}$$

where $n, m = 0, 1, 2, \ldots$, and $|q^{1-n-m}/d| < 1$. Moreover, by iterating this procedure, it follows that (5.7) extends to the family of summation formulas

$$\begin{aligned} &{}_{6+2k}W_{5+2k}(a; b, a/b, d, e_1, \ldots, e_k, aq^{n_1+1}/e_1, \ldots, aq^{n_k+1}/e_k; q, q^{1-(n_1+\cdots+n_k)}/d) \\ &= \frac{(q, aq, aq/bd, bq/d; q)_\infty}{(bq, aq/b, aq/d, q/d; q)_\infty} \prod_{j=1}^{k} \frac{(aq/be_j, bq/e_j; q)_{n_j}}{(aq/e_j, q/e_j; q)_{n_j}}, \qquad k = 1, 2, \ldots, \end{aligned} \tag{5.13}$$

where $n_1, \ldots, n_k$ are nonnegative integers and $|q^{1-(n_1+\cdots+n_k)}/d| < 1$.

For applications of (5.6) and additional summation, transformation, and expansion formulas, see BHS. The latest list of errata for BHS is available over the World Wide Web at: http://www.math.nwu.edu/preprints/gasper/index.html

References

Andrews, G. E. [1969], *On a calculus of partition functions,* Pacific J. Math., **31**, 555–562.

Andrews, G. E. and Askey, R. [1977], *Enumeration of partitions: the role of Eulerian series and q-orthogonal polynomials,* Higher Combinatorics (M. Aigner, ed.), Reidel, Boston, Mass., 3–26.

Andrews, G. E. and Askey, R. [1978], *A simple proof of Ramanujan's summation of the* ${}_1\psi_1$, Aequationes Math., **18**, 333–337.

Andrews, G. E. and Bressoud, D. M. [1984], *Identities in combinatorics III: Further aspects of ordered set sorting,* Discrete Math., **49**, 223-236.

Askey, R. [1980], *Ramanujan's extensions of the gamma and beta functions*, Amer. Math. Monthly, **87**, 346–359.

Askey, R. and Ismail, M. E. H. [1979], *A very-well-poised* ${}_6\psi_6$, Proc. Amer. Math. Soc., **77**, 218–222.

Bailey, W.N. [1935], *Generalized Hypergeometric Series,* Cambridge University Press, Cambridge, reprinted by Stechert-Hafner, New York, 1964.

Bailey, W. N. [1936], *Series of hypergeometric type which are infinite in both directions,* Quart. J. Math. (Oxford), **7**, 105-115.

Bailey, W. N. [1941], *A note on certain q-identities,* Quart. J. Math. (Oxford), **12**, 173-175.

Berndt, B. C. [1993], *Ramanujan's theory of theta-functions*, Theta Functions From the Classical to the Modern (M. Ram Murty, ed.), CRM Proceedings & Lecture Notes, **1**, Amer. Math. Soc., Providence, R.I, 1–63.

Bhatnagar, G. and Milne, S. C. [1996], *Generalized bibasic hypergeometric series and their* $U(n)$ *extensions*, Adv. in Math., to appear.

Cauchy, A.-L. [1843], *Mémoire sur les fonctions dont plusieurs valeurs sont liées entre elles par une équation linéaire, et sur diverses transformations de produits composés d'un nombre indéfini de facteurs,* C. R. Acad. Sci. Paris, T. XVII, p. 523, *Oeuvres de Cauchy,* 1re série, T. VIII, Gauthier-Villars, Paris, 1893, 42–50.

Daum, J. A. [1942], *The basic analog of Kummer's theorem,* Bull. Amer. Math. Soc., **48**, 711–713.

Dougall, J. [1907], *On Vandermonde's theorem and some more general expansions,* Proc. Edin. Math. Soc., **25**, 114–132.

Gasper, G. [1989], *Summation, transformation, and expansion formulas for bibasic series,* Trans. Amer. Math. Soc., **312**, 257–277.

Gasper, G. and Rahman, M. [1990], *Basic Hypergeometric Series*, Encyclopedia of Mathematics and Its Applications, **35**, Cambridge University Press, Cambridge and New York.

Gasper, G. and Rahman, M. [1990a], *An indefinite bibasic summation formula and some quadratic, cubic, and quartic summation and transformation formulas,* Canad. J. Math., **42**, 1–27.

Gauss, C. F. [1813], *Disquisitiones generales circa seriem infinitam ...*, Comm. soc. reg. sci. Gött. rec., Vol. II; reprinted in *Werke* **3** (1876), 123–162.

Goulden, I. [1985], *A bijective proof of the q-Saalschütz theorem,* Discrete Math., **57**, 39–44.

Hahn, W. [1949], *Über Polynome, die gleichzeitig zwei verschiedenen Orthogonalsystemen angehören,* Math. Nachr., **2**, 263-278.

Heine, E. [1847], *Untersuchungen über die Reihe ...*, J. reine angew. Math., **34**, 285–328.

Heine, E. [1878], *Handbuch der Kugelfunctionen, Theorie und Anwendungen*, Vol. 1, Reimer, Berlin.

Ismail, M. E. H. [1977], *A simple proof of Ramanujan's* $_1\psi_1$ *sum,* Proc. Amer. Math. Soc., **63**, 185–186.

Ismail, M. E. H. [1995], *The Askey–Wilson operator and summation Theorems,* Mathematical Analysis, Wavelets, and Signal Processing (M.E. Ismail, M.Z. Nashed, A.I. Zayed, and A.F. Ghaleb, eds.), Contempory Mathematics, **190**, Amer. Math. Soc., Providence, R.I, 171–178.

Jackson, F. H. [1910], *Transformations of q-series,* Messenger of Math., **39**, 145–153.

Jackson, F. H. [1921], *Summation of q-hypergeometric series,* Messenger of Math., **50**, 101–112.

Jacobi, C. G. J. [1829], *Fundamenta Nova Theoriae Functionum Ellipticarum,* Regiomonti. Sumptibus fratrum Bornträger; reprinted in Gesammelte Werke, **1** (1881), 49–239, Reimer, Berlin.

Jacobi, C. G. J. [1846], *Über einige der Binomialreihe analoge Reihen,* J. reine angew. Math., **32**, 197-204; reprinted in Gesammelte Werke, **6** (1881), 163-173, Reimer, Berlin.

Pfaff, J. F. [1797], *Observationes analyticae ad L. Euler Institutiones Calculi Integralis,* Vol. IV, Supplem. II et IV, Historia de 1793, Nova acta acad. sci. Petropolitanae, **11** (1797), 38–57.

Rahman, M. [1990], *Some extensions of the beta integral and the hypergeometric function,* Orthogonal Polynomials (P. Nevai, ed.), Kluwer Academic Publishers, 1–27.

Saalschütz, L. [1890], *Eine Summationsformel,* Zeitschr. Math. Phys., **35**, 186–188.

Slater, L. J. [1966], *Generalized Hypergeometric Functions*, Cambridge University Press, Cambridge.

Watson, G. N. [1929], *A new proof of the Rogers-Ramanujan identities,* J. London Math. Soc., **4**, 4–9.

Wilf, H. S. and Zeilberger, D. [1992], *An algorithmic proof theory for hypergeometric (ordinary and "q") multisum/integral identities*, **108**, 575–633.

Zeilberger, D. [1987], *A q-Foata proof of the q-Saalschütz identity*, Europ. J. Combinatorics, **8**, 461–463.

Fields Institute Communications
Volume **14**, 1997

$\int_{n/4}^{m/6} \ln\Gamma(z)dz$

R. Wm. Gosper, Jr.
Macsyma Inc
20 Academy Street
Arlington, Massachusetts 02174 USA

Mann muss immer unkern — Felix Knuth

Abstract. We evaluate $\int \ln\Gamma(z)dz$, which could be notated $\psi_{-2}(z)$, between rational limits with denominators 1, 2, 3, 4, and 6. The simplest interesting result is (using $t!$ for $\Gamma(t+1)$)

$$\int_0^{1/2} \ln t!dt = \frac{\gamma+3\ln\pi}{8} - \frac{\ln 2}{6} - \frac{3\zeta'(2)}{4\pi^2} - \frac{1}{2} \approx -.04285374065.$$

We note that all integrals of the form $\int z^n \ln z!dz$ are expressible in terms of $\zeta'(-n,z)$, the first derivative of the Hurwitz zeta function.

1 Introduction

We establish

$$\int_0^{1/2} \ln t!dt = \frac{\gamma+3\ln\pi}{8} - \frac{\ln 2}{6} - \frac{3\zeta'(2)}{4\pi^2} - \frac{1}{2} \approx -.04285374065,$$

$$\int_0^{1/3} \ln t!dt = \frac{2\gamma+5\ln 2\pi}{18} - \frac{25\ln 3}{72} - \frac{\sqrt{3}\pi}{54} + \frac{\sqrt{3}\psi_1(1/3)}{36\pi} - \frac{2\zeta'(2)}{3\pi^2}$$
$$-\frac{1}{3} \approx -.02296612858214,$$

$$\int_0^{1/4} \ln t!dt = \frac{3\gamma+7\ln\pi-9\ln 2}{32} - \frac{9\zeta'(2)}{16\pi^2} + \tfrac{\lambda}{4\pi} - \tfrac{1}{4} \approx -.014099944308,$$

$$\int_0^{1/6} \ln t!dt = \frac{10\gamma-23\ln 3+22\ln\pi}{144} - \frac{\sqrt{3}\pi}{36} + \frac{\sqrt{3}\psi_1(1/3)}{24\pi} - \frac{5\zeta'(2)}{12\pi^2}$$
$$-\frac{1}{6} \approx -.006818642158464,$$

1991 *Mathematics Subject Classification.* Primary: 33B15, 11M35; Secondary: 42A16.

where

$$\lambda := \text{Catalan's constant} \quad = \tfrac{1}{8}\left(\psi_1\left(\tfrac{1}{4}\right) - \pi^2\right)$$

$$= \tfrac{1}{16}\left(\psi_1\left(\tfrac{1}{4}\right) - \psi_1\left(\tfrac{3}{4}\right)\right) \approx .9159655941772,$$

which, with the Γ multiplication formula and its limit, the Barnes-Alexeiewsky recurrence,

$$\int_{z-1}^{z} \ln t!dt = \ln\sqrt{2\pi} - (1 - \ln z)z,$$

suffice for $\int_{n/4}^{m/6}$, e.g.,

$$\int_0^{2/3} \ln t!dt = \frac{\gamma + 10\ln 2 + 4\ln\pi}{9} - \frac{49\ln 3}{72} + \frac{\sqrt{3}\pi}{54} - \frac{\sqrt{3}\psi_1(1/3)}{36\pi} - \frac{2\zeta'(2)}{3\pi^2} - \frac{2}{3}$$
$$\approx -.06178124243647,$$
$$\int_0^{3/4} \ln t!dt = \frac{3\gamma + 15\ln\pi - 33\ln 2}{32} + \frac{3\ln 3}{4} - \frac{9\zeta'(2)}{16\pi^2} - \frac{\lambda}{4\pi} - \frac{3}{4} \approx -.06959909378,$$
$$\int_0^{5/6} \ln t!dt = \frac{10\gamma - 119\ln 3 + 70\ln\pi}{144} - \frac{5\ln 5 - 2\ln 2}{6} + \frac{\sqrt{3}\pi}{36} - \frac{\sqrt{3}\psi_1(1/3)}{24\pi}$$
$$- \frac{5\zeta'(2)}{12\pi^2} - \frac{5}{6} \approx -.07570064608869.$$

The key lemma is an expansion of the general polylogarithm Li_k near 1.

2 Lemmas

[1] gives, without proof,

$$\mathrm{Li}_k(e^{-z}) = \begin{cases} (-k)!z^{k-1} + \sum_{n\geq 0} \frac{\zeta(k-n)}{n!}(-z)^n, & \text{if } k \neq \lfloor k \rfloor \text{ or } k \leq 0; \text{ else} \\ (H_{k-1} - \ln z)\frac{(-z)^{k-1}}{(k-1)!} + \sum_{n\geq 0, \neq k-1} \frac{\zeta(k-n)}{n!}(-z)^n, & H_{k-1} := \psi_0(k) + \gamma; \end{cases} \quad \text{(Li1)}$$

where

$$\mathrm{Li}_k(z) := \sum_{n\geq 1} \frac{z^n}{n^k}.$$

(Earlier [3] gave (Li1) for negative integer k.) Differentiating ∂k at 1,

$$\sum_{n\geq 1} \frac{(\ln n)e^{-nz}}{n} = -\left.\frac{\partial \mathrm{Li}_k(e^{-z})}{\partial k}\right|_{k=1} \quad \text{(Li1')}$$

$$= -\sum_{n\geq 1} \frac{\zeta'(1-n)}{n!}(-z)^n + \frac{(\gamma + \ln z)^2}{2} + \gamma_1 + \frac{\pi^2}{12},$$

where

$$\gamma_1 = \frac{(\ln 2\pi)^2 - \gamma^2}{2} + \zeta''(0) + \frac{\pi^2}{24}$$

$$= \sum_{n\geq 1} \frac{\ln n}{n} - \frac{(\ln(n+1))^2 - (\ln n)^2}{2} \approx -.07281584548367672,$$

the first Stieltjes constant, [5]. Differentiating the ζ reflection formula at a negative, even integer:

$$\zeta'(-2n) = (-)^n \frac{(2n)!\zeta(2n+1)}{2^{2n+1}\pi^{2n}}.$$

We can apply this to formula (Li1′) if we first eliminate $\zeta'(1-2n)$ by setting $z = 2\pi i t$ and taking the imaginary part. The resulting infinite series coincides with the Taylor expansion of $\ln\Gamma(t)$ at 0:

$$\begin{aligned}\ln\Gamma(t) &= \tfrac{1}{2}\left(\ln\Gamma(t) - \ln(-\Gamma(-t)) + \ln\tfrac{\pi}{t\sin\pi t}\right)\\ &= \tfrac{1}{2}\ln\tfrac{\pi}{t\sin\pi t} - \gamma t - \sum_{n\geq 1}\frac{\zeta(2n+1)}{2n+1}t^{2n+1}\\ &= \sum_{n\geq 1}\frac{\ln n}{\pi n}\sin 2\pi n t + \frac{1}{2}\left(\ln\frac{2\pi^2}{\sin\pi t} + \gamma\right) - (\gamma + \ln 2\pi)t, \qquad 0 < t < 1,\end{aligned}$$

which is (the hard part of) the Fourier expansion of $\ln\Gamma(t)$. Finishing the expansion, via

$$\ln\sin\pi t = -\sum_{n\geq 1}\frac{\cos 2\pi n t}{n} - \ln 2, \qquad 0 < t < 1, \qquad \text{(ln sin)}$$

we get

$$\ln\Gamma(t) = \frac{\ln 2\pi}{2} + \sum_{n\geq 1}\frac{2(\gamma + \ln 2\pi n)\sin(2\pi n t) + \pi\cos(2\pi n t)}{2\pi n}, \qquad 0 < t < 1. \quad \text{(Fou)}$$

Differentiating (Li1) ∂k at 2 produces the same series as integrating equation (Li1′) dz:

$$\begin{aligned}\sum_{n\geq 1}\frac{(\ln n)e^{-nz}}{n^2} &= -\left.\frac{\partial \mathrm{Li}_k(e^{-z})}{\partial k}\right|_{k=2} = -\sum_{n\geq 2}\frac{\zeta'(2-n)}{n!}(-z)^n\\ &\quad - \left(\tfrac{(\gamma+\ln z)(\gamma-2+\ln z)}{2} + \gamma_1 + \tfrac{\pi^2}{12} + 1\right)z - \zeta'(2),\end{aligned}$$

Thus

$$\begin{aligned}\sum_{n\geq 1}\frac{\pi\sin(2\pi n t) - 2(\ln n)\cos(2\pi n t)}{4\pi^2 n^2} &= \ln f_1(t-1) \qquad (\psi_{-2})\\ &\quad + (\gamma - 1 + \ln(2\pi))\binom{t}{2} + \frac{\zeta'(2)}{2\pi^2},\end{aligned}$$

where

$$f_1(n) = 1^1 2^2 \ldots n^n, \qquad f_1(z) := (2\pi)^{-z/2}\exp\left(\binom{z+1}{2} + \int_0^z \ln t!\,dt\right),$$

which could equally well come from integrating (Fou) dt.

This series comes out in terms of $\zeta(2)$ and $\zeta'(2)$ for $t = j/4$ and $k/6$, yielding the identities asserted in the Introduction.

[2] denotes $\int \ln \Gamma(t)dt$ by ilg, but, if we can decide on the right constant of integration, it would seem logical to simply call it $\psi_{-2}(t)$, the first "negapolygamma". By parts, it is equivalent to $\int t\psi_0(t)dt$. f_1 is sometimes called Barnes's G function. $\zeta'(2)$ or simple functions thereof are sometimes called Glaisher's or Porter's constant ([4], [6]).

It seems strange that, at rationals like $1/3, 1/4$, and $1/6$, the function which interpolates $1^1 2^2 \ldots n^n$, *i.e.*, $\int \ln \Gamma$, comes out in simpler constants, while the function which interpolates $1 \cdot 2 \cdots n$, *i.e.*, Γ, does not. However, if we denote $\ln \Gamma$ by ψ_{-1}, this extends for subscripts -1 and -2 a pattern where $\psi_n(j/4$ or $k/6)$ simplifies iff n is even. What seems stranger is that the Fourier coefficients of $\ln \Gamma(t)$ should be elementary, while those of $\ln t$ are not.

3 Further results

(ψ_{-2}) can be repeatedly integrated by parts to give $\int_0^u t^n \ln \Gamma(t)dt$, but for $n > 0$, the only u for which the series has an obvious closed form are integers and half integers, e.g.,

$$\int_0^{1/2} t \ln dt! = \frac{\gamma}{48} + \frac{\ln \pi}{12} - \frac{4\zeta'(2) + 7\zeta(3)}{32\pi^2} - \frac{1 + \ln 2}{16} \approx -.01317042784872,$$

$$\int_0^{1/2} t^2 \ln dt! = \frac{\gamma + 5 \ln \pi}{192} - \frac{2\zeta'(2) + 3\zeta(3)}{32\pi^2} + \frac{15\zeta'(4)}{32\pi^4} - \frac{37 \ln 2}{1440} - \frac{1}{72} \approx -.00469460599089.$$

The earlier ψ_{-2} integrals, combined with the Γ and ψ reflection formulæ, yield $\int_{n/4}^{m/6} \ln \sin \pi t dt$. A superior alternative is to integrate equation (ln sin),

$$\int_0^u \ln \sin \pi t dt = -u \ln 2 - \frac{1}{2\pi} \sum_{n \geq 1} \frac{\sin 2n\pi u}{n^2}, \qquad 0 \leq u \leq 1,$$

which gives trigammas for *all* rational u, e.g.,

$$\int_0^{1/5} \ln \sin \pi t dt = \frac{1}{25\pi} \left(\frac{4}{\sqrt{5}} \phi \pi^2 - \phi \psi_1 \left(\frac{1}{5} \right) - \psi_1 \left(\frac{2}{5} \right) \right) \sin \frac{\pi}{5} - \frac{\ln 2}{5} \approx -.2973633652639,$$

where

$$\phi := \frac{1 + \sqrt{5}}{2} \approx 1.6180339887499.$$

This, with the Γ reflection formula, gives $\ln \Gamma$ integrals over intervals symmetric about integers or half integers:

$$\int_{1-z}^{z} \ln t! dt = z \ln z - (1 - z) \ln(1 - z) + (1 - 2z)(1 - \ln \sqrt{\pi}) - \int_{1-z}^{1/2} \ln \sin \pi t dt,$$

e.g.,

$$\int_{2/5}^{3/5} \ln t!dt = \frac{\ln \pi - 3\ln 2}{10} + \frac{1}{25\pi}\left(\phi\psi_1\left(\frac{2}{5}\right) - \psi_1\left(\frac{1}{5}\right) + \frac{2\pi^2}{\phi\sqrt{5}}\right)\sin\frac{\pi}{5}$$
$$+\frac{3\ln 3 - \ln 5 - 1}{5} \approx -.0238446115029.$$

With repeated integration of (ln sin) by parts, we get $\int_0^u t^n \ln \sin \pi t dt$ in terms of polygammas for all positive integer n and rational u, e.g.,

$$\int_0^{1/8} t \ln \sin \pi t dt = \tfrac{\sqrt{2}}{2048\pi^2}\left(\psi_2\left(\tfrac{1}{8}\right) - 4\psi_1\left(\tfrac{1}{8}\right)\right)$$
$$+\tfrac{5+4\sqrt{2}}{512}\pi + \tfrac{2\sqrt{2}-1}{64\pi}\lambda + 7\tfrac{37+16\sqrt{2}}{1024\pi^2}\zeta(3) - \tfrac{\ln 2}{128} \approx -.01130943023.$$

The ψ_{-3} integral (equivalent to $\int t^2\psi_0(t)dt$) over the general unit interval appears to require the ψ_{-2} integral over the general interval:

$$\int_{z-1}^{z} t \ln t!dt = \int_0^z \ln t!dt + \frac{z-2}{2} z\ln z - \frac{z^2}{4} + z - \frac{\ln 2\pi}{3} + \frac{\zeta'(2)}{2\pi^2}.$$

E.g., employing the $\psi_{-2}(1/6)$ result from the Introduction,

$$\int_{-5/6}^{1/6} t\ln t!dt = \frac{\sqrt{3}}{24\pi}\psi_1\left(\frac{1}{3}\right) - \frac{\gamma + 13\ln 2\pi}{72} - \frac{1+\ln 3}{144} + \frac{\zeta'(2)}{12\pi^2} \approx -.2815791335514.$$

4 The Hurwitz zeta function

All of these ψ_{-n} integrals are expressible via the first derivative of the Hurwitz zeta function $\zeta(s,a)$. Differentiate ∂s the reflection formula [7]:

$$\frac{\partial}{\partial s}\frac{(2\pi)^s\zeta(1-s,a)}{\Gamma(s)} = -\sum_{n\geq 1}\frac{\pi\sin(s/2-2an)\pi + 2(\ln n)\cos(s/2-2an)\pi}{n^s},$$

which, for integer s, can be written

$$\zeta'(1-s,a) = B_s(a)\frac{\psi_0(s) - \ln 2\pi}{s}$$
$$-(-)^s\frac{(s-1)!}{(2\pi)^{2s}}\frac{\partial^s}{\partial a^s}\sum_{n\geq 1}\frac{\pi\sin(2\pi an) - 2\ln(n)\cos(2\pi an)}{n^{2s}},$$

where B_s is the Bernoulli polynomial

$$B_s(a) = -s\zeta(1-s,a).$$

With this we eliminate the series from equation (ψ_{-2}),

$$\int_0^a \ln\Gamma(t)dt = a\frac{1-a+\ln 2\pi}{2} + \zeta'(-1,a) - \zeta'(-1),$$

and from its integration by parts,

$$\int_0^a t\ln\Gamma(t)dt = \frac{a}{24} + a^2\frac{1+2\ln 2\pi}{8} - \frac{a^3}{4} + a\zeta'(-1,a) - \frac{\zeta'(-2,a)}{2} + \frac{\zeta(3)}{8\pi^2},$$

etc.

We also note that ζ' provides a concise, analytic extension of the "higher factorials", e.g.:

$$1^{1^s}2^{2^s}\ldots n^{n^s} = e^{\zeta'(-s,n+1)-\zeta'(-s)},$$

$$1!2!\ldots n! \quad = n!^{n+1}e^{\zeta'(-1)-\zeta'(-1,n+1)} = e^{(n+1)\left(\zeta'(0,n+1)-\zeta'(0)\right)+\zeta'(-1)-\zeta'(-1,n+1)},$$

and reflection in the second argument provides polylogarithms, e.g.,

$$\mathrm{L}i_2(e^{2i\pi z}) = \pi^2 B_2(z) + 2i\pi\left(\zeta'(-1,z) - \zeta'(-1,1-z)\right).$$

With $z = 1/6$ and our earlier result for $\zeta'(-1,1/6)$,

$$\mathrm{L}i_2(e^{i\pi/3}) = \frac{\pi^2}{36} + \frac{i}{\sqrt{3}}\left(\frac{\psi_1(1/3)}{2} - \frac{\pi^2}{3}\right) \approx .2741556778080377 + 1.01494160640965i,$$

but this merely reminds us that we can express $\mathrm{L}i_k(e^{ip\pi/q})$ in terms of $\psi_{k-1}(j/q), 0 < j < q$, simply by grouping the terms of the $\mathrm{L}i$ series q at a time, leading to a rational function summand.

References

[1] Cohen, H., Lewin, L. and ZagieR, D. [1992], *A Sixteenth-order Polylogarithm Ladder*, Experimental Mathematics, **1** no. 1, 26.

[2] Glaisher, J. W. L. [1895], *Products and Series involving Prime Numbers Only II*, Quarterly Journal of Pure and Applied Mathematics, **XXVIII** 4.

[3] Gosper, R. Wm. [1990], *Strip Mining in the Abandoned Orefields of Nineteenth Century Mathematics*, in Computers in Mathematics, (D. Chudnovsky & R. Jenks, eds.), Lecture Notes in Pure and Applied Mathematics, Vol. 125, Marcel Dekker, New York City, NY, 282.

[4] Knuth, D. E. [1976], *An evaluation of Porter's constant*, Computers and Mathematics, with Applications, **2**, 137–139.

[5] Liang, J. J. Y., and Todd, J. [1972], *The Stieltjes Constants*, J. Res. Nat. Bur. Standards, **76**, 161–178.

[6] Pitteway, M. L .V. and Castle, C. M. A. [1988], *On the subtractive version of Euclid's algorithm*, Bull. of the Inst. of Math. and its Appl., **24**, No. 1-2, 17–21.

[7] Whittaker, E. T. and Watson, G. N. [1927], Modern Analysis, Fourth edition, Cambridge University Press, Chap XIII, 269.

Fields Institute Communications
Volume **14**, 1997

On a q-Analogue of Gauss Equation and Some q-Riccati Equations

F. Alberto Grünbaum
Department of Mathematics
University of California
Berkeley, California 94720-3840 USA

Luc Haine
Department of Mathematics
Université Catholique de Louvain
1348 Louvain-la-Neuve, Belgium

The second order linear hypergeometric differential equation of Gauss

$$k(1-k)y'' + [\gamma - (\alpha+\beta+1)k]y' - \alpha\beta y = 0 \, , \tag{1}$$

plays a central role in many areas of mathematics. Its q-analogue

$$\begin{aligned} &k(\gamma - \alpha\beta q k)\; D_q^2 y + \left[\tfrac{1-\gamma}{1-q} + \tfrac{(1-\alpha)(1-\beta)-(1-\alpha\beta q)}{1-q}\; k\right] D_q y \\ &-\tfrac{(1-\alpha)(1-\beta)}{(1-q)^2}\; y = 0 \, , \end{aligned} \tag{2}$$

where D_q denotes the standard q-derivative operator

$$D_q y(k) = \frac{y(kq) - y(k)}{k(q-1)} \, ,$$

or equivalently

$$\frac{\gamma - \alpha\beta k}{k}(y(qk) - y(k)) + \frac{q-k}{k}(y(k/q) - y(k)) = (\alpha-1)(\beta-1)y(k) \, , \tag{2'}$$

has a long and venerable history too, going back to Heine [6], see also [2],[5],[8]. Another very important q-difference equation

$$A(k)(y(qk) - y(k)) + A(1/k)(y(k/q) - y(k)) = (1 - q^{-n})(abcdq^{n-1} - 1)y(k) \tag{3}$$

where

$$A(k) = \frac{(1-ka)(1-kb)(1-kc)(1-kd)}{(1-k^2)(1-k^2 q)} \, ,$$

1991 *Mathematics Subject Classification.* Primary: 39A10; Secondary: 33D15.

The second author is a Research Associate for FNRS

was put forward by Askey and Wilson [1] (see also [7]) in their generalization of the classical Jacobi polynomials. In this note we want to call attention to a q-difference equation depending on six parameters a, b, c, d and *two new parameters* z, ε which contains the two equations above as special cases, namely

$$A(k)(y(qk)-y(k)) + A(\varepsilon/k)(y(k/q)-y(k)) = (1-z)(\frac{abcd}{qz}-1)y(k) \tag{4}$$

with

$$A(k) = \frac{(1-ka)(1-kb)(\varepsilon-kc)(\varepsilon-kd)}{(\varepsilon-k^2)(\varepsilon-k^2q)}$$

and its consequence

$$A(\varepsilon/k) = \frac{(k-\varepsilon a)(k-\varepsilon b)(k-c)(k-d)}{(k^2-\varepsilon)(k^2-\varepsilon q)} \ ,$$

with ε arbitrary. By rescaling k and renaming a, b, c, d one can always assume that $\varepsilon = 0$ or $\varepsilon = 1$. Thus, in what follows, we shall restrict attention to these two cases and think of equation (4) as depending on five parameters for ε fixed. When $\varepsilon = 1$, for the special choice $z = q^{-n}$, one obtains equation (3) above, and when $\varepsilon = 0$ if one takes

$$a = \frac{\alpha\beta}{\gamma} \ , \qquad c = q \ , \qquad d = \frac{\gamma}{b} \ , \qquad z = \alpha \tag{5}$$

and let $b \to \infty$, one gets equation (2′).

Introduce, for a general function $s(k)$, the operator D_s by means of

$$D_s y(k) = \frac{y(qk)-y(k)}{s(qk)-s(k)} \ . \tag{6}$$

Put

$$\begin{aligned} t_1 &= \varepsilon(a+b)+(c+d) \\ t_2 &= \varepsilon^2 ab + \varepsilon(a+b)(c+d) + cd \\ (q-1)t_3 + t_1 &= \varepsilon ab(c+d) + cd(a+b) \\ (q-1)t_4 + 1 &= abcd \\ (q-1)^2 t_5 &= (1-z)\left(\tfrac{abcd}{qz}-1\right) \ . \end{aligned}$$

Replacing k by kq in equation (4) and choosing

$$s(k) = \delta(k+\frac{\varepsilon}{k}) \ , \qquad \delta \text{ arbitrary,}$$

one can rewrite this equation in terms of D_s^2 and D_s as

$$\begin{aligned} &\delta^2(k^2q-\varepsilon) \begin{bmatrix} ((q-1)t_4+1)q^4k^4 - ((q-1)t_3+t_1)q^3k^3 \\ +t_2q^2k^2 - \varepsilon t_1 qk + \varepsilon^2 \end{bmatrix} D_s^2 y(k) \\ &\qquad -\delta kq(k^2q^2-\varepsilon) \begin{bmatrix} ((q-1)t_5-t_4)q^2k^2 + t_3qk \\ -\varepsilon(q(q-1)t_5+t_4) \end{bmatrix} D_s y(k) \\ &\qquad -k^2q^3(k^2q^2-\varepsilon)t_5 y(k) = 0 \ . \end{aligned}$$

Observe that in the limit $q = 1$, we get

$$(s(k)^2 - \delta t_1 s(k) + \delta^2(t_2-2\varepsilon))\frac{d^2}{ds^2}y(k) + (t_4 s(k) - \delta t_3)\frac{d}{ds}\, y(k) - t_5 y(k) = 0 \ , \tag{7}$$

and thus in the limit $y(k)$ can be considered as a function of $s(k)$. By translation and scaling in s, (7) can be brought to the standard form (1) of Gauss' hypergeometric equation. For the reasons explained above we propose to call equation (4) the Gauss-Askey-Wilson equation. Further justification for this terminology can be gleaned from the discussion below.

If one adds the term

$$\frac{\left(z, \frac{abcd}{qz}, ak, \frac{\varepsilon a}{k}; q\right)_\infty}{(ac, ad, ab\varepsilon, q; q)_\infty}$$

to the right-hand side of (4), then a solution of this inhomogeneous equation is given, using the notation of [2], by

$$y(k) = {}_4\phi_3\left(\begin{matrix} z, \ \frac{abcd}{qz}, \ ak, \ a\varepsilon/k \\ ac, \ ad, \ ab\varepsilon \end{matrix}; q, q\right) . \tag{8}$$

Notice that, of course, when $\varepsilon = 1$ and $z = q^{-n}$ this reduces to the expression for the Askey-Wilson polynomials in terms of basic hypergeometric series and that the inhomogeneous term vanishes for other special choices. For details, see [4].

There is a well-known relation between the Riccati nonlinear first order differential equation and a general second order linear equation.

In our work on the "bispectral problem", see [3], we run into the following specific Riccati equation

$$(c_1 + c_2 k - (c_3 + 2c_4)k^2)\frac{df}{dk} + c_3 f^2 + (2c_4 k + c_5)f + c_6 = 0 , \tag{9}$$

with c_i, $1 \leq i \leq 6$, arbitrary constants. Generically they can be scaled down to just five parameters. This equation gets converted by the usual "logarithmic derivative trick" and further standard manipulations into a very special equation, namely Gauss' hypergeometric equation. In particular one gets an explicit expression for $f(k)$ depending on five parameters, to wit: the three parameters of Gauss' equation and the two roots of the coefficient of df/dk in (9). Our recent work (see [4]) on a q-version of the "bispectral problem" led us to the equation

$$f(qk) = \frac{P(k)f(k) + Q(k)}{R(k)f(k) + S(k)} , \tag{10}$$

with $P(k)$, $Q(k)$, $R(k)$, $S(k)$ given by the expressions

$$\begin{aligned} S(k) &= c_1 + c_2 s(k) - (c_3 q + c_4(q+1))s(k)^2 \\ &\quad + \tfrac{\varepsilon\delta}{qk}(q^2 - 1)\left((c_3 + c_4)s(k) - \tfrac{c_2 + c_5}{q+1}\right) \\ R(k) &= c_3(s(kq) - s(k)) \\ Q(k) &= c_6(s(k) - s(kq)) \\ P(k) &= S(\varepsilon/kq) , \end{aligned} \tag{11}$$

and the c_i, $1 \leq i \leq 6$, arbitrary constants which can again be scaled to just five parameters. In order to compare equation (10) with (9), we rewrite it in terms of the operator D_s introduced in (6), namely

$$(R(k)f(k) + S(k))D_s f(k) + c_3 f(k)^2 + (c_4(s(k) + s(kq)) + c_5)f(k) + c_6 = 0 .$$

We propose to call equation (10), for arbitrary choices of the functions $P(k)$, $Q(k)$, $R(k)$ and $S(k)$, the q-analogue of the Riccati equation. In this case the "logarithmic derivative trick" can be replaced by the transformation

$$f(k)=\frac{S(k)}{R(k)}\left(\frac{w(kq)-w(k)}{w(k)}\right),$$

which converts (10) into a second order linear q-difference equation

$$\begin{aligned}R(k)S(k)S(kq)w(kq^2)\quad &-(R(kq)P(k)+R(k)S(kq))S(k)w(kq)\\ &+(P(k)S(k)-Q(k)R(k))R(kq)w(k)=0\,.\end{aligned}$$

In the special case where $P(k)$, $Q(k)$, $R(k)$ and $S(k)$ are given by (11), see [4] for details, for generic choices of the c_i, $1\le i\le 6$, one arrives at

$$\begin{aligned}f(k)\quad &=s(k)-\frac{\delta z[q(y_3+qy_1)z^2-(q+1)(y_1y_4+qy_3)z+(y_3+qy_1)y_4]}{(y_4-z^2)(y_4-q^2z^2)}\\ &+\frac{\delta qz}{y_4-qz^2}\,A(k)(k-\frac{\varepsilon}{k})\left(\frac{y(kq)}{y(k)}-1\right)\\ &+\frac{\delta(z-1)}{y_4-qz^2}\left[\frac{qz(qzy_3-y_1y_4)}{y_4-q^2z^2}+y_4k+\frac{\varepsilon qz}{k}\right],\end{aligned}$$

with $y(k)$ an arbitrary solution of the Gauss-Askey-Wilson equation (4), for an appropriate choice of the parameters a,b,c,d and z and y_i, $1\le i\le 4$, defined by

$$y_4k^4-y_3k^3+y_2k^2-\varepsilon y_1k+\varepsilon^2=(1-ka)(1-kb)(\varepsilon-kc)(\varepsilon-kd)\,.$$

These five parameters a, b, c, d and z are projectively equivalent to the c_i's above.

Acknowledgements

The first author was supported in part by NSF Grant #DMS94–00097 and by AFOSR under contract AFO F49629–92.

We thank the referees for pointing out an oversight in an earlier version of this paper.

References

[1] Askey, R. and Wilson, J. [1985], *Some basic hypergeometric orthogonal polynomials that generalize Jacobi polynomials,* Mem. Amer. Math. Soc., **319**.

[2] Gasper, G. and Rahman, M. [1990], *Basic hypergeometric series*, Encyclopedia of Mathematics & Its Applications, **35**, Cambridge University Press.

[3] Grünbaum, F. A. and Haine, L. [1996], *A theorem of Bochner, revisited,* in Algebraic Aspects of Integrable Systems (A. S. Fokas and I. M. Gel'fand, ed.), in memory of Irene Dorfman (Birkhauser, Boston), to appear.

[4] Grünbaum, F. A. and Haine, L. *Some functions that generalize the Askey-Wilson polynomials*, preprint.

[5] Hahn, W. [1949], *Beiträge zur Theorie der Heineschen Reihen, Die 24 Integrale der hypergeometrischen q-Differenzengleichung, das q-Analogon der Laplace-Transformation,* Math. Nachr., **2**, 340–379.

[6] Heine, E. [1847], *Untersuchungen über die Reihe* $1+\frac{(q^\alpha-1)(q^\beta-1)}{(q-1)(q^\gamma-1)}\,x+\frac{(q^{\alpha+1}-1)(q^\alpha-1)(q^{\beta+1}-1)(q^\beta-1)}{(q^2-1)(q-1)(q^{\gamma+1}-1)(q^\gamma-1)}\,x^2+\ldots$, J. reine angew. Math., **34**, 285–328.

[7] Koekoek, R. and Swarttouw, R. F. [1994], *The Askey-scheme of hypergeometric orthogonal polynomials and its q-analogue*, Report 94-05, Technical University Delft, Faculty of Technical Mathematics & Informatics.

[8] Koornwinder, T. H. [1994], *Compact quantum groups and q-special functions*, in Representations of Lie Groups and Quantum Groups (V. Baldoni and M. A. Picardello, ed.), Pitman Research Notes in Mathematics Series 311, (Longman Scientific & Technical), 46–128.

Fields Institute Communications
Volume **14**, 1997

Determinant Evaluations and $U(n)$ Extensions of Heine's ${}_2\phi_1$-Transformations

Robert A. Gustafson
Department of Mathematics
Texas A&M University
College Station, TX 77843 USA

Christian Krattenthaler
Institut für Mathematik der Universität Wien
Strudlhofgasse 4
A-1090 Wien, Austria

Abstract. We give new proofs for $U(n)$ extensions of Heine's three ${}_2\phi_1$-transformation formulas that were recently discovered in another paper by the authors. These new proofs proceed by combining Heine's original transformation formulas with certain determinant evaluations. As a by-product we are able to generalize two of the $U(n)$ transformation formulas.

1 Introduction

In the recent paper [3] we discovered $U(n)$ extensions of Heine's ${}_2\phi_1$-transformation formulas (see (2.1)–(2.3) for Heine's formulas) which involve $U(n)$ basic hypergeometric series of a new type. With the q-notation $(a)_n := \prod_{i=1}^{n}(1 - aq^{i-1})$ (note that this differs from the standard notation $(a;q)_n$ in Gasper and Rahman's book [1] in that we drop the base q for the sake of being able to write our formulas more compactly), these transformation formulas are the following. Our

---

1991 *Mathematics Subject Classification.* Primary: 33D20; Secondary: 33C20, 33C70.
The first author was supported in part by NSF Grant DMS-9304290.
The second author was supported in part by the Austrian Science Foundation FWF, grant P10191-MAT.

$U(n)$ extension ([3], (1.7)) of the first Heine transformation reads

$$\sum_{k_1,\dots,k_n\ge 0}\left(\prod_{1\le i<j\le n}\frac{1-\frac{X_j}{X_i}q^{k_j-k_i}}{1-\frac{X_j}{X_i}}\prod_{i=1}^n\frac{(A)_{k_i}\,(BX_i)_{k_i}}{(q)_{k_i}\,(CX_i)_{k_i}}q^{k_i(1-i)}Z^{k_i}\right)$$

$$=\prod_{i=1}^n\frac{(Aq^{1-i})_\infty\,(BZX_i)_\infty}{(Zq^{1-i})_\infty\,(CX_i)_\infty}\sum_{k_1,\dots,k_n\ge 0}\left(\prod_{1\le i<j\le n}\frac{1-\frac{X_j}{X_i}q^{k_j-k_i}}{1-\frac{X_j}{X_i}}\right. \tag{1.1}$$

$$\left.\times\prod_{i=1}^n\frac{(Z)_{k_i}\,(CX_i/A)_{k_i}}{(q)_{k_i}\,(BZX_i)_{k_i}}q^{k_i(1-i)}A^{k_i}\right),$$

provided the series terminate or $|q|<1$, $|Zq^{1-n}|<1$, $|Aq^{1-n}|<1$. Our $U(n)$ extension ([3], (1.8)) of the second Heine transformation reads

$$\sum_{k_1,\dots,k_n\ge 0}\left(\prod_{1\le i<j\le n}\frac{1-\frac{X_j}{X_i}q^{k_j-k_i}}{1-\frac{X_j}{X_i}}\prod_{i=1}^n\frac{(A)_{k_i}\,(BX_i)_{k_i}}{(q)_{k_i}\,(CX_i)_{k_i}}q^{k_i(1-i)}Z^{k_i}\right)$$

$$=\prod_{i=1}^n\frac{(Cq^{1-i}/B)_\infty\,(BZX_i)_\infty}{(Zq^{1-i})_\infty\,(CX_i)_\infty}\sum_{k_1,\dots,k_n\ge 0}\left(\prod_{1\le i<j\le n}\frac{1-\frac{X_j}{X_i}q^{k_j-k_i}}{1-\frac{X_j}{X_i}}\right. \tag{1.2}$$

$$\left.\times\prod_{i=1}^n\frac{(ABZ/C)_{k_i}\,(BX_i)_{k_i}}{(q)_{k_i}\,(BZX_i)_{k_i}}q^{k_i(1-i)}\left(\frac{C}{B}\right)^{k_i}\right),$$

provided the series terminate or $|q|<1$, $|Zq^{1-n}|<1$, $|Cq^{1-n}/B|<1$. Finally, our $U(n)$ extension ([3], (1.9)) of the third Heine transformation reads

$$\sum_{k_1,\dots,k_n\ge 0}\left(\prod_{1\le i<j\le n}\frac{1-\frac{X_j}{X_i}q^{k_j-k_i}}{1-\frac{X_j}{X_i}}\prod_{i=1}^n\frac{(A)_{k_i}\,(BX_i)_{k_i}}{(q)_{k_i}\,(CX_i)_{k_i}}q^{k_i(1-i)}Z^{k_i}\right)$$

$$=\prod_{i=1}^n\frac{(ABZq^{1-i}/C)_\infty}{(Zq^{1-i})_\infty}\sum_{k_1,\dots,k_n\ge 0}\left(\prod_{1\le i<j\le n}\frac{1-\frac{X_j}{X_i}q^{k_j-k_i}}{1-\frac{X_j}{X_i}}\right. \tag{1.3}$$

$$\left.\times\prod_{i=1}^n\frac{(C/B)_{k_i}\,(CX_i/A)_{k_i}}{(q)_{k_i}\,(CX_i)_{k_i}}q^{k_i(1-i)}\left(\frac{ABZ}{C}\right)^{k_i}\right),$$

provided the series terminate or $|q|<1$, $|Zq^{1-n}|<1$, $|ABZq^{1-n}/C|<1$. In particular, from these identities there also result $U(n)$ extensions of the q-Gauss summation formula and the q-binomial theorem, cf. [3].

We proved (1.1)–(1.3) in [3] by induction with respect to n. The purpose of this paper is to give new, direct proofs of these identities, avoiding induction. These new proofs, which are described in Section 2 and are shorter than the original ones, proceed by a combination of Heine's original ${}_2\phi_1$-transformations and certain determinant evaluations. One determinant evaluation is simply the Vandermonde determinant. The other determinant evaluation, Lemma 1 in Section 2, comes from a determinant lemma ([4], Lemma 2.2) (see Lemma 2 in Section 2) that has been already used successfully in a number of cases by the second author ([2], [4], [5],

[6]) in the computation of generating functions for tableaux and plane partitions. In [2] this also lead to find $SU(n)$ basic hypergeometric identities. However, the particular technique that we use in the present paper seems to be new in the subject of multidimensional basic hypergeometric series. Hopefully, it will prove to be useful in other cases, too.

As a by-product of our new proofs, we are able to generalize the identities (1.2) and (1.3). We summarize the new (and old) results in the Theorem in Section 2 (identities (2.4)–(2.6)), and subsequently prove them. Interestingly, in view of our new technique, the easiest to prove is the third identity, (2.6). It just requires the Vandermonde determinant and Heine's third ${}_2\phi_1$-transformation (which is displayed in (2.3)). Next in the hierarchy comes the second identity, (2.5). It requires a fancier determinant evaluation, Lemma 1, and Heine's second ${}_2\phi_1$-transformation (which is displayed in (2.2)). The most difficult one, again in view of our technique, is the first identity, (2.4). In fact, we do not know how to prove (2.4) in a way similar to the proofs of the other identities. However, once (2.5) and (2.6) are established, (2.4) follows by simply applying (2.5) and (2.6) in succession.

2 The $U(n)$ extensions of Heine's ${}_2\phi_1$-transformations and their proofs

Before we state the $U(n)$ extensions of Heine's ${}_2\phi_1$-transformation formulas that we are going to prove, it might be convenient to recall Heine's original ${}_2\phi_1$-transformations. These are (cf. [1], (1.4.1), (1.4.5), (1.4.6))

$$\sum_{k\ge 0}\frac{(a)_k\,(b)_k}{(q)_k\,(c)_k}z^k=\frac{(a)_\infty\,(bz)_\infty}{(c)_\infty\,(z)_\infty}\sum_{k\ge 0}\frac{(c/a)_k\,(z)_k}{(q)_k\,(bz)_k}a^k, \tag{2.1}$$

$$\sum_{k\ge 0}\frac{(a)_k\,(b)_k}{(q)_k\,(c)_k}z^k=\frac{(c/b)_\infty\,(bz)_\infty}{(c)_\infty\,(z)_\infty}\sum_{k\ge 0}\frac{(abz/c)_k\,(b)_k}{(q)_k\,(bz)_k}\left(\frac{c}{b}\right)^k, \tag{2.2}$$

$$\sum_{k\ge 0}\frac{(a)_k\,(b)_k}{(q)_k\,(c)_k}z^k=\frac{(abz/c)_\infty}{(z)_\infty}\sum_{k\ge 0}\frac{(c/a)_k\,(c/b)_k}{(q)_k\,(c)_k}\left(\frac{abz}{c}\right)^k. \tag{2.3}$$

Our $U(n)$ extensions of these identities are the following.

Theorem 2.1 *There hold*

$$\sum_{k_1,\dots,k_n\ge 0}\left(\prod_{1\le i<j\le n}\frac{1-\frac{X_j}{X_i}q^{k_j-k_i}}{1-\frac{X_j}{X_i}}\prod_{i=1}^{n}\frac{(A)_{k_i}\,(BX_i)_{k_i}}{(q)_{k_i}\,(CX_i)_{k_i}}q^{k_i(1-i)}Z^{k_i}\right)$$
$$=\prod_{i=1}^{n}\frac{(Aq^{1-i})_\infty\,(BZX_i)_\infty}{(Zq^{1-i})_\infty\,(CX_i)_\infty}\sum_{k_1,\dots,k_n\ge 0}\left(\prod_{1\le i<j\le n}\frac{1-\frac{X_j}{X_i}q^{k_j-k_i}}{1-\frac{X_j}{X_i}}\right. \tag{2.4}$$
$$\left.\times\prod_{i=1}^{n}\frac{(Z)_{k_i}\,(CX_i/A)_{k_i}}{(q)_{k_i}\,(BZX_i)_{k_i}}q^{k_i(1-i)}A^{k_i}\right),$$

provided the series terminate or $|q|<1$, $|Zq^{1-n}|<1$, $|Aq^{1-n}|<1$,

$$\sum_{k_1,\dots,k_n\ge 0}\left(\prod_{1\le i<j\le n}\frac{1-\frac{X_j}{X_i}q^{k_j-k_i}}{1-\frac{X_j}{X_i}}\prod_{i=1}^{n}\frac{(A_i)_{k_i}\,(BX_i)_{k_i}}{(q)_{k_i}\,(CX_i)_{k_i}}q^{k_i(1-i)}Z^{k_i}\right)$$

$$=\prod_{i=1}^{n}\frac{(Cq^{1-i}/B)_\infty\,(BZX_i)_\infty}{(Zq^{1-i})_\infty\,(CX_i)_\infty}\sum_{k_1,\dots,k_n\ge 0}\left(\prod_{1\le i<j\le n}\frac{1-\frac{X_j}{X_i}q^{k_j-k_i}}{1-\frac{X_j}{X_i}}\right.\tag{2.5}$$

$$\left.\times\prod_{i=1}^{n}\frac{(A_iBZ/C)_{k_i}\,(BX_i)_{k_i}}{(q)_{k_i}\,(BZX_i)_{k_i}}q^{k_i(1-i)}\left(\frac{C}{B}\right)^{k_i}\right),$$

provided the series terminate or $|q|<1$, $|Zq^{1-n}|<1$, $|Cq^{1-n}/B|<1$,

$$\sum_{k_1,\dots,k_n\ge 0}\left(\prod_{1\le i<j\le n}\frac{1-\frac{X_j}{X_i}q^{k_j-k_i}}{1-\frac{X_j}{X_i}}\prod_{i=1}^{n}\frac{(A_i)_{k_i}\,(BX_i)_{k_i}}{(q)_{k_i}\,(CA_iX_i)_{k_i}}q^{k_i(1-i)}Z^{k_i}\right)$$

$$=\prod_{i=1}^{n}\frac{(BZq^{1-i}/C)_\infty}{(Zq^{1-i})_\infty}\sum_{k_1,\dots,k_n\ge 0}\left(\prod_{1\le i<j\le n}\frac{1-\frac{X_j}{X_i}q^{k_j-k_i}}{1-\frac{X_j}{X_i}}\right.\tag{2.6}$$

$$\left.\times\prod_{i=1}^{n}\frac{(CA_i/B)_{k_i}\,(CX_i)_{k_i}}{(q)_{k_i}\,(CA_iX_i)_{k_i}}q^{k_i(1-i)}\left(\frac{BZ}{C}\right)^{k_i}\right),$$

provided the series terminate or $|q|<1$, $|Zq^{1-n}|<1$, $|BZq^{1-n}/C|<1$.

Remark It should be noted that (2.4) is identical with (1.1), while (2.5) is more general than (1.2), and (2.6) is more general than (1.3). In fact, (1.2) comes out of (2.5) by setting $A_i\equiv A$, and (1.3) comes out of (2.6) by setting $A_i\equiv A$ and replacing C by C/A.

We start with the proof of the third identity.

Proof of (2.6) We have

$$\prod_{i=1}^{n}\left(X_iq^{k_i}\right)^{1-i}\times\prod_{1\le i<j\le n}\left(1-\frac{X_j}{X_i}q^{k_j-k_i}\right)=\prod_{1\le i<j\le n}\left(\frac{1}{X_jq^{k_j}}-\frac{1}{X_iq^{k_i}}\right)\tag{2.7}$$

$$=\det_{1\le s,t\le n}\left(\left(\frac{1}{X_sq^{k_s}}\right)^{t-1}\right),\tag{2.8}$$

the second line being due to Vandermonde's determinant. Hence we may write the left-hand side of (2.6), when multiplied by $\prod_{1\le i<j\le n}(1/X_j-1/X_i)$, as

$$\det_{1\le s,t\le n}\left(X_s^{1-t}\sum_{k_s\ge 0}\frac{(A_s)_{k_s}\,(BX_s)_{k_s}}{(q)_{k_s}\,(CA_sX_s)_{k_s}}\left(Zq^{1-t}\right)^{k_s}\right).$$

Now, to the sum inside the determinant we apply the third Heine transformation, (2.3), with $a=A_s$, $b=BX_s$, $c=CA_sX_s$, $z=Zq^{1-t}$. Thus we obtain

$$\det_{1\le s,t\le n}\left(X_s^{1-t}\frac{(BZq^{1-t}/C)_\infty}{(Zq^{1-t})_\infty}\sum_{k_s\ge 0}\frac{(CX_s)_{k_s}\,(CA_s/B)_{k_s}}{(q)_{k_s}\,(CA_sX_s)_{k_s}}\left(\frac{BZq^{1-t}}{C}\right)^{k_s}\right),$$

which, by taking factors out of the determinant and by using linearity of the determinant with respect to rows, equals

$$\prod_{i=1}^{n} \frac{(BZq^{1-i}/C)_\infty}{(Zq^{1-i})_\infty}$$

$$\times \sum_{k_1,\dots,k_n\ge 0} \left(\prod_{i=1}^{n} \frac{(CX_i)_{k_i}\,(CA_i/B)_{k_i}}{(q)_{k_i}\,(CA_iX_i)_{k_i}} \left(\frac{BZ}{C}\right)^{k_i} \det_{1\le s,t\le n}\left((X_sq^{k_s})^{1-t}\right)\right).$$

By using (2.8), this last expression is turned into the right-hand side of (2.6), when multiplied by $\prod_{1\le i<j\le n}(1/X_j - 1/X_i)$.

□

For the proof of (2.5) we need the following determinant evaluation.

Lemma 1 *Let $B, C, X_1, \dots, X_n$ be indeterminates. Then there holds*

$$\det_{1\le s,t\le n}\left(\frac{(BX_s)_{k_s+t-1}}{(CX_s)_{k_s+t}}\right) = B^{\binom{n}{2}} q^{\sum_{i=1}^{n}\binom{i-1}{2}}$$
$$\times \prod_{i=1}^{n}\left((X_iq^{k_i})^{n-1} \frac{(BX_i)_{k_i}\,(Cq/B)_{i-1}}{(CX_i)_{k_i+n}}\right) \prod_{1\le i<j\le n}\left(\frac{1}{X_jq^{k_j}} - \frac{1}{X_iq^{k_i}}\right). \tag{2.9}$$

This determinant evaluation follows easily from a determinant lemma ([4], Lemma 2.2), which we state here without proof.

Lemma 2 *Let $X_1, \dots, X_n$, $A_2, \dots, A_n$, and $B_2, \dots B_n$ be indeterminates. Then there holds*

$$\det_{1\le s,t\le n}\Big((X_s + A_n)\cdots(X_s + A_{t+1})(X_s + B_t)\cdots(X_s + B_2)\Big)$$
$$= \prod_{1\le i<j\le n}(X_i - X_j) \prod_{2\le i\le j\le n}(B_i - A_j). \tag{2.10}$$

□

Proof of Lemma 1 By taking some factors out of the determinant, we write the left-hand side of (2.9) as

$$\prod_{i=1}^{n} \frac{(BX_i)_{k_i}}{(CX_i)_{k_i+n}} \det_{1\le s,t\le n}\Big((1 - BX_sq^{k_s})\cdots(1 - BX_sq^{k_s}q^{t-2})$$
$$\times(1 - CX_sq^{k_s}q^t)\cdots(1 - CX_sq^{k_s}q^{n-1})\Big)$$

$$= \prod_{i=1}^{n} \frac{(BX_i)_{k_i}}{(CX_i)_{k_i+n}} (X_iq^{k_i})^{n-1} \det_{1\le s,t\le n}\left(\left(\frac{1}{X_sq^{k_s}} - B\right)\cdots\left(\frac{1}{X_sq^{k_s}} - Bq^{t-2}\right)\right.$$
$$\left.\times\left(\frac{1}{X_sq^{k_s}} - Cq^t\right)\cdots\left(\frac{1}{X_sq^{k_s}} - Cq^{n-1}\right)\right).$$

Now we apply Lemma 2 with the replacements $X_s \to 1/X_s q^{k_s}$, $A_t \to -Cq^{t-1}$, $B_t \to -Bq^{t-2}$. A few simplifications then give the right-hand side of (2.9).

□

Now we are in the position to prove our second identity.

Proof of (2.5) By (2.7) and Lemma 1 with C replaced by Cq^{-n}, the left-hand side of (2.5), when multiplied by $\prod_{1\le i<j\le n}(1/X_j - 1/X_i)$, equals

$$B^{-\binom{n}{2}} q^{-\sum_{i=1}^{n}\binom{i-1}{2}} \prod_{i=1}^{n} \frac{(CX_i q^{-n})_n}{(Cq^{1-n}/B)_{i-1}}$$

$$\times \det_{1\le s,t\le n}\left(\sum_{k_s\ge 0} \frac{(A_s)_{k_s}(BX_s)_{k_s+t-1}}{(q)_{k_s}(CX_s q^{-n})_{k_s+t}} \left(X_s q^{k_s}\right)^{1-n} Z^{k_s}\right).$$

Now, to the sum inside the determinant we apply the second Heine transformation, (2.2), with $a = A_s$, $b = BX_s q^{t-1}$, $c = CX_s q^{t-n}$, $z = Zq^{1-n}$. Thus we obtain

$$B^{-\binom{n}{2}} q^{-\sum_{i=1}^{n}\binom{i-1}{2}} \prod_{i=1}^{n} \frac{(CX_i q^{-n})_n}{(Cq^{1-n}/B)_{i-1}} \det_{1\le s,t\le n}\left(\frac{(BX_s)_{t-1}}{(CX_s q^{-n})_t} X_s^{1-n}\right.$$

$$\left.\times \frac{(Cq^{1-n}/B)_\infty (BZX_s q^{t-n})_\infty}{(CX_s q^{t-n})_\infty (Zq^{1-n})_\infty} \sum_{k_s\ge 0} \frac{(A_s BZ/C)_{k_s}(BX_s q^{t-1})_{k_s}}{(q)_{k_s}(BZX_s q^{t-n})_{k_s}} \left(\frac{Cq^{1-n}}{B}\right)^{k_s}\right)$$

$$= B^{\binom{n}{2}} {}_{-}\sum_{i=1}^{n}\binom{i-1}{2} \prod_{i=1}^{n} \frac{(Cq^{i-n}/B)_\infty (BZX_i q^{-n})_\infty}{(CX_i)_\infty (Zq^{1-n})_\infty}$$

$$\times \sum_{k_1,\dots,k_n\ge 0}\left(\prod_{i=1}^{n} \frac{(A_i BZ/C)_{k_i}}{(q)_{k_i}} \left(\frac{C}{B}\right)^{k_i} \left(X_i q^{k_i}\right)^{1-n} \det_{1\le s,t\le n}\left(\frac{(BX_s)_{k_s+t-1}}{(BZX_s q^{-n})_{k_s+t}}\right)\right).$$

The determinant in the last line can be evaluated by means of Lemma 1, with C replaced by BZq^{-n}. Subsequently, a few simplifications lead to the right-hand side of (2.5), when multiplied by $\prod_{1\le i<j\le n}(1/X_j - 1/X_i)$.

□

Finally, we are able to prove our first identity.

Proof of (2.4) To the left-hand side of (2.4) we apply (1.2) (which is (2.5) with $A_i \equiv A$). To the resulting series we apply (1.3) (which is (2.6) with $A_i \equiv A$ and C replaced by C/A), with the replacements $A \to ABZ/C$, $C \to BZ$, $Z \to C/B$. Thus we arrive at the right-hand side of (2.4).

□

References

[1] Gasper, G. and Rahman, M. [1990], *Basic hypergeometric series*, Encyclopedia of Mathematics And Its Applications 35, Cambridge University Press, Cambridge.

[2] Gessel, I. M. and Krattenthaler, C. [1995], *Cylindric partitions*, Trans. Amer. Math. Soc., to appear.

[3] Gustafson, R. A. and Krattenthaler, C. [1994], *Heine transformations for a new kind of basic hypergeometric series in $U(n)$*, Integral Transforms and Special Functions, to appear.

[4] Krattenthaler, C. [1990], *Generating functions for plane partitions of a given shape*, Manuscripta Math., **69**, 173–202.

[5] Krattenthaler, C. [1990], *A determinant evaluation and some enumeration results for plane partitions*, in Number-Theoretic Analysis (E. Hlawka, R. F. Tichy, eds.), Lect. Notes in Math. 1452, Springer-Verlag, Berlin, New York, 121–131.

[6] Krattenthaler, C. [1995], The major counting of nonintersecting lattice paths and generating functions for tableaux, Mem. Amer. Math. Soc. 115, no. 552, Amer. Math. Soc., Providence, R. I.

Fields Institute Communications
Volume **14**, 1997

Some Generating Functions for q-Polynomials

Mourad E. H. Ismail
Department of Mathematics
University of South Florida
Tampa, FL 33620 USA

David R. Masson
Department of Mathematics
University of Toronto
Toronto, Ontario, Canada M5S 3G3

Sergei K. Suslov
Department of Mathematics
Arizona State University
Tempe, AZ 85287-1804 USA

Abstract. We use the product formula for the continuous q-ultraspherical polynomials and the connection relation for the continuous q-Jacobi polynomials to derive several bilinear generating functions for the continuous q-ultraspherical polynomials and linear generating functions for continuous q-Jacobi polynomials.

1 Introduction

The separation of variables of the wave equation in three space dimensions using polar coordinates leads to the expansions

$$e^{ir\cos\theta} = \left(\frac{2}{r}\right)^{\nu} \Gamma(\nu) \sum_{n=0}^{\infty} i^n (\nu+n)\, J_{\nu+n}(r)\, C_n^{\nu}(\cos\theta), \tag{1.1}$$

and

$$\begin{aligned}&(\sin\theta\sin\phi)^{-\nu+1/2}\, J_{\nu-1/2}(r\sin\theta\sin\phi)\, e^{ir\cos\theta\, \cos\phi} \\ &= \frac{2^{2\nu}}{\sqrt{2\pi r}}\Gamma^2(\nu) \sum_{n=0}^{\infty} i^n \frac{n!\,(\nu+n)}{\Gamma(2\nu+n)} J_{\nu+n}(r)\, C_n^{\nu}(\cos\theta)\, C_n^{\nu}(\cos\phi).\end{aligned} \tag{1.2}$$

The C_n's are the ultraspherical polynomials (spherical harmonics). Formula (1.1) is the special case $\phi = 0$ of (1.2). Formulas (1.1) and (1.2) can be thought of as linear

1991 *Mathematics Subject Classification*. Primary: 33C45, 33D45; Secondary: 33D15.
Research partially supported by NSF grant #DMS 9203659 and NSERC grant #A5384.

and bilinear generating functions for the ultraspherical polynomials that involve Bessel functions. Ismail and Zhang [1994] found a q-analogue of (1.1). Their result is

$$\mathcal{E}_q(z;-i,b) = \frac{(q;q)_\infty b^{-\nu}}{(-b^2;q^2)_\infty (q^\nu;q)_\infty} \times \sum_{n=0}^{\infty} i^n (1-q^{\nu+n}) q^{n^2/4} J_{\nu+n}^{(2)}(2b;q) C_n(z;q^\nu|q), \tag{1.3}$$

where $\mathcal{E}_q$, $J_\nu^{(2)}$ and C_n are the q-exponential function (Ismail and Zhang [1994]), the q-Bessel function (Ismail [1982]) and the continuous q-ultraspherical polynomials (Askey and Ismail [1983], Askey and Wilson [1985]) defined by

$$\mathcal{E}_q(\cos\theta;a,b) := \sum_{n=0}^{\infty} \frac{q^{n^2/4}\, b^n}{(q;q)_n} (aq^{(1-n)/2}e^{i\theta}, aq^{(1-n)/2}e^{-i\theta};q)_n, \tag{1.4}$$

$$J_\nu^{(2)}(z;q) := \frac{(q^{\nu+1};q)_\infty}{(q;q)_\infty} \sum_{n=0}^{\infty} \frac{(-1)^n (z/2)^{\nu+2n}}{(q^{\nu+1},q;q)_n} q^{n(n+\nu)}, \tag{1.5}$$

$$C_n(\cos\theta;\beta|q) = \sum_{k=0}^{n} \frac{(\beta;q)_k(\beta;q)_{n-k}}{(q;q)_k(q;q)_{n-k}} e^{i\theta(n-2k)}, \tag{1.6}$$

respectively. Here we followed the notation of (Gasper and Rahman [1990]) for q-shifted factorials and q-functions. Watson [1944] describes the special case $\phi = 0$ of (1.2) as a degenerate form of the addition theorem for Bessel functions. This makes it desirable to look for a full q-analogue of (1.2), that is we look for a bilinear generating function of the type

$$\sum_{n=0}^{\infty} c_n\, C_n(\cos\theta;q^\nu|q)\, C_n(\cos\phi;q^\nu|q). \tag{1.7}$$

This point was raised in (Ismail and Zhang [1994]).

Although the continuous q-ultraspherical polynomials appeared first in Rogers' work in 1893, it was not until the late 1970's that their weight function was found (Askey and Ismail [1983], Askey and Wilson [1985]), and the important role they play as orthogonal polynomials was realized. Rogers found a generating function for the continuous q-ultraspherical polynomials. Two additional generating functions were established in (Askey and Ismail [1983]). Gasper and Rahman [1983] derived the Poisson kernel and a related bilinear generating function. Recently Ismail and Stanton [1986] found several new linear and multilinear generating functions for the continuous q-ultraspherical polynomials and the more general polynomials of Al-Salam and Chihara.

In this paper we discuss bilinear generating functions for the continuous q-ultraspherical polynomials of the form (1.7) and other generating functions involving the continuous q-Jacobi polynomials and q-Bessel functions. This is a modest step towards the goal of finding an addition theorem for q-Bessel functions and continuous q-Jacobi polynomials. Our attempts led to another q-analogue of Bessel functions which we introduce in Section 4. In Section 2 we discuss the bilinear generating function (1.7). In Section 3 we extend and use methods similar to those employed in Section 2 to obtain linear generating functions for the more general case of continuous q-Jacobi polynomials. In Section 4 we discuss some noteworthy special and limiting cases of the results of Sections 2 and 3.

One integral representation which will be used repeatedly is the Nassrallah–Rahman integral (Gasper and Rahman [1990], (6.3.8))

$$\int_0^\pi \frac{(e^{2i\theta}, e^{-2i\theta}, Be^{i\theta}, Be^{-i\theta}; q)_\infty}{\prod_{k=1}^{5}(A_k e^{i\theta}, A_k e^{-i\theta}; q)_\infty} \, d\theta \tag{1.8}$$

$$= \frac{2\pi(A_4A_5, A_1B, A_2B, A_3B, A_1A_2A_3A_4, A_1A_2A_3A_5; q)_\infty}{(q, A_1A_2A_3B; q)_\infty \prod_{1\le j<k\le 5}(A_jA_k; q)_\infty}$$

$$\times {}_8W_7(A_1A_2A_3B/q; A_1A_2, A_1A_3, A_2A_3, B/A_4, B/A_5; q, A_4A_5),$$

where ${}_8W_7$ is a very well-poised ${}_8\phi_7$ series, that is

$${}_8W_7(a; b, c, d, e, f; q, z) \tag{1.9}$$

$$:= {}_8\phi_7\left(\begin{matrix} a, q\sqrt{a}, -q\sqrt{a}, b, c, d, e, f \\ \sqrt{a}, -\sqrt{a}, aq/b, aq/c, aq/d, aq/e, q/f \end{matrix} \,\middle|\, q, z\right).$$

The normalization for continuous q-Jacobi polynomials used here is

$$P_n^{(\alpha,\beta)}(\cos\theta|q) \tag{1.10}$$

$$:= \frac{(q^{\alpha+1}; q)_n}{(q; q)_n} \, {}_4\phi_3\left(\begin{matrix} q^{-n}, q^{n+\alpha+\beta+1}, q^{(2\alpha+1)/4}e^{i\theta}, q^{(2\alpha+1)/4}e^{-i\theta} \\ q^{\alpha+1}, -q^{(\alpha+\beta+1)/2}, -q^{(\alpha+\beta+2)/2} \end{matrix} \,\middle|\, q, q\right).$$

This is the same normalization used in (Askey and Wilson [1985]).

To use (1.4), (1.6) and (1.10) we need to recover $e^{\pm i\theta}$ from knowing $\cos\theta$. Throughout this work we choose

$$e^{\pm i\theta} = x \pm \sqrt{x^2 - 1}, \quad x = \cos\theta, \tag{1.11}$$

so that $|e^{-i\theta}| \le |e^{i\theta}|$, that is the branch of the square root is the one that makes

$$\sqrt{x^2-1} \approx x, \quad \text{as } x \to \infty.$$

Generating functions are power series in a variable and its radius of convergence can be found explicitly from knowing the large n asymptotics of the coefficients. Since we deal primarily with continuous q-ultraspherical and continuous q-Jacobi polynomials we will now state their large n asymptotics. They are

$$C_n(x; \beta|q) \approx \frac{(\beta, \beta e^{-2i\theta}; q)_\infty}{(q, e^{-2i\theta}; q)_\infty} e^{in\theta}, \quad x = \cos\theta \notin [-1, 1], \tag{1.12}$$

$$C_n(x; \beta|q) \approx \left|\frac{(\beta, \beta e^{-2i\theta}; q)_\infty}{(q, e^{-2i\theta}; q)_\infty}\right| \cos(n\theta + \phi), \quad x = \cos\theta \in (-1, 1), \tag{1.13}$$

with

$$\phi := arg\left\{\frac{(\beta, \beta e^{-2i\theta}; q)_\infty}{(q, e^{-2i\theta}; q)_\infty}\right\}, \tag{1.14}$$

and

$$P_n^{(\alpha,\beta)}(x|q) \approx \frac{q^{n(2\alpha+1)/4}\, e^{in\theta}\, A(e^{i\theta})}{(q, -q^{(\alpha+\beta+1)/2}, -q^{(\alpha+\beta+2)/2}; q)_\infty}, \tag{1.15}$$
$$x = \cos\theta \notin [-1, 1],$$

$$P_n^{(\alpha,\beta)}(x|q) \approx \frac{q^{n(2\alpha+1)/4}|A(e^{i\theta})|}{(q, -q^{(\alpha+\beta+1)/2}, -q^{(\alpha+\beta+2)/2}; q)_\infty}\cos(n\theta - \psi), \tag{1.16}$$
$$x = \cos\theta \in (-1, 1),$$

where

$$A(z) := \frac{(q^{(2\alpha+1)/4}z, q^{(2\alpha+3)/4}z, -q^{(2\beta+1)/4}z, -q^{(2\beta+3)/4}z; q)_\infty}{(z^2; q)_\infty}, \tag{1.17}$$

and

$$\psi := arg\, A(e^{i\theta}). \tag{1.18}$$

The above asymptotic relationships follow from (Ismail and Wilson [1982]) and (7.5.2), (7.5.23), (7.5.24) in (Gasper and Rahman [1990]).

The reader can easily apply (1.12)–(1.16) to determine the domain of analyticity of the generating functions in the following sections, so we may not specify these domains in the statements of our results.

2 Continuous q-Ultraspherical Polynomials

The function $w(x; a_1, a_2, a_3, a_4)$,

$$w(\cos\theta; a_1, a_2, a_3, a_4) := \frac{(e^{2i\theta}, e^{-2i\theta}; q)_\infty}{\prod_{j=1}^{4}(a_j e^{i\theta}, a_j e^{-i\theta}; q)_\infty}, \tag{2.1}$$

will appear repeatedly throughout this work. The weight function for the Askey–Wilson polynomials is $w(x; a_1, a_2, a_3, a_4)/\sqrt{1-x^2}$, (Askey and Wilson [1985]). We will also use the kernel

$$K(\cos\theta, \cos\phi, \cos\psi) \tag{2.2}$$

$$:= \frac{(q, q^\nu, q^\nu; q)_\infty}{2\pi(q^{2\nu}; q)_\infty}(q^\nu e^{2i\theta}, q^\nu e^{-2i\theta}, q^\nu e^{2i\phi}, q^\nu e^{-2i\phi}; q)_\infty$$

$$\times w\left(\cos\psi; q^{\nu/2}e^{i(\theta+\phi)}, q^{\nu/2}e^{i(\theta-\phi)}, q^{\nu/2}e^{i(\phi-\theta)}, q^{\nu/2}e^{-i(\theta+\phi)}\right).$$

The continuous q-ultraspherical polynomials satisfy the product formula, (8.4.1) in (Gasper and Rahman [1990])

$$C_n(x;q^\nu|q)C_n(y;q^\nu|q) = \frac{(q^{2\nu};q)_n}{(q;q)_n} q^{-\nu n/2} \int_0^\pi K(x,y,\cos\psi)C_n(\cos\psi;q^\nu|q)\,d\psi. \quad (2.3)$$

Inserting the right-hand side of (2.3) into (1.7) we see that the generating function we seek is

$$\int_0^\pi K(x,y,\cos\psi)\left[\sum_{n=0}^\infty c_n \frac{(q^{2\nu};q)_n}{(q;q)_n} q^{-\nu n/2} C_n(\cos\psi;q^\nu|q)\right] d\psi. \quad (2.4)$$

Since we wish to extend (1.3) we may choose

$$c_n = \frac{(q;q)_n}{(q^{2\nu};q)_n} \frac{(q;q)_\infty (2/b)^\nu}{(-b^2/4;q^2)_\infty (q^\nu;q)_\infty} i^n (1-q^{n+\nu}) q^{n(n+2\nu)/4} J^{(2)}_{\nu+n}(b;q). \quad (2.5)$$

The series inside (2.4) is now summable by (1.3) and the result leads to the following theorem.

Theorem 2.1 *The continuous q-ultraspherical polynomials have the bilinear generating function*

$$\frac{(q;q)_\infty (2/b)^\nu}{(-b^2/4;q^2)_\infty (q^\nu;q)_\infty} \sum_{n=0}^\infty i^n \frac{(1-q^{n+\nu})(q;q)_n}{(q^{2\nu};q)_n} q^{n(n+2\nu)/4} \quad (2.6)$$
$$\times J^{(2)}_{\nu+n}(b;q) C_n(\cos\theta;q^\nu|q) C_n(\cos\phi;q^\nu|q)$$
$$= \sum_{n=0}^\infty \frac{q^{n^2/4}}{(q;q)_n}\left(\frac{b}{2}\right)^n (-iq^{(1+\nu-n)/2}e^{i(\theta+\phi)}, -iq^{(1-\nu-n)/2}e^{-i(\theta+\phi)};q)_n$$
$$\times {}_4\phi_3\left(\begin{matrix} q^{-n}, q^\nu e^{2i\theta}, q^\nu e^{2i\phi}, q^\nu \\ iq^{(1+\nu-n)/2}e^{i(\theta+\phi)}, -iq^{(1+\nu-n)/2}e^{i(\theta+\phi)}, q^{2\nu} \end{matrix}\,\middle|\, q,q\right).$$

Proof Let G denote the left-hand side of (2.6). The use of the formulas (2.4), (2.5), (1.3) and (1.4) shows that the bilinear generating function G has the following integral representation

$$G = \int_0^\pi K(\cos\theta,\cos\phi,\cos\psi)\mathcal{E}_q(\cos\psi;-i,b/2)\,d\psi \quad (2.7)$$
$$= \sum_{n=0}^\infty \frac{q^{n^2/4}}{(q;q)_n}\left(\frac{b}{2}\right)^n \int_0^\pi K(\cos\theta,\cos\phi,\cos\psi)$$
$$\times \frac{(-iq^{(1-n)/2}e^{i\psi}, -iq^{(1-n)/2}e^{-i\psi};q)_\infty}{(-iq^{(1+n)/2}e^{i\psi}, -iq^{(1+n)/2}e^{-i\psi};q)_\infty}\,d\psi.$$

The integral in (2.7) can now be evaluated by the Nassrallah–Rahman integral of (1.8). The result is

$$\frac{2\pi(q^{2\nu}, -iq^{(1+\nu-n)/2}e^{i(\theta+\phi)}, -iq^{(1+\nu-n)/2}e^{i(\theta-\phi)}; q)_\infty}{(q, q^\nu, q^\nu, q^\nu e^{2i\theta}, q^\nu e^{-2i\theta}, q^\nu e^{2i\phi}, q^\nu e^{-2i\phi}; q)_\infty} \tag{2.8}$$

$$\times \frac{(-iq^{(1+\nu-n)/2}e^{i(\phi-\theta)}, -iq^{(3\nu+n+1)/2}e^{i(\theta+\phi)}; q)_\infty}{(-iq^{(1+\nu+n)/2}e^{i(\theta+\phi)}, -iq^{(1+\nu+n)/2}e^{i(\theta-\phi)}; q)_\infty}$$

$$\times \left((-iq^{(1+\nu+n)/2}e^{i(\phi-\theta)}, -iq^{(3\nu-n-1)/2}e^{i(\theta+\phi)}; q)_\infty\right)^{-1}$$

$$\times {}_8W_7(-iq^{(3\nu-n-1)/2}e^{i(\theta+\phi)}; q^\nu, q^\nu e^{2i\theta}, q^\nu e^{2i\phi},$$

$$-iq^{(1-\nu-n)/2}e^{i(\theta+\phi)}, q^{-n}; q, -iq^{(1+\nu+n)/2}e^{-i(\theta+\phi)}).$$

Interchanging the sum and integral in (2.7) can be easily justified by the Lebesgue dominated convergence theorem. The terminating ${}_8\phi_7$ in the expression (2.8) can be reduced to a terminating balanced ${}_4\phi_3$ series

$$\frac{(-iq^{(3\nu-n+1)/2}e^{i(\theta+\phi)}, -iq^{(1-n-\nu)/2}e^{-i(\theta+\phi)}; q)_n}{(-iq^{(1+\nu-n)/2}e^{-i(\theta-\phi)}, -iq^{(1-n+\nu)/2}e^{i(\theta-\phi)}; q)_n}$$

$$\times {}_4\phi_3\left(\begin{matrix} q^{-n}, q^\nu, q^\nu e^{2i\theta}, q^\nu e^{2i\phi} \\ q^{2\nu}, iq^{(1+\nu-n)/2}e^{i(\theta+\phi)}, -iq^{(1+\nu-n)/2}e^{i(\theta+\phi)} \end{matrix} \middle| q, q\right)$$

by Watson's transformation (Gasper and Rahman [1990], (III.18)). This completes the proof of our theorem.

We now state an alternate formulation of (2.6) which makes the special cases of Section 4 more explicit by the evaluation of an inner sum. The alternate form also more clearly exhibits the structure of singularities from the right-hand side because the asymptotic behavior of its summand is easy to obtain.

Theorem 2.2 *An alternate form of (2.6) is*

$$\sum_{n=0}^{\infty} i^n \frac{(1-q^{n+\nu})(q;q)_n}{(q^{2\nu};q)_n} q^{n(n+2\nu)/4} \tag{2.9}$$

$$\times J^{(2)}_{\nu+n}(b;q)C_n(\cos\theta;q^\nu|q)C_n(\cos\phi;q^\nu|q)$$

$$= \frac{(-b^2/4;q^2)_\infty (q^\nu, q^\nu, -q;q)_\infty}{(q, q^{2\nu}, -q^{1+\nu};q)_\infty} \sum_{n=0}^{\infty} \frac{q^{n(\nu+n/4)}}{(q;q)_n} \left(\frac{b}{2}\right)^{n+\nu}$$

$$\times(-iq^{(1-\nu-n)/2}e^{i(\theta+\phi)}, -iq^{(1-\nu-n)/2}e^{-i(\theta+\phi)};q)_n$$

$$\times \frac{(iq^{(1+3\nu+n)/2}e^{i(\theta-\phi)}, iq^{(1+3\nu+n)/2}e^{i(\phi-\theta)};q)_\infty}{(iq^{(1+\nu+n)/2}e^{i(\theta-\phi)}, iq^{(1+\nu+n)/2}e^{i(\phi-\theta)};q)_\infty}$$

$$\times {}_8W_7(-q^\nu;q^\nu, iq^{(1+\nu+n)/2}e^{i(\theta+\phi)}, iq^{(1+\nu+n)/2}e^{-i(\theta+\phi)},$$

$$iq^{(1-\nu-n)/2}e^{i(\phi-\theta)}, iq^{(1-\nu-n)/2}e^{i(\theta-\phi)};q,q^\nu).$$

Proof The application of Sears' transformation (Gasper and Rahman [1990], (III.15)) and Watson's transformation (Gasper and Rahman [1990], (III.19)) enable us to transform the ${}_4\phi_3$ in (2.6) to an ${}_8\phi_7$ function in the following manner

$$ {}_4\phi_3\left(\begin{matrix} q^{-n}, & q^\nu e^{2i\theta}, & q^\nu e^{2i\phi}, & q^\nu \\ \multicolumn{4}{c}{iq^{(1+\nu-n)/2}e^{i(\theta+\phi)}, -iq^{(1+\nu-n)/2}e^{i(\theta+\phi)}, q^{2\nu}} \end{matrix} \middle| q,q\right) \tag{2.10}$$

$$= q^{\nu n}\frac{(-iq^{(1-\nu-n)/2}e^{i(\theta+\phi)}, q^\nu;q)_n}{(-iq^{(1+\nu-n)/2}e^{i(\theta+\phi)}, q^{2\nu};q)_n}$$

$$\times {}_4\phi_3\left(\begin{matrix} q^{-n}, q^\nu, iq^{(1-\nu-n)/2}e^{i(\phi-\theta)}, iq^{(1-\nu-n)/2}e^{i(\theta-\phi)} \\ q^{1-\nu-n}, iq^{(1+\nu-n)/2}e^{i(\theta+\phi)}, iq^{(1+\nu-n)/2}e^{-i(\theta+\phi)} \end{matrix} \middle| q,q\right)$$

$$= q^{n\nu}\frac{(q^\nu, -q, -iq^{(1-\nu-n)/2}e^{i(\theta+\phi)}, -iq^{(1+\nu+n)/2}e^{i(\theta+\phi)};q)_\infty}{(q^{2\nu}, -q^{1+\nu}, -iq^{(1+\nu-n)/2}e^{i(\theta+\phi)}, -iq^{(1-\nu+n)/2}e^{i(\theta+\phi)};q)_\infty}$$

$$\times\frac{(iq^{(1+3\nu+n)/2}e^{i(\theta-\phi)}, iq^{(1+3\nu+n)/2}e^{i(\phi-\theta)};q)_\infty}{(iq^{(1+\nu+n)/2}e^{i(\theta-\phi)}, iq^{(1+\nu+n)/2}e^{i(\phi-\theta)};q)_\infty}$$

$$\times {}_8W_7(-q^\nu;q^\nu, iq^{(1+\nu+n)/2}e^{i(\theta+\phi)}, iq^{(1+\nu+n)/2}e^{-i(\theta+\phi)},$$
$$iq^{(1-\nu-n)/2}e^{i(\phi-\theta)}, iq^{(1-\nu-n)/2}e^{i(\theta-\phi)};q,q^\nu).$$

Substituting the last line into (2.6) establishes (2.9).

Note that the left-hand side of (2.9) is analytic in b for $|b| < 2$. On the other hand the series on the right-hand side of (2.9) converges for $|b| < 2$. To see this we first expand the ${}_8W_7$ as

$$\sum_{k=0}^{n-1} + \sum_{k=n}^{\infty},$$

then observe that the dominant terms in the first and second sums correspond to $k = n-1$ and $k = n$, respectively and make comparable contributions to their sum. After determining the asymptotic behavior of the remaining factors we find that the series on the right-hand side converges for $|b| < 2$.

Before we proceed to establish a second bilinear generating function of the type (1.7) we give a lemma needed in the sequel.

Lemma 2.3 *The continuous q-ultraspherical polynomials have the generating functions*

$$\begin{aligned}
&\sum_{n=0}^{\infty} C_n(\cos\theta;\beta|q)\,\frac{(\gamma;q)_n}{(\beta^2;q)_n}\,t^n \qquad (2.11)\\
&= \frac{(\gamma t e^{-i\theta};q)_\infty}{(te^{-i\theta};q)_\infty}\,{}_3\phi_2\left(\begin{matrix}\beta,\gamma,\beta e^{-2i\theta}\\ \beta^2,\gamma t e^{-i\theta}\end{matrix}\,\middle|\, q, te^{i\theta}\right)\\
&= \frac{(\beta t e^{i\theta},\gamma t e^{-i\theta}/\beta;q)_\infty}{(te^{i\theta},te^{-i\theta};q)_\infty}\,{}_3\phi_2\left(\begin{matrix}\beta,\beta^2/\gamma,\beta e^{2i\theta}\\ \beta^2,\beta t e^{i\theta}\end{matrix}\,\middle|\, q, \frac{\gamma t}{\beta}e^{-i\theta}\right)\\
&= \frac{(\gamma,\beta t e^{i\theta},\beta t e^{-i\theta};q)_\infty}{(\beta^2,te^{i\theta},te^{-i\theta};q)_\infty}\,{}_3\phi_2\left(\begin{matrix}\beta^2/\gamma,te^{i\theta},te^{-i\theta}\\ \beta t e^{i\theta},\beta t e^{-i\theta}\end{matrix}\,\middle|\, q, \gamma\right).
\end{aligned}$$

Proof Exercise (7.16) in (Gasper and Rahman [1990]) shows that the left-hand side of (2.11) has the q-integral representation

$$\frac{2i\sin\theta\,(\beta,\beta,\beta e^{2i\theta},\beta e^{-2i\theta};q)_\infty}{(1-q)\,(q,\beta^2,e^{2i\theta},e^{-2i\theta};q)_\infty} \times \int_{e^{i\theta}}^{e^{-i\theta}} \frac{(que^{i\theta},que^{-i\theta},\gamma tu;q)_\infty}{(\beta ue^{i\theta},\beta ue^{-i\theta},tu;q)_\infty}\,d_qu.$$

The above q-integral can be evaluated by (Askey, Rahman, and Suslov [1996], (15.7)), namely

$$\begin{aligned}
&\int_a^b \frac{(qu/a,qu/b,cu;q)_\infty}{(du,eu,fu;q)_\infty}\,d_qu\\
&= b(1-q)\frac{(q,a/b,qb/a,bc,abde;q)_\infty}{(ad,ae,bd,be,bf;q)_\infty}\\
&\times {}_3\phi_2\left(\begin{matrix}bd,be,c/f\\ bc,abde\end{matrix}\,\middle|\, q, af\right),
\end{aligned}$$

which is also a special case of (Gasper and Rahman [1990], (2.10.19)). One can then transform the first ${}_3\phi_2$ in (2.11) to multiples of the second and third ${}_3\phi_2$'s through the use of (III.9) and (III.10) in (Gasper and Rahman [1990]). This completes the proof.

The different right-hand sides in (2.11) give different analytic continuations of the left-hand side.

Note that the generating function (7.4.1) in (Gasper and Rahman [1990]) is the special case $\gamma = \beta^2$ of (2.11). The generating functions in (Askey and Ismail [1983]) correspond to the special case $\gamma = 0$ and the limiting case $\gamma \to \infty$ after replacing t by $-\beta t/\gamma$ in (2.11). Letting $\gamma = \beta$ in (2.11) we obtain a q-version of the generating function (Rainville [1971], (144.31)). Thus our generating function (2.11) provides a common generalization of several older results.

Theorem 2.4 *The continuous q-ultraspherical polynomials have a bilinear generating function of the form*

$$\sum_{n=0}^{\infty} \frac{(q,\gamma;q)_n}{(\beta^2,\beta^2;q)_n} C_n(\cos\theta;\beta|q)\, C_n(\cos\phi;\beta|q) t^n \tag{2.12}$$

$$= \frac{(\beta,\gamma,t^2,\beta te^{i(\theta+\phi)},\beta te^{i(\theta-\phi)},\beta te^{i(\phi-\theta)},\beta te^{-i(\theta+\phi)};q)_\infty}{(\beta^2,\beta^2,\beta t^2,te^{i(\theta+\phi)},te^{i(\theta-\phi)},te^{i(\phi-\theta)},te^{-i(\theta+\phi)};q)_\infty}$$

$$\times \sum_{n=0}^{\infty} \gamma^n \frac{(\beta t^2;q)_{2n}}{(t^2;q)_{2n}} \frac{(\beta^2/\gamma,te^{i(\theta+\phi)},te^{i(\theta-\phi)},te^{i(\phi-\theta)},te^{-i(\theta+\phi)};q)_n}{(q,\beta te^{i(\theta+\phi)},\beta te^{i(\theta-\phi)},\beta te^{i(\phi-\theta)},\beta te^{-i(\theta+\phi)};q)_n}$$

$$\times {}_8W_7(\beta t^2 q^{2n-1}; tq^n e^{i(\theta+\phi)}, tq^n e^{i(\theta-\phi)}, tq^n e^{i(\phi-\theta)}, tq^n e^{-i(\theta+\phi)},\beta;q,\beta)$$

provided that $\max\{|t|,|\beta|,|\gamma|\} < 1$ *and* $0 \le \theta,\phi \le \pi$.

Proof We again use the product formula (2.3), change q^ν to β in (2.2)–(2.3), and obtain

$$\sum_{n=0}^{\infty} c_n C_n(x;\beta|q) C_n(y;\beta|q) \tag{2.13}$$

$$= \int_0^\pi K(x,y,\cos\psi)\left[\sum_{n=0}^{\infty} c_n \frac{(\beta^2;q)_n}{(q;q)_n}\beta^{-n/2} C_n(\cos\psi;\beta|q)\right] d\psi.$$

The choice

$$c_n = \frac{(q,\gamma;q)_n}{(\beta^2,\beta^2;q)_n}(t\sqrt{\beta})^n$$

and the generating function (2.11) enable us to reduce the right-hand side of (2.13) to

$$\frac{(q,\beta,\beta,\gamma,\beta e^{2i\theta},\beta e^{-2i\theta},\beta e^{2i\phi},\beta e^{-2i\phi};q)_\infty}{2\pi(\beta^2,\beta^2;q)_\infty}$$

$$\times \sum_{n=0}^{\infty} \gamma^n \frac{(\beta^2/\gamma;q)_n}{(q;q)_n}$$

$$\times \int_0^\pi w\left(\cos\psi;\sqrt{\beta}e^{i(\theta+\phi)},\sqrt{\beta}e^{i(\theta-\phi)},\sqrt{\beta}e^{i(\phi-\theta)},\sqrt{\beta}e^{-i(\theta+\phi)}\right)$$

$$\times \frac{(\beta tq^n e^{i\psi},\beta tq^n e^{-i\psi};q)_\infty}{(tq^n e^{i\psi},tq^n e^{-i\psi};q)_\infty}\, d\psi.$$

The integral in the above formula can be evaluated by (1.8) and we arrive at (2.12) after t is replaced by $t/\sqrt{\beta}$; and the proof is complete.

The bilinear generating function (2.12) is an extension of the kernel L originally due to Gasper and Rahman. The L kernel,(Gasper and Rahman [1990], (8.6.4)), is the special case $\gamma = \beta^2$.

3 Continuous q-Jacobi Polynomials

Motivated by the generating function (1.3) we now proceed to extend it from the continuous q-ultraspherical polynomials to the more general continuous q-Jacobi polynomials $P_n^{(\alpha,\beta)}(x|q)$ of (1.10). The extension given here is different from the one given by Ismail, Rahman and Zhang in (Ismail, Rahman, and Zhang [1996]) where the q-Bessel function is replaced by a q-analogue of the Coulomb wave function. The proof given here relies on the following connection relation

$$
\begin{aligned}
q^{n(\beta-3\alpha-1)/4} & \frac{(q^{\alpha+\beta+1};q)_n}{(q^{\beta+1};q)_n} P_n^{(\alpha,\beta)}(\cos\theta|q) \qquad (3.1)\\
&= \frac{(q,q^{(\beta-\alpha)/2},q^{(\alpha+\beta+2)/2};q)_\infty}{2\pi\,(q^{\beta+1};q)_\infty}\\
&\times(-q^{(2\beta+1)/4}e^{i\theta},-q^{(2\beta+1)/4}e^{-i\theta};q^{1/2})_\infty\\
&\times\int_0^\pi C_n(\cos\phi;q^{(\alpha+\beta+1)/2}|q)\\
&\times w(\cos\phi;q^{(\beta-\alpha)/4}e^{i\theta},q^{(\beta-\alpha)/4}e^{-i\theta},-q^{(\alpha+\beta+1)/4},-q^{(\alpha+\beta+3)/4})\,d\phi,
\end{aligned}
$$

where $\beta > \alpha$ and the function w is as in (2.1), (cf. (Gasper and Rahman [1990], Ex 7.35)).

Consider the generating function

$$
\begin{aligned}
G := & \sum_{n=0}^{\infty} i^n(1-q^{n+(\alpha+\beta+1)/2})q^{n(n+\beta-3\alpha-1)/4} \qquad (3.2)\\
&\times\frac{(q^{\alpha+\beta+1};q)_n}{(q^{\beta+1};q)_n} J^{(2)}_{n+(\alpha+\beta+1)/2}(z;q)\,P_n^{(\alpha,\beta)}(\cos\theta|q).
\end{aligned}
$$

Using (1.3) we represent G in the form

$$G = \frac{(q^{(\beta-\alpha)/2}, q^{(\alpha+\beta+1)/2}, q^{(\alpha+\beta+2)/2}; q)_\infty}{2\pi\ (q^{\beta+1}; q)_\infty} \tag{3.3}$$

$$\times(-z^2/4; q^2)_\infty(-q^{(2\beta+1)/4}e^{i\theta}, -q^{(2\beta+1)/4}e^{-i\theta}; q^{1/2})_\infty$$

$$\times \sum_{n=0}^{\infty} \frac{q^{n^2/4}}{(q;q)_n} \left(\frac{z}{2}\right)^{n+(\alpha+\beta+1)/2}$$

$$\times \int_0^\pi w(\cos\phi; q^{(\beta-\alpha)/4}e^{i\theta}, q^{(\beta-\alpha)/4}e^{-i\theta}, -q^{(\alpha+\beta+1)/4}, -q^{(\alpha+\beta+3)/4})$$

$$\times \frac{(-iq^{(1-n)/2}e^{i\phi}, -iq^{(1-n)/2}e^{-i\phi}; q)_\infty}{(-iq^{(1+n)/2}e^{i\phi}, -iq^{(1+n)/2}e^{-i\phi}; q)_\infty}\, d\phi.$$

The integral in (3.3) can be evaluated by (1.8) and its value is a multiple of a terminating very well-poised ${}_8\phi_7$. We then reduce the latter function to a multiple of a terminating balanced ${}_4\phi_3$ by Watson's transformation (Gasper and Rahman [1990], (III.17)). The result is the following theorem.

Theorem 3.1 *The continuous q-Jacobi polynomials have the generating function*

$$\sum_{n=0}^{\infty} i^n(1 - q^{n+(\alpha+\beta+1)/2})q^{n(n+\beta-3\alpha-1)/4}\frac{(q^{\alpha+\beta+1}; q)_n}{(q^{\beta+1}; q)_n} \tag{3.4}$$

$$\times J^{(2)}_{n+(\alpha+\beta+1)/2}(z; q)P_n^{(\alpha,\,\beta)}(\cos\theta|q)$$

$$= \frac{(q^{(\alpha+\beta+1)/2}; q)_\infty\ (-z^2/4; q^2)_\infty}{(q;q)_\infty}$$

$$\times \sum_{n=0}^{\infty} \frac{q^{n^2/4}}{(q;q)_n} \left(\frac{z}{2}\right)^{n+(\alpha+\beta+1)/2}$$

$$\times(-iq^{(\beta-\alpha+2-2n)/4}e^{i\theta}, -iq^{(\alpha-\beta+2-2n)/4}e^{-i\theta}; q)_n$$

$$\times {}_4\phi_3\left(\begin{matrix} q^{-n}, -q^{(2\beta+1)/4}e^{i\theta}, -q^{(2\beta+3)/4}e^{i\theta}, q^{(\beta-\alpha)/2} \\ iq^{(\beta-\alpha-2n+2)/4}e^{i\theta}, -iq^{(\beta-\alpha-2n+2)/4}e^{i\theta}, q^{\beta+1} \end{matrix} \middle| q, q\right).$$

It is clear that (1.3) is the special case $\beta = \alpha$ of (3.4).

Theorem 3.2 *An alternate form of (3.4) is*

$$\sum_{n=0}^{\infty} i^n (1-q^{n+(\alpha+\beta+1)/2}) q^{n(n+\beta-3\alpha-1)/4} \frac{(q^{\alpha+\beta+1};q)_n}{(q^{\beta+1};q)_n} \tag{3.5}$$

$$\times J^{(2)}_{n+(\alpha+\beta+1)/2}(z;q) P_n^{(\alpha,\,\beta)}(\cos\theta|q)$$

$$= \frac{(-z^2/4;q^2)_\infty(-q, q^{(\alpha+\beta+1)/2}, q^{(\alpha+\beta+2)/2};q)_\infty}{(q, q^{\beta+1}, -q^{(\beta-\alpha+2)/2};q)_\infty}$$

$$\times \sum_{n=0}^{\infty} \frac{q^{n(n-2\alpha+2\beta)/4}}{(q;q)_n} \left(\frac{z}{2}\right)^{n+(\alpha+\beta+1)/2}$$

$$\times(-iq^{(\alpha-\beta+2-2n)/4}e^{i\theta}, -iq^{(\alpha-\beta+2-2n)/4}e^{-i\theta};q)_n$$

$$\times \frac{(-iq^{(3\beta-\alpha+2n+3)/4}, -iq^{(3\beta-\alpha+2n+5)/4};q)_\infty}{(-iq^{(\alpha+\beta+2n+3)/4}, -iq^{(\alpha+\beta+2n+5)/4};q)_\infty}$$

$$\times {}_8W_7\Big(-q^{(\beta-\alpha)/2}; q^{(\beta-\alpha)/2}, iq^{-(\alpha+\beta+2n-1)/4}, -iq^{-(\alpha+\beta+2n+1)/4},$$

$$iq^{(\beta-\alpha+2n+2)/4}e^{i\theta}, iq^{(\beta-\alpha+2n+2)/4}e^{-i\theta}; q, q^{(\alpha+\beta+2)/2}\Big).$$

Proof The application of Sears' transformation (Gasper and Rahman [1990], (III.15)) and Watson's transformation (Gasper and Rahman [1990], (III.19)) enable us to transform the terminating ${}_4\phi_3$ series in (3.4) to a nonterminating ${}_8\phi_7$ series of the form

$${}_4\phi_3\left(\begin{matrix} q^{-n}, -q^{(2\beta+1)/4}e^{i\theta}, -q^{(2\beta+3)/4}e^{i\theta}, q^{(\beta-\alpha)/2} \\ iq^{(\beta-\alpha-2n+2)/4}e^{i\theta}, -iq^{(\beta-\alpha-2n+2)/4}e^{i\theta}, q^{\beta+1} \end{matrix} \middle| q, q\right)$$

$$= q^{n(\beta-\alpha)/2} \frac{(-q, -iq^{(3\beta-\alpha+2n+3)/4}, -iq^{(3\beta-\alpha+2n+5)/4};q)_\infty}{(-q^{(\beta-\alpha+2)/2}, -iq^{(\alpha+\beta+2n+3)/4}, -iq^{(\alpha+\beta+2n+5)/4};q)_\infty}$$

$$\times \frac{(q^{(\alpha+\beta+2)/2}, -iq^{(\alpha-\beta-2n+2)/4}e^{i\theta}, -iq^{(\beta-\alpha+2n+5)/4}e^{i\theta};q)_\infty}{(q^{\beta+1}, -iq^{(\alpha-\beta+2n+2)/4}e^{i\theta}, -iq^{(\beta-\alpha-2n+2)/4}e^{i\theta};q)_\infty}$$

$$\times {}_8W_7\Big(-q^{(\beta-\alpha)/2}; q^{(\beta-\alpha)/2}, iq^{-(\alpha+\beta+2n-1)/4}, -iq^{-(\alpha+\beta+2n+1)/4},$$

$$iq^{(\beta-\alpha+2n+2)/4}e^{i\theta}, iq^{(\beta-\alpha+2n+2)/4}e^{-i\theta}; q, q^{(\alpha+\beta+2)/2}\Big).$$

Substituting this into (3.4) we establish the generating function (3.5). This completes the proof.

It is always interesting to compare a q-series identity with the case $q = 1$. The limiting case $q \to 1$ of (3.4) will be discussed in Section 4 and it involves a confluent hypergeometric function (see (4.7)).

We next establish two generating functions for the continuous q-Jacobi polynomials based on the connecting relation (3.1). The idea is to start with a generating relation for the continuous q-ultraspherical polynomials and then apply (3.1) to lift it to the level of continuous q-Jacobi polynomials. Both our generating functions extend known generating functions of Jacobi polynomials and are of independent interest.

Theorem 3.3 *The continuous q-Jacobi polynomials have the generating relation*

$$\begin{aligned}
&\sum_{n=0}^{\infty} \frac{(\gamma;q)_n}{(q^{\alpha+1};q)_n} P_n^{(\alpha,\beta)}(\cos\theta|q)t^n \qquad (3.6)\\
&= \frac{(\gamma, tq^{\alpha+1/2}, tq^{\alpha+\beta+1};q)_\infty}{(q^{\alpha+\beta+1}, tq^{(\alpha+\beta+1)/2}, tq^{(3\alpha+\beta+2)/2};q)_\infty}\\
&\times \frac{(tq^{\alpha+(2\beta+3)/4}e^{i\theta}, tq^{\alpha+(2\beta+3)/4}e^{-i\theta};q)_\infty}{(tq^{(2\alpha+1)/4}e^{i\theta}, tq^{(2\alpha+1)/4}e^{-i\theta};q)_\infty}\\
&\times \sum_{n=0}^{\infty} \gamma^n \frac{(q^{\alpha+\beta+1}/\gamma, tq^{(\alpha+\beta+1)/2}, tq^{(3\alpha+\beta+2)/2};q)_n}{(q, tq^{\alpha+1/2}, tq^{\alpha+\beta+1};q)_n}\\
&\times \frac{(tq^{(2\alpha+1)/4}e^{i\theta}, tq^{(2\alpha+1)/4}e^{-i\theta};q)_n}{(tq^{\alpha+(2\beta+3)/4}e^{i\theta}, tq^{\alpha+(2\beta+3)/4}e^{-i\theta};q)_n}\\
&\times {}_8W_7\Big(tq^{n+(3\alpha+\beta)/2}; q^{(\alpha-\beta)/2}, q^{(\alpha+\beta+1)/2}, tq^{n+(\alpha+\beta)/2},\\
&\qquad q^{(2\alpha+1)/4}e^{i\theta}, q^{(2\alpha+1)/4}e^{-i\theta}; q, tq^{n+(\alpha+\beta+2)/2}\Big)\\
&= \frac{(\gamma, q^{(\alpha+\beta+2)/2}, tq^{\alpha+1/2}, tq^{\alpha+1}, t^2q^{\alpha+\beta+1};q)_\infty}{(q^{\alpha+\beta+1}, q^{\alpha+1}, tq^{(\alpha+\beta+1)/2}, tq^{(\alpha+\beta+2)/2}, t^2q^{(3\alpha+\beta+2)/2};q)_\infty}\\
&\times \frac{(tq^{\alpha+(2\beta+3)/4}e^{i\theta}, tq^{\alpha+(2\beta+3)/4}e^{-i\theta};q)_\infty}{(tq^{(2\alpha+1)/4}e^{i\theta}, tq^{(2\alpha+1)/4}e^{-i\theta};q)_\infty}\\
&\times \sum_{n=0}^{\infty} \gamma^n \frac{(t^2q^{(3\alpha+\beta+2)/2};q)_{2n}}{(t^2q^{\alpha+\beta+1};q)_{2n}}\\
&\times \frac{(q^{\alpha+\beta+1}/\gamma, tq^{(\alpha+\beta+1)/2}, tq^{(\alpha+\beta+2)/2};q)_n}{(q, tq^{\alpha+1/2}, tq^{\alpha+1};q)_n}\\
&\times \frac{(tq^{(2\alpha+1)/4}e^{i\theta}, tq^{(2\alpha+1)/4}e^{-i\theta};q)_n}{(tq^{\alpha+(2\beta+3)/4}e^{i\theta}, tq^{\alpha+(2\beta+3)/4}e^{-i\theta};q)_n}\\
&\times {}_8W_7\Big(t^2q^{2n+(3\alpha+\beta)/2}; q^{(\alpha-\beta)/2}, tq^{n+(\alpha+\beta+1)/2}, tq^{n+(\alpha+\beta)/2},\\
&\qquad tq^{n+(2\alpha+1)/4}e^{i\theta}, tq^{n+(2\alpha+1)/4}e^{-i\theta}; q, q^{(\alpha+\beta+2)/2}\Big).
\end{aligned}$$

Proof Use the connection relation (3.1), the generating function (2.11) and the integral representation (1.8) in the left-hand side of (3.6) to reduce it to the right-hand side in (3.6). In this process one needs to apply the symmetry relation

$$P_n^{(\beta,\alpha)}(-x|q) = (-1)^n q^{n(\beta-\alpha)/2} P_n^{(\alpha,\beta)}(x|q). \tag{3.7}$$

We then transform the first ${}_8W_7$ in (3.6) to a multiple of the second ${}_8W_7$ through the use of (Gasper and Rahman [1990], (III.23)).

The reader may note that (3.7) is the corrected version of (ii) in Exercise 7.32 of (Gasper and Rahman [1990]), where the factor $q^{n(\beta-\alpha)/2}$ was missing.

When $\alpha \to \beta$ the continuous q-Jacobi polynomials reduce to the continuous q-ultraspherical polynomials and (3.6) reduces to the last equality in (2.11). On the other hand the special case $\gamma = q^{\alpha+\beta+1}$ of (3.6) can be thought of as a q-analogue of the generating function, (Rainville [1971], (132.10))

$$\begin{aligned}\sum_{n=0}^{\infty} \frac{(\alpha+\beta+1)_n}{(\alpha+1)_n} P_n^{(\alpha,\beta)}(x) t^n &\\ = (1-2xt+t^2)^{-(\alpha+\beta+1)/2} &\\ \times {}_2F_1\left(\begin{matrix}(\alpha-\beta)/2, (\alpha+\beta+1)/2\\ \alpha+1\end{matrix}\,\middle|\, \frac{2t(1-x)}{1-2xt+t^2}\right).\end{aligned} \tag{3.8}$$

The special case $\gamma = q^{\alpha+1}$ of (3.6) gives a q-extension of the generating function, (Rainville [1971], (140.6)),

$$\sum_{n=0}^{\infty} P_n^{(\alpha,\beta)}(x)\, t^n = 2^{\alpha+\beta} R^{-1} (1-t+R)^{-\alpha} (1+t+R)^{-\beta}, \tag{3.9}$$

where $R = \sqrt{1-2xt+t^2}$.

Our next theorem provides a q-extension of the following generating function, (Askey [1975], (3.25)),

$$\begin{aligned}\sum_{n=0}^{\infty} \frac{(\alpha+\beta+2n+1)(\alpha+\beta+1)_n}{(\alpha+\beta+1)(\alpha+1)_n} P_n^{(\alpha,\beta)}(x)\, t^n &\\ = \frac{(1-t^2)}{(1-2xt+t^2)^{(\alpha+\beta+3)/2}} &\\ \times {}_2F_1\left(\begin{matrix}(\alpha-\beta)/2, (\alpha+\beta+3)/2\\ \alpha+1\end{matrix}\,\middle|\, \frac{2t(1-x)}{1-2xt+t^2}\right).\end{aligned} \tag{3.10}$$

Theorem 3.4 *We have*

$$\sum_{n=0}^{\infty} \frac{(1-q^{(\alpha+\beta+2n+1)/2})(q^{\alpha+\beta+1};q)_n}{(1-q^{(\alpha+\beta+1)/2})(q^{\alpha+1};q)_n} P_n^{(\alpha,\beta)}(\cos\theta|q)t^n \tag{3.11}$$

$$= (1-t^2q^{\alpha+\beta+1}) \frac{(tq^{\alpha+1/2}, tq^{\alpha+\beta+2};q)_\infty}{(tq^{(3\alpha+\beta+4)/2}, tq^{(\alpha+\beta+1)/2};q)_\infty}$$

$$\times \frac{(tq^{\alpha+(2\beta+7)/4}e^{i\theta}, tq^{\alpha+(2\beta+7)/4}e^{-i\theta};q)_\infty}{(tq^{(2\alpha+1)/4}e^{i\theta}, tq^{(2\alpha+1)/4}e^{-i\theta};q)_\infty}$$

$$\times {}_8W_7\Big(tq^{(3\alpha+\beta+2)/2}; q^{(\alpha-\beta)/2}, q^{(\alpha+\beta+3)/2}, tq^{(\alpha+\beta+2)/2}, q^{(2\alpha+1)/4}e^{i\theta}, q^{(2\alpha+1)/4}e^{-i\theta}; q, tq^{(\alpha+\beta+2)/2}\Big)$$

$$= \frac{(q^{(\alpha+\beta+2)/2}, tq^{\alpha+1/2}, tq^{\alpha+1}, t^2q^{\alpha+\beta+1};q)_\infty}{(q^{\alpha+1}, tq^{(\alpha+\beta+1)/2}, tq^{(\alpha+\beta+2)/2}, t^2q^{(3\alpha+\beta+4)/2};q)_\infty}$$

$$\times \frac{(tq^{\alpha+(2\beta+7)/4}e^{i\theta}, tq^{\alpha+(2\beta+7)/4}e^{-i\theta};q)_\infty}{(tq^{(2\alpha+1)/4}e^{i\theta}, tq^{(2\alpha+1)/4}e^{-i\theta};q)_\infty}$$

$$\times {}_8W_7\Big(t^2q^{(3\alpha+\beta+2)/2}; q^{(\alpha-\beta)/2}, tq^{(\alpha+\beta+2)/2}, tq^{(\alpha+\beta+3)/2}, tq^{(2\alpha+1)/4}e^{i\theta}, tq^{(2\alpha+1)/4}e^{-i\theta}; q, q^{(\alpha+\beta+2)/2}\Big).$$

Proof We apply the connection relation (3.1), the generating function (8.6.7) in (Gasper and Rahman [1990]) and the integral representation (1.8) along the same lines as in the proof of Theorem 3.3.

4 Special Cases

In this section we consider special and limiting cases of formulas (2.6) and (2.9) when the right-hand sides reduce to a single sum.

We first show how (2.6) tends to (1.2) as $q \to 1$. Replace b by $(1-q)r$ and use the fact that the q-gamma function $\Gamma_q(z)$,

$$\Gamma_q(z) := \frac{(1-q)^{1-z}(q;q)_\infty}{(q^z;q)_\infty} \tag{4.1}$$

tends to $\Gamma(z)$ as $q \to 1$. The ${}_4\phi_3$ in (2.6) tends to

$${}_2F_1(-n, \nu; 2\nu; -2\sin\theta\sin\phi/\cos(\theta+\phi)).$$

Thus the right-hand side of (2.6) tends to

$$\frac{(r/2)^\nu}{\Gamma(\nu+1)} \sum_{n=0}^{\infty} \frac{(ir\cos(\theta+\phi))^n}{n!}\, {}_2F_1(-n, \nu; 2\nu; -2\sin\theta\sin\phi/\cos(\theta+\phi)).$$

The last sum can be evaluated by the method on page 125 of (Rainville [1971]) and the above expression becomes

$$\frac{(r/2)^\nu}{\Gamma(\nu+1)}\, e^{ir\cos(\theta+\phi)}\, {}_1F_1(\nu; 2\nu; 2ir\sin\theta\sin\phi).$$

Now the above ${}_1F_1$ reduces to a Bessel function through the second Kummer transformation (Rainville [1971], (70.1)).

We now obtain (1.3) as a special case of (2.9). Let $e^{i\phi} = q^{-\nu/2}$ in (2.9), then the ${}_8W_7$ on the right-hand side can be summed by the ${}_6\phi_5$ summation formula (Gasper and Rahman [1990], (II.20)). The result is the Ismail–Zhang formula (1.3).

Another interesting special case is the case $\phi = \pi/2$ in (2.9). In this case the ${}_8W_7$ function in (2.9) can be summed by the q-analogue of the Whipple summation formula for ${}_3F_2$, (Gasper and Rahman [1990], (II.18))

$$\begin{aligned}&{}_8W_7(-q^\nu; q^\nu, -q^{(1+\nu+n)/2}e^{i\theta}, q^{(1+\nu+n)/2}e^{-i\theta}, \\ &\qquad q^{(1-\nu-n)/2}e^{i\theta}, -q^{(1-\nu-n)/2}e^{-i\theta}; q, q^\nu) \\ &= \frac{(q^{1-n}, q^{1+2\nu+n}, q^{\nu+1}e^{-2i\theta}, q^{\nu+1}e^{-2i\theta}; q^2)_\infty}{(q^{(\nu-n+1)/2}e^{-i\theta}, q^{(3\nu+n+1)/2}e^{i\theta}, -q^{(3\nu+n+1)/2}e^{-i\theta}, -q^{(\nu-n+1)/2}e^{i\theta}; q)_\infty} \\ &\quad \times (-q^\nu, -q^{\nu+1}; q)_\infty.\end{aligned} \tag{4.2}$$

Thus the ${}_8W_7$ vanishes for odd n. We now replace the ${}_8W_7$ in (2.9) by the right-hand side of (4.2) and get

$$\begin{aligned}&\sum_{n=0}^\infty q^{n(n+\nu)} \frac{(1-q^{\nu+2n})(q;q^2)_n}{(q^{1+2\nu};q^2)_n} J^{(2)}_{\nu-2n}(z;q)\, C_{2n}(\cos\theta; q^\nu|q) \\ &= \frac{(q^\nu;q)_\infty(-z^2/4;q^2)_\infty}{(q;q)_\infty} \left(\frac{z}{2}\right)^\nu {}_2\phi_1\left(\begin{matrix} q^{\nu+1}e^{2i\theta}, q^{\nu+1}e^{-2i\theta} \\ q^{2\nu+1}\end{matrix} \middle| q^2, -z^2/4\right).\end{aligned} \tag{4.3}$$

The relationship (4.3) is a q-generalization of the special case $\phi = \pi/2$ of (1.2). This suggests that the function

$$(-z^2/4;q^2)_\infty \left(\frac{z}{2}\right)^\nu {}_2\phi_1\left(\begin{matrix} q^{\nu+1}e^{2i\theta}, q^{\nu+1}e^{-2i\theta} \\ q^{2\nu+1}\end{matrix} \middle| q^2, -z^2/4\right) \tag{4.4}$$

on the right-hand side of (4.3) is a constant multiple of a q-analogue of $(\sin\theta)^{-\nu+1/2}$ $J_{\nu-1/2}(z\sin\theta)$. A similar q-analogue of a Bessel function has appeared in Rahman's work (Rahman [1988]).

Formula (4.3) has a companion formula involving continuous q-ultraspherical polynomials of odd degrees. To see this apply the Askey–Wilson divided difference operator (Askey and Wilson [1985]) to (4.3). The result is

$$\begin{aligned}&\sum_{n=0}^\infty q^{n(n+\nu)} \frac{(1-q^{\nu+2n+1})(q^3;q^2)_n}{(q^{1+2\nu};q^2)_n} J^{(2)}_{\nu+2n+1}(z;q)\, C_{2n+1}(\cos\theta; q^\nu|q) \\ &= 2\,\frac{(q^\nu;q)_\infty(-z^2/4;q^2)_\infty}{(1-q)(q;q)_\infty} \\ &\quad \times \left(\frac{z}{2}\right)^{\nu+1} \cos\theta\, {}_2\phi_1\left(\begin{matrix} q^{\nu+1}e^{2i\theta}, q^{\nu+1}e^{-2i\theta} \\ q^{2\nu+1}\end{matrix} \middle| q^2, -z^2/4\right).\end{aligned} \tag{4.5}$$

We now proceed to establish the case $q = 1$ of (3.4) and (3.5). It is easy to see that the limiting case $q \to 1$ of (3.5) is

$$\sum_{n=0}^{\infty} i^n \frac{(\alpha+\beta+2n+1)\,(\alpha+\beta+1)_n}{(\alpha+\beta+1)\,(\beta+1)_n}\, J_{n+(\alpha+\beta+1)/2}(z)\, P_n^{(\alpha,\,\beta)}(\cos\theta) \tag{4.6}$$

$$= \frac{(z/2)^{(\alpha+\beta+3)/2}}{\Gamma((\alpha+\beta+1)/2)} \sum_{n=0}^{\infty} \frac{(iz\cos\theta)^n}{n!}\, {}_2F_1\left(\begin{matrix} -n, (\beta-\alpha)/2 \\ \beta+1 \end{matrix} \,\middle|\, 1+\frac{1}{\cos\theta} \right).$$

From Section 69 in (Rainville [1971]) it is easy to get

$$\sum_{n=0}^{\infty} {}_2F_1(-n, u; v; x)\, \frac{t^n}{n!} = e^t\, {}_1F_1(u; v; -xt),$$

which reduces (4.6) to the generating function

$$\sum_{n=0}^{\infty} i^n \frac{(\alpha+\beta+2n+1)\,(\alpha+\beta+1)_n}{(\alpha+\beta+1)\,(\beta+1)_n}\, J_{n+(\alpha+\beta+1)/2}(t)\, P_n^{(\alpha,\,\beta)}(x) \tag{4.7}$$

$$= \frac{(t/2)^{(\alpha+\beta+3)/2}}{\Gamma((\alpha+\beta+1)/2)}\, e^{ixt}\, {}_1F_1\left((\beta-\alpha)/2; \beta+1; -it(1+x)\right).$$

Formula (4.7) is well-known and was generalized by Fields and Wimp [1961]. The Fields and Wimp formulas generated considerable interest. Some time later Fields and Ismail [1975] generalized and reformulated these expansions in terms of pairs of inverse relations. Later Gessel and Stanton [1983] showed that the correct setup for these identities is the Lagrange inversion and proceeded to derive many interesting q-analogues of these expansions. Gessel and Stanton [1996] gave a second family of q-Lagrange inversions. The generating functions obtained here do not seem to follow from the Gessel–Stanton work, so there are more families of q-Lagrange inversions remaining to be discovered. These problems are closely related to polynomial expansions in the so-called generalized powers (Suslov [1989]).

Acknowledgments

We wish to thank Mizan Rahman for valuable discussions and comments on this paper. Two of us (M. I. and S. S.) greatly acknowledge the hospitality of the Department of Mathematics at the University of Toronto where this work was started.

References

Askey, R. A. [1975], *Orthogonal Polynomials and Special Functions*, Regional Conference Series in Applied Mathematics, Vol. 21, SIAM, Philadelphia.

Askey, R. A. and Ismail, M. E. H. [1983], *A generalization of ultraspherical polynomials*, in "Studies in Pure Mathematics", (Erdös, P., ed.), Birkhauser, Basel, 55–78.

Askey, R. A., Rahman, M., and Suslov, S. K. [1996], *On a general q-Fourier transformation with nonsymmetric kernels*, J. Comp. Appl. Math. **68**, 25–55.

Askey, R. A. and Wilson, J. A. [1985], *Some basic hypergeometric orthogonal polynomials that generalize Jacobi polynomials*, Memoirs Amer. Math. Soc. **319**.

Fields, J. L. and Ismail, M. E. H. [1975], *Polynomial expansions*, Math. Comp. **29**, 894–902.

Fields, J. L. and Wimp, J. [1961], *Expansions of hypergeometric functions in hypergeometric functions*, Math. Comp. **15**, 390–395.

Gasper, G. and Rahman, M. [1983], *Positivity of the Poisson kernel for the continuous q-ultraspherical polynomials*, SIAM J. Math. Anal. **14**, 409–420.

Gasper, G. and Rahman, M. [1990], *Basic Hypergeometric Series*, Cambridge University Press, Cambridge.

Gessel, I. and Stanton, D. [1983], *Applications of q-Lagrange inversion to basic hypergeometric series*, Trans. Amer. Math. Soc. **277**, 173–201.

Gessel, I. and Stanton, D. [1996], *Another family of q-Lagrange inversion formulas*, Rocky Mountain J. Math. **16**, 373–384.

Ismail, M. E. H. [1982], *The zeros of basic Bessel functions, the functions $J_{v+ax}(x)$, and associated orthogonal polynomials*, J. Math. Anal. Appl. **82**, 1–19.

Ismail, M. E. H., Rahman, M. and Zhang, R. [1996], *Diagonalization of certain integral operators II*, J. Comp. Appl. Math., **68**, 163-196.

Ismail, M. E. H. and Stanton, D. [1996], *Classical orthogonal polynomials as moments*, Can. J. Math., to appear.

Ismail, M. E. H. and Wilson, J. A. [1982], *Asymptotic and generating relations for q-Jacobi and ${}_4\phi_3$ polynomials*, J. Approx. Theory **36**, 43–54.

Ismail, M. E. H. and Zhang, R. [1994], *Diagonalization of certain integral operators*, Advances in Math. **109**, 1–33.

Rahman, M. [1988], *An addition theorem and some product formulas for q-Bessel functions*, Can. J. Math. **40**, 1203–1221.

Rainville, E. D. [1971], *Special Functions*, Chelsea, Bronx.

Suslov, S. K. [1989], *On the theory of difference analogues of special functions of hypergeometric type*, Russian Math. Surveys **44**, 227–278, London Math. Soc.

Watson, G. N. [1944], *A Treatise on the Theory of Bessel Functions*, second edition, Cambridge University Press, Cambridge.

Fields Institute Communications
Volume **14**, 1997

Addition Formulas for q-Special Functions

Erik Koelink
Vakgroep Wiskunde
Universiteit van Amsterdam,
Plantage Muidergracht 24
1018 TV Amsterdam, the Netherlands

Abstract. A general addition formula for a two-parameter family of Askey-Wilson polynomials is derived from the quantum $SU(2)$ group theoretic interpretation. This formula contains most of the previously known addition formulas for q-Legendre polynomials as special or limiting cases. A survey of the literature on addition formulas for q-special functions using quantum groups and quantum algebras is given.

1 Survey and Introduction

Many of the well-known special functions, such as the Jacobi polynomials and Bessel functions, satisfy addition formulas, which can be found in e.g., Askey [1975], Erdélyi et al. [1953-1955], Watson [1944]. Often there exists a group theoretic interpretation of such an addition formula. This means that there exists a group G and a representation t of G in a Hilbert space V such that for a suitable basis $\{e_n\}$ of V the matrix elements $t_{n,m}: G \to \mathbb{C}$ defined by $t_{n,m}(g) = \langle t(g)e_m, e_n\rangle$ are known in terms of special functions. Then the homomorphism property

$$t_{n,m}(gh) = \sum_p t_{n,p}(g)t_{p,m}(h), \qquad g,h \in G \tag{1.1}$$

gives an addition formula for a suitably chosen basis of V and for certain elements $g,h \in G$. Usually we number the basis such that $t_{0,0}$ is left and right invariant with respect to a certain subgroup K (a spherical function) and we use (1.1) for $n = m = 0$. See Vilenkin [1968] and Vilenkin and Klimyk [1991-1993] for more information.

1991 *Mathematics Subject Classification.* Primary: 33D80, 33D45; Secondary: 33D55, 17B37.

Research supported by a Fellowship of the Research Council of the Katholieke Universiteit Leuven and by the Netherlands Organization for Scientific Research (NWO) under project number 610.06.100.

Before the introduction of quantum groups and quantum algebras (also known as q-algebras) only a few addition formulas were known for q-special functions. Addition formulas for q-Krawtchouk polynomials and q-Hahn polynomials were proved by Dunkl [1977] and Stanton [1981] using the interpretation of these polynomials on finite groups of Lie type. See also Stanton [1980] for a partial result (a product formula). This method is limited to q-Krawtchouk and q-Hahn polynomials, cf. the list in Stanton [1984]. Also, by a purely analytic method Rahman and Verma [1986] have shown that the continuous q-ultraspherical polynomials satisfy an addition formula similar to Gegenbauer's addition formula for the ultraspherical polynomials; see Section 5 how this addition formula for the Legendre case follows from a general result given in this paper.

With the introduction of quantum groups and quantum algebras (or quantised universal enveloping algebras) a new setting for various q-special functions emerged, see e.g., the survey papers by Koornwinder [1990], Noumi [1991] and Koelink [1995c] for the quantum group approach and Floreanini and Vinet [1993a], [1993c] and references given later in this section for the quantum algebra approach. For a good introduction to quantum groups, quantised universal enveloping algebras and their applications we refer to Chari and Pressley [1994], which also contains a wealth of references on the subject.

From this setting two approaches for deriving addition formulas for q-special functions can be obtained. To understand the quantum group approach we recall that a quantum group $\mathcal{A}_q(G)$ is a deformation (with deformation parameter q) of a function algebra $\mathcal{A}(G)$ on the group G. For a compact group G this would mean the polynomials on G or, in Woronowicz's [1987] approach, the continuous functions on G. We assume that the comultiplication

$$\Delta\colon \mathcal{A}(G) \to \mathcal{A}(G) \otimes \mathcal{A}(G) \cong \mathcal{A}(G\times G), \qquad (\Delta f)(g,h) = f(gh),$$

is well-defined and we demand that $\Delta\colon \mathcal{A}_q(G) \to \mathcal{A}_q(G)\otimes\mathcal{A}_q(G)$ survives the deformation unchanged (in a suitable sense). In the deformation the representations of G correspond to corepresentations of $\mathcal{A}_q(G)$ and for a suitable basis of the representation space we get matrix elements $t_{n,m} \in \mathcal{A}_q(G)$ satisfying the cohomomorphism property

$$\Delta(t_{n,m}) = \sum_p t_{n,p} \otimes t_{p,m}, \tag{1.2}$$

which must be considered as the analogue of (1.1). If the matrix elements can be expressed in terms of q-special functions we consider (1.2) as an (implicit) addition formula. Usually $\mathcal{A}_q(G)$ is a non-commutative algebra and thus we need to investigate the representations of $\mathcal{A}_q(G)$ in order to convert (1.2) into an addition formula involving q-special functions in commuting variables. Usually, $n = m = 0$ in (1.2) and $t_{0,0}$ can be considered as a spherical function. In this paper we show that in the case of the quantum $SU(2)$ group we can obtain a general addition formula for Askey-Wilson polynomials from (1.2) from which various known examples can be obtained as special or limiting cases.

Let us now briefly consider the quantum algebra approach. A quantum algebra is a deformation $\mathcal{U}_q\mathfrak{g}$ of the universal enveloping algebra $\mathcal{U}\mathfrak{g}$ of the Lie algebra $\mathfrak{g}$. The representation theory of $\mathcal{U}_q\mathfrak{g}$ is usually similar to the representation theory of $\mathfrak{g}$. Classically we can obtain elements of the corresponding group G by exponentiating Lie algebra elements. In the quantum algebra approach the action of

$exp_q(\alpha_1 X_1)\dots exp_q(\alpha_n X_n)$ is calculated in a representation of $\mathcal{U}_q\mathfrak{g}$. Here exp_q can be one of the q-analogues of the exponential function, α_i are scalars and X_i are generators of $\mathcal{U}_q\mathfrak{g}$. For a suitable basis $\{f_m\}$ of the representation space we get

$$exp_q(\alpha_1 X_1)\dots exp_q(\alpha_n X_n)\, f_m = \sum_k U_{m,k}(\alpha_1,\dots,\alpha_n)\, f_k. \tag{1.3}$$

The matrix coefficients $U_{m,k}(\alpha_1,\dots,\alpha_n)$ can be calculated in terms of special functions in $\alpha_1,\dots,\alpha_n$. By working with explicit realisations, in which the f_m correspond to certain special functions, addition formulas can be derived from (1.3). This method is motivated by the classical relation between Lie algebras and special functions as described in Miller [1968] and references for the q-algebra approach are given later on. In special cases the approaches from the quantum algebra and the quantum group can be matched, see Floreanini and Vinet [1993a].

It should be observed that in the quantum group case the coalgebra structure, and in particular the comultiplication Δ, is needed to find the addition formula (1.2) in non-commuting variables. The algebra structure of $\mathcal{A}_q(G)$ is needed to transform (1.2) into an addition formula for q-special functions in commuting variables, although identities for q-special functions in non-commuting variables are of interest in their own right, see Koornwinder's contribution to these proceedings. In contrast, in the quantum algebra approach only the algebra structure is needed, and there are examples of quantum algebras used in relation with q-special functions which do not carry a bialgebra structure.

Let us now give some references to the literature for several addition formulas using one of these methods. For addition formulas for q-Legendre polynomials from the quantum group theoretic point of view see Koornwinder [1989], Masuda et al. [1991], Noumi and Mimachi [1990a], [1990b], [1991], [1992a], Vaksman and Soibelman [1988] for (implicit) addition formulas in non-commuting variables and Koelink [1994a], [1995a], Koornwinder [1991] (analytically proved by Rahman [1989]) and Noumi and Mimachi [1990a] for addition formulas involving only commuting variables. For addition formulas for q-Bessel functions from the quantum group theoretic point of view see Koelink [1994b], [1995b]. For the q-algebra point of view see Floreanini and Vinet [1992a], [1994], [1995b], Kalnins and Miller [1994], Kalnins et al. [1994]. For analytic proofs of related addition formulas for q-Bessel functions see Koelink [1993], [1995d], Koelink and Swarttouw [1995], Koornwinder and Swarttouw [1992], Rahman [1988], Swarttouw [1992]. Of these references Koelink [1993], Koornwinder and Swarttouw [1992] are closely related to the q-algebra approach, since they use generating functions as a starting point, which can be interpreted as the left hand side of (1.3), and Koelink [1995d] is related to the quantum group approach, since it can be proved formally from an identity similar to (1.2). Using the quantum group approach it is possible to derive an addition formula for q-disk polynomial involving little q-Jacobi polynomials, cf. Floris [1994], Floris and Koelink [1995]. The announcement by Dijkhuizen and Noumi [1995] suggests that generalisations to Askey-Wilson polynomials might be possible in this context. Using other quantum algebras it is possible to obtain addition formulas for general ${}_r\varphi_s$ (Floreanini and Vinet [1993b]), for basic Lauricella φ_D (Floreanini et al. [1994]), for continuous analogues of addition formulas (Floreanini and Vinet [1995a], Kalnins and Miller [1994]), for q-Laguerre polynomials (Kalnins et al. [1992], Kalnins and Miller [1994], Kalnins et al. [1993]). See Groza and Kachurik [1990] for addition and product formulas from the quantum $SU(2)$ group interpretation of q-Krawtchouk, q-Hahn and q-Racah polynomials.

In this paper we show how to derive an explicit addition formula for Askey-Wilson polynomials involving 3 parameters from the implicit addition formula, i.e., involving non-commuting variables. So we start off with an identity of the type as in (1.2) and we convert it into an identity for Askey-Wilson polynomials. In Section 2 we recall results on the quantum $SU(2)$ group and its relation with Askey-Wilson polynomials. In Section 3 we consider suitable vectors in the representation space of an irreducible representation of $\mathcal{A}_q(SU(2))$ in which all the elements of $\mathcal{A}_q(SU(2))$ under consideration act as a multiplication operator or as a shift operator. From this we obtain in Section 4 an addition formula for Askey-Wilson polynomials involving a very-well-poised ${}_8\varphi_7$-series, which is not a polynomial identity. In Section 5 we consider two special cases in which a polynomial addition formula can be obtained.

The notation for q-shifted factorials and q-hypergeometric series follows the excellent book by Gasper and Rahman [1990].

2 Generalised matrix elements on the quantum $SU(2)$ group

In this section we recall the relation between the quantum $SU(2)$ group and the Askey-Wilson polynomials. We also give the appropriate version of (1.2) in this case, which is the starting point for the derivation of the addition formula. References for this section are Noumi and Mimachi [1990a], [1992b] and Koelink [1995c], from which the notation has been taken and where further references can be found.

$\mathcal{A}_q(SU(2))$ is the complex unital associative algebra generated by α, β, γ, δ subject to the relations

$$\begin{gathered}\alpha\beta = q\beta\alpha, \quad \alpha\gamma = q\gamma\alpha, \quad \beta\delta = q\delta\beta, \quad \gamma\delta = q\delta\gamma,\\ \beta\gamma = \gamma\beta, \quad \alpha\delta - q\beta\gamma = \delta\alpha - q^{-1}\beta\gamma = 1\end{gathered} \tag{2.1}$$

for some constant $q \in \mathbb{C}$. With a $*$-operator given by

$$\alpha^* = \delta, \quad \beta^* = -q\gamma, \quad \gamma^* = -q^{-1}\beta, \quad \delta^* = \alpha$$

the algebra $\mathcal{A}_q(SU(2))$ becomes a $*$-algebra for real q, and from now on we fix $0 < q < 1$.

We use the following realisation of $\mathcal{A}_q(SU(2))$ by a non-faithful representation. An irreducible infinite dimensional $*$-representation π of $\mathcal{A}_q(SU(2))$ in the Hilbert space $\ell^2(\mathbb{Z}_+)$ with orthonormal basis $\{e_n \mid n \in \mathbb{Z}_+\}$ is given by

$$\pi(\alpha)\, e_n = \sqrt{1-q^{2n}}\, e_{n-1}, \qquad \pi(\gamma)\, e_n = q^n\, e_n, \tag{2.2}$$

where we use the convention $e_{-p} = 0$ for $p \in \mathbb{N}$. Note that $-q\pi(\gamma) = \pi(\beta)$. For all $\xi \in \mathcal{A}_q(SU(2))$, $\pi(\xi)$ is a bounded operator on $\ell^2(\mathbb{Z}_+)$. The one-dimensional $*$-representation $\tau_\phi\colon \mathcal{A}_q(SU(2)) \to \mathbb{C}$ is defined by

$$\tau_\phi(\alpha) = e^{i\phi}, \qquad \tau_\phi(\gamma) = 0. \tag{2.3}$$

The algebra $\mathcal{A}_q(SU(2))$ is an example of a Hopf $*$-algebra. The comultiplication Δ, which is a $*$-homomorphism of $\mathcal{A}_q(SU(2)) \to \mathcal{A}_q(SU(2)) \otimes \mathcal{A}_q(SU(2))$, is given on the generators by

$$\Delta(\alpha) = \alpha\otimes\alpha + \beta\otimes\gamma, \quad \Delta(\gamma) = \gamma\otimes\alpha + \delta\otimes\gamma. \tag{2.4}$$

The irreducible corepresentations of $\mathcal{A}_q(SU(2))$ are parametrised by $l \in \frac{1}{2}\mathbb{Z}_+$. For each $\sigma \in \mathbb{R} \cup \{\infty\}$ there exists a canonical basis of the $2l+1$-dimensional representation space and the elements $b^l_{i,j}(\tau,\sigma) \in \mathcal{A}_q(SU(2))$, $l \in \frac{1}{2}\mathbb{Z}_+$, $i,j \in \{-l,-l+1,\ldots,l-1,l\}$, $\sigma,\tau \in \mathbb{R} \cup \{\infty\}$, are (up to a simple algebra isomorphism) the mixed basis matrix elements with respect to two such canonical bases, cf. Noumi and Mimachi [1990a], [1992b], Koelink [1995c]. These generalised matrix elements satisfy the following cohomomorphism property;

$$\Delta\big(b^l_{i,j}(\tau,\sigma)\big) = \sum_{n=-l}^{l} \big(D.b^l_{i,n}(\tau,\mu)\big) \otimes b^l_{n,j}(\mu,\sigma), \qquad \forall\, \mu \in \mathbb{R} \cup \{\infty\}. \tag{2.5}$$

Here $D.\colon \mathcal{A}_q(SU(2)) \to \mathcal{A}_q(SU(2))$ is the bijective algebra homomorphism given by $D.\alpha = q^{-\frac{1}{2}}\alpha$, $D.\beta = q^{\frac{1}{2}}\beta$, $D.\gamma = q^{-\frac{1}{2}}\gamma$ and $D.\delta = q^{\frac{1}{2}}\delta$. The case of most interest to us is $l \in \mathbb{Z}_+$, $i=j=0$.

In order to describe the generalised matrix elements $b^l_{i,j}(\tau,\sigma)$ in more detail we introduce certain simple elements of $\mathcal{A}_q(SU(2))$. In case $l=\frac{1}{2}$ the 2×2-matrix $b^{\frac{1}{2}}(\tau,\sigma)$ is given by

$$\frac{1}{\sqrt{(1+q^{2\sigma})(1+q^{2\tau})}} \begin{pmatrix} \alpha_{\tau,\sigma}, & \beta_{\tau,\sigma} \\ \gamma_{\tau,\sigma}, & \delta_{\tau,\sigma} \end{pmatrix} = b^{1/2}(\tau,\sigma) = \begin{pmatrix} b^{1/2}_{-1/2,-1/2}, & b^{1/2}_{-1/2,1/2} \\ b^{1/2}_{1/2,-1/2}, & b^{1/2}_{1/2,1/2} \end{pmatrix}$$

with

$$\begin{aligned} \alpha_{\tau,\sigma} &= q^{1/2}\alpha - iq^{\sigma-1/2}\beta + iq^{\tau+1/2}\gamma + q^{\sigma+\tau-1/2}\delta, \\ \beta_{\tau,\sigma} &= -q^{\sigma+1/2}\alpha - iq^{-1/2}\beta - iq^{\sigma+\tau+1/2}\gamma + q^{\tau-1/2}\delta, \\ \gamma_{\tau,\sigma} &= -q^{\tau+1/2}\alpha + iq^{\tau+\sigma-1/2}\beta + iq^{1/2}\gamma + q^{\sigma-1/2}\delta, \\ \delta_{\tau,\sigma} &= q^{\tau+\sigma+1/2}\alpha + iq^{\tau-1/2}\beta - iq^{\sigma+1/2}\gamma + q^{-1/2}\delta. \end{aligned} \tag{2.6}$$

We let $\sigma,\tau \in \mathbb{R} \cup \{\infty\}$, e.g., $\alpha_{\infty,\sigma} = q^{1/2}\alpha - iq^{\sigma-1/2}\beta$ and $\alpha_{\infty,\infty} = q^{1/2}\alpha$. Then it follows that

$$\begin{aligned} \alpha_{\tau,\sigma} &= \alpha_{\tau,\infty} + q^{\sigma}\beta_{\tau,\infty}, \quad \beta_{\tau,\sigma} = \beta_{\tau,\infty} - q^{\sigma}\alpha_{\tau,\infty}, \\ \gamma_{\tau,\sigma} &= \gamma_{\tau,\infty} + q^{\sigma}\delta_{\tau,\infty}, \quad \delta_{\tau,\sigma} = \delta_{\tau,\infty} - q^{\sigma}\gamma_{\tau,\infty}. \end{aligned} \tag{2.7}$$

Next we define

$$\rho_{\tau,\sigma} = \frac{1}{2}\big(q^{-\tau-\sigma-1}\alpha_{\tau+1,\sigma+1}\delta_{\tau,\sigma} - q^{-\tau-\sigma-1} - q^{\tau+\sigma+1}\big). \tag{2.8}$$

The limit case $\sigma \to \infty$ is defined by

$$\rho_{\tau,\infty} = \lim_{\sigma\to\infty} 2q^{\sigma+\tau-1}\rho_{\tau,\sigma} = q^{-2}\alpha_{\tau+1,\infty}\delta_{\tau,\infty} - q^{-2},$$

and $\rho_{\infty,\sigma}$ is defined by letting $\tau \to \infty$ in the same expression. Note that $\rho_{\tau,\sigma}$ is self-adjoint; $\rho^*_{\tau,\sigma} = \rho_{\tau,\sigma}$.

We have the following factorisation and commutation relations.

Proposition 2.1 *For $\sigma,\tau \in \mathbb{R}\cup\{\infty\}$ we have in $\mathcal{A}_q(SU(2))$*

$$\beta_{\tau+1,\sigma-1}\gamma_{\tau,\sigma} = 2q^{\tau+\sigma}\rho_{\tau,\sigma} - q^{2\sigma-1} - q^{2\tau+1},$$

$$\gamma_{\tau-1,\sigma+1}\beta_{\tau,\sigma} = 2q^{\tau+\sigma}\rho_{\tau,\sigma} - q^{2\sigma+1} - q^{2\tau-1},$$

$$\alpha_{\tau+1,\sigma+1}\delta_{\tau,\sigma} = 2q^{\tau+\sigma+1}\rho_{\tau,\sigma} + 1 + q^{2\sigma+2\tau+2},$$

$$\delta_{\tau-1,\sigma-1}\alpha_{\tau,\sigma} = 2q^{\tau+\sigma-1}\rho_{\tau,\sigma} + 1 + q^{2\sigma+2\tau-2}.$$

Proof This can be proved straightforwardly from (2.1), (2.6) and (2.8). A more conceptual proof uses Koelink [1995c], Proposition 6.5, Theorem 5.1, which implies that each product on the left-hand side is a polynomial in $\rho_{\tau,\sigma}$. The Clebsch-Gordan series, cf. e.g., Koelink [1995c], Section 4.7, imply that this polynomial is of degree 1. A simple calculation determines the coefficients in this polynomial.

□

Corollary 2.2 *The following relations hold in $\mathcal{A}_q(SU(2))$;*

$$\alpha_{\tau,\sigma}\rho_{\tau,\sigma} = \rho_{\tau-1,\sigma-1}\alpha_{\tau,\sigma}, \qquad \beta_{\tau,\sigma}\rho_{\tau,\sigma} = \rho_{\tau-1,\sigma+1}\beta_{\tau,\sigma},$$

$$\gamma_{\tau,\sigma}\rho_{\tau,\sigma} = \rho_{\tau+1,\sigma-1}\gamma_{\tau,\sigma}, \qquad \delta_{\tau,\sigma}\rho_{\tau,\sigma} = \rho_{\tau+1,\sigma+1}\delta_{\tau,\sigma}.$$

Proof The proofs of these statements are all similar. To prove the first, multiply the last equation of Proposition 2.1 by $\alpha_{\tau,\sigma}$ and use the third equation in the left hand side. Cancelling terms proves the first statement of the corollary.

□

In order to be able to express $b^l_{i,j}(\tau,\sigma)$ in these simple terms we need Askey-Wilson polynomials $p_n(x;a,b,c,d|q)$, where we follow the normalisation as in Askey and Wilson [1985], (1.15), see also Gasper and Rahman [1990], (7.5.2). We use the following notation for the Askey-Wilson polynomials as q-analogues of the Jacobi polynomials;

$$p_n^{(\alpha,\beta)}(x;s,t|q) = p_n(x;q^{1/2}t/s, q^{1/2+\alpha}s/t, -q^{1/2}/(st), -stq^{1/2+\beta}|q). \tag{2.9}$$

Observe that $\alpha,\beta\to\infty$ in (2.9) gives Askey-Wilson polynomials with two parameters set to zero, which are the Al-Salam–Chihara polynomials. So the Al-Salam–Chihara polynomials can be considered as the Hermite case of the q-Jacobi polynomials (2.9), and in this form they play a role in the sequel.

The generalised matrix elements $b^l_{i,j}(\tau,\sigma)$ can be expressed using these simple elements and the Askey-Wilson polynomials. Here we only need the cases where $i=0$ or $j=0$. For $n\in\mathbb{Z}_+$ we have

$$b^l_{n,0}(\tau,\sigma) = d^{l,n}_{\tau,\sigma}\, c^n_{n,0}(\tau,\sigma)\, p^{(n,n)}_{l-n}(\rho_{\tau,\sigma};q^\tau,q^\sigma|q^2),$$

$$b^l_{0,n}(\tau,\sigma) = d^{l,n}_{\sigma,\tau}\, c^n_{0,n}(\tau,\sigma)\, p^{(n,n)}_{l-n}(\rho_{\tau,\sigma};q^\sigma,q^\tau|q^2),$$

$$b^l_{-n,0}(\tau,\sigma) = d^{l,n}_{-\tau,-\sigma}\, c^n_{-n,0}(\tau,\sigma)\, p^{(n,n)}_{l-n}(\rho_{\tau,\sigma};q^{-\tau},q^{-\sigma}|q^2),$$

$$b^l_{0,-n}(\tau,\sigma) = d^{l,n}_{-\sigma,-\tau}\, c^n_{0,-n}(\tau,\sigma)\, p^{(n,n)}_{l-n}(\rho_{\tau,\sigma};q^{-\sigma},q^{-\tau}|q^2)$$

with the so-called minimal elements in $\mathcal{A}_q(SU(2))$ defined by

$$
\begin{aligned}
c^n_{n,0}(\tau,\sigma) &= q^{n(1-\sigma)}\prod_{j=0}^{n-1}\delta_{\tau+2n-1-2j,\sigma-1}\,\gamma_{\tau+2n-2-2j,\sigma},\\
c^n_{0,n}(\tau,\sigma) &= q^{n(1-\tau)}\prod_{j=0}^{n-1}\delta_{\tau-1,\sigma+2n-1-2j}\,\beta_{\tau,\sigma+2n-2-2j},\\
c^n_{-n,0}(\tau,\sigma) &= q^{-n(\sigma+2\tau-2n)}\prod_{j=0}^{n-1}\beta_{\tau-2n+1+2j,\sigma-1}\,\alpha_{\tau-2n+2+2j,\sigma},\\
c^n_{0,-n}(\tau,\sigma) &= q^{-n(2\sigma+\tau-2n)}\prod_{j=0}^{n-1}\gamma_{\tau-1,\sigma+1-2n+2j}\,\alpha_{\tau,\sigma+2-2n+2j},
\end{aligned}
\tag{2.10}
$$

and with the constant given by

$$
\begin{aligned}
d^{l,n}_{\tau,\sigma} &= \frac{C^{l,0}(\sigma)C^{l,n}(\tau)q^{-l}}{(q^{2l+2n+2};q^2)_{l-n}} = d^{l,n}_{\tau,-\sigma},\\
C^{l,j}(\sigma) &= q^{l+j}\begin{bmatrix}2l\\ l-j\end{bmatrix}_{q^2}^{1/2}\left(\frac{1+q^{-4j-2\sigma}}{(1+q^{-2\sigma})(-q^{2-2\sigma};q^2)_{l-j}(-q^{2+2\sigma};q^2)_{l+j}}\right)^{1/2}.
\end{aligned}
\tag{2.11}
$$

Note that $C^{l,j}(\sigma) = C^{l,-j}(-\sigma)$. Since we work in a non-commutative algebra, we have to be careful about the ordering in the product. We use the convention that $\prod_{i=0}^k \psi_i = \psi_0\psi_1\dots\psi_k$.

The following identity in the $\mathcal{A}_q(SU(2))$ is the starting point for the 'addition formula'. The identity is obtained from (2.5) with $i=j=0$ after applying $id\otimes\tau_{\phi/2}$. We assume from now on that $\tau,\sigma,\mu\in\mathbb{R}$. Explicitly,

$$
\begin{aligned}
d^{l,0}_{\tau,\sigma}p_l^{(0,0)}(\rho^\phi_{\tau,\sigma};q^\tau,q^\sigma|q^2) &= d^{l,0}_{\mu,\tau}d^{l,0}_{\mu,\sigma}p_l^{(0,0)}(\cos\phi;q^\mu,q^\sigma|q^2)p_l^{(0,0)}(D.\rho_{\tau,\mu};q^\mu,q^\sigma|q^2)\\
&+\sum_{n=1}^{l}d^{l,n}_{\mu,\tau}d^{l,n}_{\mu,\sigma}e^{-in\phi}(-q^{1+\sigma+\mu}e^{i\phi},q^{1-\sigma+\mu}e^{i\phi};q^2)_n p^{(n,n)}_{l-n}(\cos\phi;q^\mu,q^\sigma|q^2)\\
&\qquad\times\big(D.c^n_{0,n}(\tau,\mu)\big)p^{(n,n)}_{l-n}(D.\rho_{\tau,\mu};q^\mu,q^\tau|q^2)\\
&+\sum_{n=1}^{l}d^{l,n}_{-\mu,-\tau}d^{l,n}_{-\mu,-\sigma}e^{-in\phi}(-q^{1-\sigma-\mu}e^{i\phi},q^{1+\sigma-\mu}e^{i\phi};q^2)_n p^{(n,n)}_{l-n}(\cos\phi;q^{-\mu},q^{-\sigma}|q^2)\\
&\qquad\times\big(D.c^n_{0,-n}(\tau,\mu)\big)p^{(n,n)}_{l-n}(D.\rho_{\tau,\mu};q^{-\mu},q^{-\tau}|q^2),
\end{aligned}
\tag{2.12}
$$

where $\rho^\phi_{\tau,\sigma} = (id\otimes\tau_{\phi/2})\Delta(\rho_{\tau,\sigma})$. To this identity in $\mathcal{A}_q(SU(2))$ we apply the $*$-representation π in order to obtain an identity in the space $\mathcal{B}(\ell^2(\mathbb{Z}_+))$ of bounded linear operators. Next we find two suitable bases of (generalised) eigenvectors, and taking mixed basis matrix elements of this operator valued identity gives the addition theorem. Note that if we apply the one-dimensional $*$-representation $\tau_{\theta/2}$ to (2.12) instead of π, we obtain the (degenerate) addition formula for the Askey-Wilson polynomials, cf. Noumi and Mimachi [1990a], Theorem 4, see also Koelink [1994a], (3.15), Koelink [1995c], (8.1).

The motivation for choosing the representation $\pi\otimes\tau_{\phi/2}$ in $\ell^2(\mathbb{Z}_+)$ with a relatively large kernel in order to represent (2.5) for $i=j=0$ instead of the representation $\pi\otimes\pi$ in $\ell^2(\mathbb{Z}_+)\otimes\ell^2(\mathbb{Z}_+)$ is twofold. Firstly, comparing the non-commutative addition formula (2.5) for $i=j=0$ in terms of the Askey-Wilson polynomials with the addition formula for the Legendre polynomials shows that it suffices to obtain an identity in commuting variables from (2.5) in which all three of the Askey-Wilson polynomials have a different argument. As is shown in this paper, this can be done using $\pi\otimes\tau_{\phi/2}$, but not $\tau_{\theta/2}\otimes\tau_{\phi/2}$ as remarked before. Secondly, although it is possible to find generalised eigenvectors for both $\pi(D.\rho_{\tau,\mu})\otimes\pi(\rho_{\mu,\sigma})$ and $(\pi\otimes\pi)\Delta(\rho_{\tau,\sigma})$, it does not seem possible to evaluate in general the matrix elements of the corresponding transition matrix explicitly. These matrix elements must occur explicitly in the corresponding addition formula.

Also $\tau_{\phi/2}(\rho_{\tau,\sigma})=0$ if and only if σ or τ equals infinity, so that we have to use $\pi\otimes\pi$ if $\mu=\infty$ or if $\sigma=\tau=\infty$. The corresponding addition formulas can then be obtained in the same way as the addition formula for little q-Legendre polynomials in Koornwinder's [1991] using Proposition 3.1. It turns out that all of these addition formulas can be obtained as limit cases of the addition formula derived in Section 4, see also Section 5.

3 Basis of the representation space

In order to turn (2.12) into an identity for q-special functions we have to study the operators $\pi(\rho^{\phi}_{\tau,\sigma})$ and $\pi(D.\rho_{\tau,\sigma})$. These operators are given as a five term recurrence operator, but it turns out that $\ell^2(\mathbb{Z}_+)$ has an orthogonal basis of eigenvectors of $\pi(\rho_{\tau,\infty})$ in which these operators are tridiagonal. This gives the opportunity to determine the eigenvectors in terms of orthogonal polynomials, and in this case the Al-Salam–Chihara polynomials are involved. The action of the minimal elements on these eigenvectors is given by certain shift operators, and thus the action of each of the generalised matrix elements can be calculated.

So we first recall the following basis of $\ell^2(\mathbb{Z}_+)$ and the action of certain operators in this basis.

Proposition 3.1 (i) *$\ell^2(\mathbb{Z}_+)$ has an orthogonal basis of the form $v_\lambda=v_\lambda(q^\tau)$, where $\lambda=-q^{2n}$, $n\in\mathbb{Z}_+$, $\lambda=q^{2\tau+2n}$, $n\in\mathbb{Z}_+$. The vector v_λ is an eigenvector of the self-adjoint operator $\pi(\rho_{\tau,\infty})$ for the eigenvalue λ. Moreover, with the normalisation $\langle v_\lambda,e_0\rangle=1$ we have*

$$\langle v_\lambda,v_\lambda\rangle=q^{-2n}(q^2;q^2)_n(-q^{2-2\tau};q^2)_n(-q^{2\tau};q^2)_\infty,\qquad \lambda=-q^{2n},$$

$$\langle v_\lambda,v_\lambda\rangle=q^{-2n}(q^2;q^2)_n(-q^{2+2\tau};q^2)_n(-q^{-2\tau};q^2)_\infty,\qquad \lambda=q^{2\tau+2n}.$$

(ii) *For $\lambda=-q^{2n}$, $\lambda=q^{2\tau+2n}$, $n\in\mathbb{Z}_+$, we have*

$$\pi(\alpha_{\tau,\infty})v_\lambda(q^\tau)=iq^{\frac12-\tau}(1+\lambda)v_{\lambda/q^2}(q^{\tau-1}),\qquad \pi(\beta_{\tau,\infty})v_\lambda(q^\tau)=iq^{\frac12}v_\lambda(q^{\tau-1}),$$

$$\pi(\gamma_{\tau,\infty})v_\lambda(q^\tau)=iq^{\frac12}(q^{2\tau}-\lambda)v_\lambda(q^{\tau+1}),\qquad \pi(\delta_{\tau,\infty})v_\lambda(q^\tau)=-iq^{\frac12+\tau}v_{\lambda q^2}(q^{\tau+1}),$$

with the convention $v_{-q^{-2}}(q^\tau)=0=v_{q^{2\tau-2}}(q^\tau)$.

Proof Part (i) follows from Koelink [1995a], Section 4, since $-q\pi(\gamma) = \pi(\beta)$ implies $\pi(\alpha_{\tau,\sigma}) = \pi(\alpha_{\sigma,\tau})$, $\pi(\delta_{\tau,\sigma}) = \pi(\delta_{\sigma,\tau})$ and thus $\pi(\rho_{\tau,\sigma}) = \pi(\rho_{\sigma,\tau})$.

For part (ii) observe that by Proposition 2.1 it suffices to prove the statements for $\beta_{\tau,\infty}$ and $\delta_{\tau,\infty}$, and these statements are proved analogously. Let us prove the last statement. From Corollary 2.2 we get $\delta_{\tau,\infty}\rho_{\tau,\infty} = q^{-2}\rho_{\tau,\infty}\delta_{\tau,\infty}$, so that by part (i) $\pi(\delta_{\tau,\infty})v_\lambda(q^\tau) = Cv_{\lambda q^2}(q^{\tau+1})$ for some constant C. Use the normalisation, π being a $*$-representation, $\delta^*_{\tau,\infty} = q^{-1}\alpha_{\tau+1,\infty}$, (2.2) and (2.6) to get

$$C = \langle \pi(\delta_{\tau,\infty})v_\lambda(q^\tau), e_0\rangle = \langle v_\lambda(q^\tau), \pi(q^{-\frac{1}{2}}\alpha + iq^{\tau+\frac{1}{2}}\gamma)e_0\rangle = -iq^{\tau+\frac{1}{2}}.$$

□

Remark 3.2 *(i) The vectors v_λ can be expressed in terms of the standard basis $\{e_n \mid n \in \mathbb{Z}_+\}$, since $\langle v_\lambda, e_n\rangle$ equals a constant times an Al-Salam–Carlitz polynomial $U_n^{(a)}$, which must be considered as a big q-Hermite polynomial in this case, see* Koelink [1995a]. *So the big q-Hermite polynomials occur as matrix elements of the transition operator in $\ell^2(\mathbb{Z}_+)$ mapping eigenvectors of $\pi(\rho_{\infty,\infty})$, i.e., the standard basis of $\ell^2(\mathbb{Z}_+)$, to eigenvectors of $\pi(\rho_{\tau,\infty})$.*

(ii) The basis of Proposition 3.1 induces the orthogonal decomposition $\ell^2(\mathbb{Z}_+) = V_1 \oplus V_2$, where V_1, respectively V_2, is spanned by $v_{-q^{2n}}$, $n \in \mathbb{Z}_+$, respectively $v_{q^{2\tau+2n}}$, $n \in \mathbb{Z}_+$. When needed we use $V_i^\tau = V_i$ to stress the dependence on τ.

From (2.4) and (2.3) we see that $(id \otimes \tau_{\phi/2}) \circ \Delta$ multiplies α and γ by $e^{i\phi/2}$ and β and δ by $e^{-i\phi/2}$. From this we see that $(id \otimes \tau_{\phi/2}) \circ \Delta$ multiplies $\alpha_{\tau,\infty}$ and $\gamma_{\tau,\infty}$ by $e^{i\phi/2}$ and $\beta_{\tau,\infty}$ and $\delta_{\tau,\infty}$ by $e^{-i\phi/2}$ and from this we find the action of $(id \otimes \tau_{\phi/2}) \circ \Delta$ on $\alpha_{\tau,\sigma}$, etcetera, cf. (2.7). Next recall Proposition 2.1 to find

$$\begin{aligned}\Big[2q^{\tau+\sigma}(\pi \otimes \tau_{\phi/2})\Delta(\rho_{\tau,\sigma}) - q^{2\sigma-1} - q^{2\tau+1}\Big]v_\lambda &= (\pi \otimes \tau_{\phi/2})\Delta(\beta_{\tau+1,\sigma-1}\gamma_{\tau,\sigma})v_\lambda \\ &= \pi\Big((e^{-i\phi/2}\beta_{\tau+1,\infty} - q^{\sigma-1}e^{i\phi/2}\alpha_{\tau+1,\infty})(e^{i\phi/2}\gamma_{\tau,\infty} + q^\sigma e^{-i\phi/2}\delta_{\tau,\infty})\Big)v_\lambda,\end{aligned}$$

and this can be calculated explicitly by Proposition 3.1(ii). We obtain

$$\begin{aligned} &2(\pi \otimes \tau_{\phi/2})\Delta(\rho_{\tau,\sigma})v_\lambda \\ &\quad = e^{-i\phi}qv_{\lambda q^2} + e^{i\phi}q^{-1}(1 - q^{-2\tau}\lambda)(1+\lambda)v_{\lambda/q^2} + \lambda q^{1-\tau}(q^{-\sigma} - q^\sigma)v_\lambda. \end{aligned} \tag{3.1}$$

For convenience we restrict our attention from now on to the subspace V_1 of $\ell^2(\mathbb{Z}_+)$. We use the notation $v_n = v_n(q^\tau)$ for the orthogonal basis $v_{-q^{2n}} = v_{-q^{2n}}(q^\tau)$, $n \in \mathbb{Z}_+$, of V_1. Put $\lambda = -q^{2n}$ in (3.1) and compare the recurrence relation with the three-term recurrence relation for the Al-Salam–Chihara polynomials $h_n(x) = h_n(x; s, t|q) = \lim_{\alpha,\beta\to\infty} p_n^{(\alpha,\beta)}(x; s, t|q)$, cf. Askey and Wilson [1985], (1.24), Gasper and Rahman [1990] , (7.5.3),

$$2xh_n(x) = h_{n+1}(x) + (t - t^{-1})q^{n+\frac{1}{2}}s^{-1}h_n(x) + (1-q^n)(1+q^ns^{-2})h_{n-1}(x), \tag{3.2}$$

to see that the operator $(\pi\otimes\tau_{\phi/2})\Delta(\rho_{\tau,\sigma})$ can be realised as the multiplication operator on a suitable weighted L^2-space. This is the content of the following proposition, which follows from the spectral theory of Jacobi matrices, cf. Berezanskiĭ[1968], Ch. VII, Section 1, Dombrowski [1990]. In this case it is also easily proved using the three-term recurrence relation (3.2) of the Al-Salam–Chihara polynomials and Remark 3.4(i).

Proposition 3.3 *Denote by dm the orthogonality measure for the Al-Salam–Chihara polynomials* $h_n(\cdot;q^\tau,q^\sigma|q^2)$ *normalised by* $\int_{\mathbb{R}} dm(x) = (-q^{2\tau};q^2)_\infty$ *and define the mapping* $\Lambda\colon V_1 \to L^2(dm)$ *by*

$$\Lambda\colon v_n(q^\tau) \longmapsto q^{-n}e^{in\phi}\, h_n(\cdot;q^\tau,q^\sigma|q^2).$$

Then Λ *extends to a unitary operator and* $\Lambda\big((\pi\otimes\tau_{\phi/2})\Delta(\rho_{\tau,\sigma})\big) = M\Lambda$, *where the multiplication operator* $M\colon L^2(dm)\to L^2(dm)$ *is defined by* $Mf\colon x\mapsto xf(x)$.

Remark 3.4 *(i) The fact that* Λ *extends to a unitary operator follows from the determinacy of the moment problem for the Al-Salam–Chihara polynomials. The multiplication operator* M *is a bounded operator on* $L^2(dm)$, *since the support of the orthogonality measure dm is compact.*

(ii) Proposition 3.3 states that

$$w_x(q^\tau,q^\sigma) = \sum_{n=0}^{\infty} \frac{q^n e^{-in\phi} h_n(x;q^\tau,q^\sigma|q^2)}{(q^2,-q^{2-2\tau};q^2)_n}\, v_n(q^\tau), \qquad x\in supp(dm)$$

is a generalised eigenvector of $(\pi\otimes\tau_{\phi/2})\Delta(\rho_{\tau,\sigma})$. *So the Al-Salam–Chihara polynomials, considered as the Hermite case of the Askey-Wilson polynomials, occur as matrix elements of the unbounded transition operator sending eigenvectors of* $\pi(\rho_{\tau,\infty})$ *to generalised eigenvectors of* $(\pi\otimes\tau_{\phi/2})\Delta(\rho_{\tau,\sigma})$, *cf. Remark 3.2(i).*

(iii) In a similar way as in Proposition 3.3 it is possible to give a complete description of the spectral properties of the element $\rho_{\tau,\sigma}$ *in such an irreducible *-representation. For this observe that* $\rho_{\tau,\sigma} = (id\otimes\tau_0)\Delta(\rho_{\tau,\sigma})$. *This can be used to prove that the Haar functional on the quantum* $SU(2)$ *group on the subalgebra generated by* $\rho_{\tau,\sigma}$ *is given by an Askey-Wilson integral, see* Koelink and Verding [1996]. *This statement is due to* Koornwinder [1993], *and is proved there using corepresentations and the Schur type orthogonality relations.*

Next we study $\pi(D.\rho_{\tau,\sigma})$. Observe that $D. = (id\otimes\tau_{(i\ln q)/2})\circ\Delta$, so we can apply the previous calculations to find in V_1 the eigenvector

$$u_y = u_y(q^\tau,q^\sigma) = \sum_{n=0}^{\infty} \frac{q^{2n} h_n(y;q^\tau,q^\sigma|q^2)}{(q^2,-q^{2-2\tau};q^2)_n}\, v_n(q^\tau), \qquad y\in \mathrm{supp}(dm). \tag{3.3}$$

To see that $u_y\in\ell^2(\mathbb{Z}_+)$ for y in the support of dm we can use the fact that the similarly defined vector with q^{2n} replaced by q^n is a generalised eigenvector of the self-adjoint operator $\pi(\rho_{\tau,\sigma})$, cf. Remark 3.4(ii), and then use Berezanskiĭ[1968], Ch. VII, Section 1.1, (1.24). Another way to see this is to use the asymptotic properties of the Al-Salam–Chihara polynomials as $n\to\infty$, cf. Askey and Ismail [1984], Section 3.1.

Proposition 3.5 *The operators* $\pi(D.\alpha_{\tau,\sigma})$ *and* $\pi(D.\beta_{\tau,\sigma})$ *map* V_1^τ *into* $V_1^{\tau-1}$ *and the operators* $\pi(D.\gamma_{\tau,\sigma})$ *and* $\pi(D.\delta_{\tau,\sigma})$ *map* V_1^τ *into* $V_1^{\tau+1}$. *Moreover,*

$$\pi(D.\alpha_{\tau,\sigma})\, u_y(q^\tau, q^\sigma) = \frac{iq^{1+\sigma}}{1+q^{2-2\tau}}(1+2yq^{1-\tau-\sigma}+q^{2-2\tau-2\sigma})\, u_y(q^{\tau-1}, q^{\sigma-1}),$$

$$\pi(D.\beta_{\tau,\sigma})\, u_y(q^\tau, q^\sigma) = \frac{iq}{1+q^{2-2\tau}}(1-2yq^{1+\sigma-\tau}+q^{2+2\sigma-2\tau})\, u_y(q^{\tau-1}, q^{\sigma+1}),$$

$$\pi(D.\gamma_{\tau,\sigma})\, u_y(q^\tau, q^\sigma) = i(1+q^{2\tau})\, u_y(q^{\tau+1}, q^{\sigma-1}),$$

$$\pi(D.\delta_{\tau,\sigma})\, u_y(q^\tau, q^\sigma) = -iq^\sigma(1+q^{2\tau})\, u_y(q^{\tau+1}, q^{\sigma+1}).$$

Proof First observe that $D.\gamma_{\tau,\infty} = q^{-\frac{1}{2}}\gamma_{\tau,\infty}$ and $D.\delta_{\tau,\infty} = q^{\frac{1}{2}}\delta_{\tau,\infty}$. Now (2.6) and Proposition 3.1(ii) show that $\pi(D.\gamma_{\tau,\sigma})$ maps V_1^τ into $V_1^{\tau+1}$. And similarly for the other operators.

To prove the second statement we note that the results for $D.\gamma_{\tau,\sigma}$ and $D.\delta_{\tau,\sigma}$ imply the results for $D.\beta_{\tau,\sigma}$ and $D.\alpha_{\tau,\sigma}$ by Proposition 2.1, since u_y is an eigenvector of $\pi(D.\rho_{\tau,\sigma})$. Since $D.$ is a homomorphism, Corollary 2.2 implies

$$(D.\gamma_{\tau,\sigma})(D.\rho_{\tau,\sigma}) = (D.\rho_{\tau+1,\sigma-1})(D.\gamma_{\tau,\sigma}).$$

Since the eigenspaces of $\pi(D.\rho_{\tau,\sigma})$ in V_1^τ are one-dimensional, this implies

$$\pi(D.\gamma_{\tau,\sigma})\, u_y(q^\tau, q^\sigma) = C\, u_y(q^{\tau+1}, q^{\sigma-1})$$

for some constant C. To calculate C we take the inner product with $v_0(q^{\tau+1})$ and use that π is a $*$-representation to get

$$C\,(-q^{2+2\tau};q^2)_\infty = \langle u_y(q^\tau,q^\sigma), \pi\big((D.\gamma_{\tau,\sigma})^*\big)v_0(q^{\tau+1})\rangle$$

$$= \langle u_y(q^\tau,q^\sigma), \pi\big(-q^{-\frac{1}{2}}\beta_{\tau+1,\infty} + q^{\sigma-\frac{1}{2}}\alpha_{\tau+1,\infty}\big)v_0(q^{\tau+1})\rangle = i(-q^{2\tau};q^2)_\infty$$

by (2.7), $\gamma^*_{\tau,\infty} = -\beta_{\tau+1,\infty}$, $\delta^*_{\tau,\infty} = q^{-1}\alpha_{\tau+1,\infty}$ and Proposition 3.1(ii). This implies the value for C. The statement for $\pi(D.\delta_{\tau,\sigma})$ is proved analogously.

□

Remark 3.6 *We can rewrite the result of Proposition 3.5 as identities for Al-Salam–Chihara polynomials by use of* (3.3), (2.7) *and Proposition 3.1(ii) to find special cases of*

$$(1-q^{2n}abcd)(1-2ay+a^2)p_n(y;aq,b,c,d|q)$$

$$= -ap_{n+1}(y;a,b,c,d|q) + (1-abq^n)(1-acq^n)(1-adq^n)p_n(y;a,b,c,d|q),$$

which can be proved from the orthogonality relations for the Askey-Wilson polynomials, and of

$$(1-q^{2n-2}abcd)p_n(y;a/q,b,c,d|q) = (1-q^{n-2}abcd)p_n(y;a,b,c,d|q)$$

$$-\tfrac{a}{q}(1-q^n)(1-bcq^{n-1})(1-bdq^{n-1})(1-cdq^{n-1})p_{n-1}(y;a,b,c,d|q)$$

which is a special case of the connection coefficients given by Askey and Wilson [1985], Section 6.

From Proposition 3.5 and (2.10) we obtain the following corollary by iteration.

Corollary 3.7 *With $y = (z + z^{-1})/2$ we have*

$$\pi\big(D.c^n_{0,n}(\tau,\sigma)\big)\, u_y(q^\tau, q^\sigma) = q^{n(\tau+\sigma+n)}(q^{1+\sigma-\tau}z, q^{1+\sigma-\tau}/z; q^2)_n\, u_y(q^\tau, q^{\sigma+2n}),$$

$$\pi\big(D.c^n_{0,-n}(\tau,\sigma)\big)\, u_y(q^\tau, q^\sigma) = (-1)^n q^{n(\tau-\sigma+n)} \times(-q^{1-\sigma-\tau}z, -q^{1-\sigma-\tau}/z; q^2)_n\, u_y(q^\tau, q^{\sigma-2n}).$$

4 Addition formula for Askey-Wilson polynomials

The results of Section 3 allow us to calculate the action of the elements in (2.12) under the representation π on suitable vectors. In this section we show how we can obtain a very general type of addition formula for the Legendre case, i.e. $\alpha = \beta = 0$, of the q-Jacobi polynomials defined in (2.9).

Apply π to (2.12) and let the resulting identity in $\mathcal{B}\big(\ell^2(\mathbb{Z}_+)\big)$ act on the eigenvector $u_y(q^\tau, q^\mu)$ of $\pi(D.\rho_{\tau,\mu})$. The action of the right hand side of (2.12) follows from Corollary 3.7. To the resulting identity we apply the unitary operator Λ of Proposition 3.3, which also shows how the left hand side of (2.12) looks under Λ. In order to find the result we need the action of Λ on the right hand side of (2.12). For this we have to calculate the following function of x in the Hilbert space $L^2(dm)$;

$$\big(\Lambda u_y(q^\tau, q^\mu)\big)(x) = \sum_{m=0}^\infty \frac{q^m e^{im\phi}}{(q^2, -q^{2-2\tau}; q^2)_m} h_m(y; q^\tau, q^\mu|q^2) h_m(x; q^\tau, q^\sigma|q^2). \quad (4.1)$$

Observe that $\big(\Lambda u_y(q^\tau, q^\mu)\big)(x)$ is the inner product of the eigenvector $u_y(q^\tau, q^\mu)$ of $\pi(D.\rho_{\tau,\mu})$ with the generalised eigenvector $w_x(q^\tau, q^\sigma)$ of $(\pi \otimes \tau_{\phi/2})\Delta(\rho_{\tau,\sigma})$.

Note that in case $\mu = \sigma$ (4.1) is the Poisson kernel for the Al-Salam–Chihara polynomials at $t = qe^{i\phi}$. This absolutely convergent expression, which can be considered as a non-symmetric Poisson kernel for the Al-Salam–Chihara polynomials, has been evaluated by Askey, Rahman and Suslov [1994], Section 14, Case II in terms of a very-well-poised ${}_8\varphi_7$-series. Explicitly, if we define for $|t| < 1$, $x = \cos\psi$, $y = \cos\theta$ the function

$$\begin{aligned} P(t; x, y; \tau; \sigma, \mu) &= \frac{(-q^{-\sigma-\mu}t, q^{1+\sigma-\tau}te^{i\theta}, q^{1+\sigma-\tau}te^{-i\theta}, q^{1+\mu-\tau}te^{i\psi}, q^{1+\mu-\tau}te^{-i\psi}; q^2)_\infty}{(q^{2+\mu+\sigma-2\tau}t, te^{i\theta+i\psi}, te^{i\theta-i\psi}, te^{i\psi-i\theta}, te^{-i\theta-i\psi}; q^2)_\infty} \\ &\quad \times {}_8W_7(q^{\mu+\sigma-2\tau}t; -q^{\sigma+\mu}t, q^{1+\mu-\tau}e^{i\theta}, q^{1+\mu-\tau}e^{-i\theta}, \\ &\qquad q^{1+\sigma-\tau}e^{i\psi}, q^{1+\sigma-\tau}e^{-i\psi}; q^2, -q^{-\sigma-\mu}t), \end{aligned} \quad (4.2)$$

then $\big(\Lambda u_y(q^\tau, q^\mu)\big)(x) = P(qe^{i\phi}; x, y; \tau; \sigma, \mu)$. Here we use the standard notation for very-well-poised basic hypergeometric series Gasper and Rahman [1990], Section 2.1;

$${}_8W_7(a; b, c, d, e, f; q, z) = {}_8\varphi_7\left(\begin{matrix} a, q\sqrt{a}, -q\sqrt{a}, b, c, d, e, f \\ \sqrt{a}, -\sqrt{a}, qa/b, qa/c, qa/d, qa/e, qa/f \end{matrix}; q, z\right).$$

These remarks prove the following addition theorem for the Askey-Wilson polynomials.

Theorem 4.1 *We have the following 'addition formula' for Askey-Wilson polynomials;*

$$d^{l,0}_{\tau,\sigma} p_l^{(0,0)}(x;q^\tau,q^\sigma|q^2) P(qe^{i\phi};x,y;\tau;\sigma,\mu)$$

$$= A_0\, p_l^{(0,0)}(z;q^\mu,q^\sigma|q^2) p_l^{(0,0)}(y;q^\mu,q^\tau|q^2) P(qe^{i\phi};x,y;\tau;\sigma,\mu)$$

$$+\sum_{n=1}^{l} A_n\, e^{-in\phi}(-q^{1+\sigma+\mu}e^{i\phi},q^{1-\sigma+\mu}e^{i\phi};q^2)_n(q^{1+\mu-\tau}e^{i\theta},q^{1+\mu-\tau}e^{-i\theta};q^2)_n$$

$$\times p_{l-n}^{(n,n)}(z;q^\mu,q^\sigma|q^2) p_{l-n}^{(n,n)}(y;q^\mu,q^\tau|q^2) P(qe^{i\phi};x,y;\tau;\sigma,\mu+2n)$$

$$+\sum_{n=1}^{l} B_n\, e^{-in\phi}(-q^{1-\sigma-\mu}e^{i\phi},q^{1+\sigma-\mu}e^{i\phi};q^2)_n(-q^{1-\mu-\tau}e^{i\theta},-q^{1-\mu-\tau}e^{-i\theta};q^2)_n$$

$$\times p_{l-n}^{(n,n)}(z;q^{-\mu},q^{-\sigma}|q^2) p_{l-n}^{(n,n)}(y;q^{-\mu},q^{-\tau}|q^2) P(qe^{i\phi};x,y;\tau;\sigma,\mu-2n)$$

with $A_n = d^{l,n}_{\mu,\tau} d^{l,n}_{\mu,\sigma} q^{n(\tau+\mu+n)}$, $B_n = (-1)^n d^{l,n}_{-\mu,-\tau} d^{l,n}_{-\mu,-\sigma} q^{n(\tau-\mu+n)}$, $x = \cos\psi$, $y=\cos\theta$, $z=\cos\phi$ *and* $d^{l,n}_{\mu,\sigma}$ *and the Poisson kernel defined by* (2.11) *and* (4.2).

Initially Theorem 4.1 only holds for $x=\cos\psi$ almost everywhere as an identity in $L^2(dm)$, but by continuity it holds everywhere.

Remark 4.2 *Theorem 4.1 has been derived from the identity* (2.12) *after applying* π *to it. We have only considered the resulting operator identity restricted to the subspace* V_1^τ *of* $\ell^2(\mathbb{Z}_+)$. *If we study the restriction to the subspace* V_2^τ, *we obtain the same formula after replacing* τ, σ *and* μ *by* $-\tau$, $-\sigma$ *and* $-\mu$. *Replacing* π *by any other irreducible infinite-dimensional representation of* $\mathcal{A}_q(SU(2))$ *does not lead to greater generality.*

Remark 4.3 *To recover the (degenerate) addition formula for the 2-parameter family of Askey-Wilson polynomials, i.e. the result of applying* $\tau_{\theta/2}$ *to* (2.12), *we formally replace* $e^{i\theta}$ *by* $q^{-1}e^{i\theta}$, $e^{i\phi}$ *by* $q^{-1}e^{i\phi}$, *and* $e^{i\psi}$ *by* $q^{-1}e^{i\theta+i\phi}$. *For this choice of parameters the very-well-poised* ${}_8\varphi_7$*-series in the non-symmetric Poisson kernel reduces to a very-well-poised* ${}_6\varphi_5$*-series, which can be summed by* Gasper and Rahman [1990], *(II.20).*

Remark 4.4 *(i) Because of the similarity between* (2.5) *and the group theoretic proof of the addition formula for the Legendre polynomials, cf.* Vilenkin [1968], *Section III.4.2,* Vilenkin and Klimyk [1991-1993], *Section 6.6.1, we want to have a limit transition of Theorem 4.1 to the addition formula for the Legendre polynomials as* $q\uparrow 1$. *A straighforward limit does not seem possible in general, but it does work for the addition formula of* Rahman and Verma [1986]. *Also the more sophisticated technique of* Van Assche and Koornwinder [1991] *does not seem to work here, since we do not have a three-term recurrence relation for the non-symmetric Poisson kernel in* μ, *i.e. a simple relation between the non-symmetric Poisson kernels with* $\mu-2$, μ *and* $\mu+2$. *However, this technique can be used for the special case* $\mu\to\infty$ *described in Section 5.*

(ii) Usually an addition formula leads to a product formula. In this case this can be done if $P(qe^{i\phi}; x, y; \tau; \sigma, \mu+2n)$ is part of a set of e.g. biorthogonal rational functions in x with respect to n. This seems not known. However, in the special cases described in the next section a product formula can be derived.

5 Polynomial addition formulas

In general it seems hard to reduce Theorem 4.1 to a polynomial identity, but for the special cases $\mu \to \infty$ and $\sigma = \tau = \mu = 0$ we can obtain polynomial identities. In case $\mu \to \infty$ we obtain a polynomial addition formula for the two parameter family of Askey-Wilson polynomials, in which the right hand side involves big q-Jacobi polynomials. In case $\sigma = \tau = \mu = 0$ we obtain the addition formula for the continuous q-Legendre polynomials, cf. Rahman and Verma [1986], (1.24), see also Koelink [1994a], Theorem. 4.1 for another quantum group theoretic derivation.

5.1 The case $\mu \to \infty$ In order to formulate the special case $\mu \to \infty$ we introduce the big q-Jacobi polynomials

$$P_n^{(\alpha,\beta)}(x; c, d; q) = {}_3\varphi_2\left(\begin{matrix} q^{-n}, q^{n+\alpha+\beta+1}, q^{1+\alpha}x/c \\ q^{\alpha+1}, -q^{1+\alpha}d/c \end{matrix}; q, q\right) \tag{5.1}$$

and the q-Laguerre polynomials corresponding to (2.9), i.e. let $\beta \to \infty$,

$$l_n^{(\alpha)}(x; s, t \mid q) = p_n(x; q^{1/2+\alpha}s/t, q^{1/2}t/s, -q^{1/2}/(st), 0 \mid q). \tag{5.2}$$

In Theorem 4.1 we replace $e^{i\phi}$ by $q^{\mu+\sigma-1}z^{-1}$ and let $\mu \to \infty$ such that $2q^{\mu+\tau-1}\cos\theta$ tends to y. Observe that then $2q^{\mu+\sigma-1}\cos\phi$ tends to z. Then we can use the limit transition of the Askey-Wilson polynomials to the big q-Jacobi polynomials as described by Koornwinder [1993], Proposition 6.1 to treat the limit case $\mu \to \infty$ of the Askey-Wilson polynomials on the right hand side of Theorem 4.1.

It remains to consider the limit case $\mu \to \infty$ of the non-symmetric Poisson kernel (4.2), which can be done directly in the very-well-poised ${}_8\varphi_7$-series. The result is a ${}_3\varphi_2$-series which can be written as a q-Laguerre polynomial when y and z are mass points of the orthogonality measure for the corresponding big q-Jacobi polynomials by simple transformations for ${}_3\varphi_2$-series. The limit case can also be evaluated directly from the infinite sum (4.1). Using the limit as described in Koornwinder [1993], Proposition 6.1 we get

$$q^{m\mu}h_m(\cos\theta; q^\tau, q^{\mu+2n}|q^2) \to y^m q^{(1-\tau)\tau}(-y^{-1}q^{-2n}; q^2)_m$$

as $\mu \to \infty$. The resulting sum is then a generating function for the Al-Salam–Chihara polynomials, which has been evaluated by Sergeĭ Suslov (private communication). Explicitly, for the Al-Salam–Chihara polynomial defined by

$$s_m(\cos\psi; a, b, |q) = a^{-m}(ab; q)_m\, {}_3\varphi_2\left(\begin{matrix} q^{-m}, ae^{i\psi}, ae^{-i\psi} \\ ab,\ 0 \end{matrix}; q, q\right)$$

we have the generating function

$$\sum_{m=0}^{\infty} \frac{(u; q)_m t^m}{(q, ab; q)_m} s_m(\cos\psi; a, b|q) = \frac{(ute^{-i\psi}; q)_\infty}{(te^{-i\psi}; q)_\infty}\, {}_3\varphi_2\left(\begin{matrix} u, ae^{i\psi}, ae^{-i\psi} \\ ab, ute^{-i\psi} \end{matrix}; q, te^{i\psi}\right).$$

For the choices of y and z as above this can be rewritten as a q-Laguerre polynomial (5.2).

Keeping track of all the constants involved and replacing the base q^2 by q and the parameters τ, σ by 2τ, 2σ we obtain the following theorem.

Theorem 5.1 *The following addition formula holds for $l, m, p \in \mathbb{Z}_+$, $\sigma, \tau \in \mathbb{R}$, $x \in \mathbb{C}$;*

$$p_l^{(0,0)}(x; q^\tau, q^\sigma \mid q) l_m^{(p)}(x; q^\tau, q^\sigma \mid q)$$

$$= \sum_{n=0}^{l} D^{n,l}(\tau,\sigma) P_{l-n}^{(n,n)}(-q^m; q^{2\tau}, 1; q) P_{l-n}^{(n,n)}(-q^{m+p}; q^{2\sigma}, 1; q) l_{m+n}^{(p)}(x; q^\tau, q^\sigma \mid q)$$

$$+ \sum_{n=1}^{l} D^{n,l}(\tau,\sigma)(q^m, q^{m+p}, -q^{m+p-2\sigma}, -q^{m-2\tau}; q^{-1})_n$$

$$\times P_{l-n}^{(n,n)}(-q^{m-n}; q^{2\tau}, 1; q) P_{l-n}^{(n,n)}(-q^{m+p-n}; q^{2\sigma}, 1; q^2) l_{m-n}^{(p)}(x; q^\tau, q^\sigma \mid q),$$

with the constant given by

$$D^{n,l}(\tau,\sigma) = (-q^{2\sigma-l}, -q^{2\tau-l}; q)_{l-n} \frac{(q^{l-n+1}; q)_n}{(q;q)_n} (q^{n+1}; q)_l q^{\frac{1}{2}(l-n)(l-n-2\sigma-2\tau)}$$

and the notation for the q-Legendre polynomials $p_l^{(0,0)}(\cdot; q^\tau, q^\sigma \mid q)$, the q-Laguerre polynomials $l_m^{(p)}(\cdot; q^\tau, q^\sigma \mid q)$, the big q-ultraspherical polynomials $P_{l-n}^{(n,n)}(\cdot; q^{2\tau}, 1; q)$ defined by (2.9), (5.2), (5.1).

Remark 5.2 *(i) This result can also be proved directly from the case $i = j = 0$, $\mu \to \infty$ of* (2.5) *by applying the representation $\pi \otimes \pi$ of $\mathcal{A}_q(SU(2)) \otimes \mathcal{A}_q(SU(2))$ instead of $\pi \otimes \tau_{\phi/2}$. In this case the q-Laguerre polynomials have an interpretation as matrix elements of the unbounded transition operator mapping eigenvectors of $\pi(\rho_{\tau,\infty}) \otimes \pi(\rho_{\infty,\sigma})$ to generalised eigenvectors of $(\pi \otimes \pi)\Delta(\rho_{\tau,\sigma})$.*

(ii) Using transformations for the big q-Jacobi polynomials like

$$P_n^{(\alpha,\beta)}(-x; c, d; q) = (-q^{\alpha-\beta} d/c)^n \frac{(q^{\beta+1}, -q^{\beta+1}c/d; q)_n}{(q^{\alpha+1}, -q^{\alpha+1}d/c; q)_n} P_n^{(\beta,\alpha)}(x; d, c; q),$$

and $P_n^{(\alpha,\beta)}(Ax; Ac, Ad; q) = P_n^{(\alpha,\beta)}(x; c, d; q)$, for $A > 0$, we can rewrite Theorem 5.1 such that the big q-Jacobi polynomials in the right hand side are evaluated at other points of the spectrum.

(iii) Theorem 5.1, and thus Theorem 4.1, contains as special cases two addition formulas for q-Legendre polynomials previously obtained from the quantum $SU(2)$ group. Firstly, we rescale $x = q^{1-\sigma-\tau}y/2$ and we take the limits $\tau \to \infty$. The limit transition of Askey-Wilson polynomials to big q-Jacobi polynomials as in Koornwinder [1993], *(6.2) shows that $p_l^{(0,0)}(x; q^\tau, q^\sigma \mid q)$ and $l_m^{(p)}(x; q^\tau, q^\sigma \mid q)$ tend to the big q-Legendre and big q-Laguerre polynomials. Moreover, the big q-ultraspherical polynomial $P_{l-n}^{(n,n)}(x; q^{2\tau}, 1; q)$ tends to a little q-Jacobi polynomial, and we obtain an addition formula for a big q-Legendre polynomial. If we next let the order p of the big q-Laguerre polynomial tend to infinity, so that the big q-Laguerre polynomials tend to the Al-Salam–Carlitz $U_n^{(a)}$, we obtain the addition formula for the big q-Legendre polynomials derived in* Koelink [1995a], *Theorem 4.3. For this we have to observe that a big q-ultraspherical polynomial of argument zero*

can be rewritten as a dual q-Krawtchouk polynomial by Gasper and Rahman [1990], *(III.5), (III.6).*

Secondly, rescaling $x = q^{1-\sigma-\tau}y/2$ *and letting both* $\sigma, \tau \to \infty$, *and using the limit transition of the Askey-Wilson polynomials to the little q-Jacobi polynomial, cf.* Koornwinder [1993], *(6.4), and the limit transition of the big q-Jacobi polynomials to the little q-Jacobi polynomials, shows that the addition formula in Theorem 5.1 contains* Koornwinder's [1991] *addition formula for the little q-Legendre polynomials as a special cas. See* Rahman [1989] *for an analytic proof of this limit case.*

Since we have orthogonality relations for the q-Laguerre polynomials we can derive a product formula. For this we need the normalised orthogonality measure $dm^{(\alpha)}(x; s, t \mid q)$ for the q-Laguerre polynomials;

$$\int_{\mathbb{R}} \left(l_m^{(\alpha)} l_n^{(\alpha)}\right)(x; s, t \mid q)\, dm^{(\alpha)}(x; s, t \mid q) = \delta_{n,m}(q, q^{1+\alpha}, -qs^{-2}, -q^{1+\alpha}t^{-2} : q)_n,$$

see Askey and Wilson [1985], Theorem 2.5 in combination with (5.2) for the explicit form of the orthogonality measure as an absolutely continuous measure on $[-1,1]$ plus a finite number, possibly zero, of discrete mass points off $[-1,1]$.

Corollary 5.3 *For* $l \in \mathbb{Z}_+$, *we have the product formula*

$$\begin{aligned} &P_l^{(0,0)}(-q^m; q^{2\tau}, 1; q) P_l^{(0,0)}(-q^{m+p}; q^{2\sigma}, 1; q) \\ &\quad = \frac{1}{C} \int_{\mathbb{R}} p_l^{(0,0)}(x; q^\tau, q^\sigma \mid q) \left(l_m^{(p)}(x; q^\tau, q^\sigma \mid q)\right)^2 dm^{(p)}(x; q^\tau, q^\sigma \mid q), \end{aligned}$$

with

$$C = D^{0,l}(\tau, \sigma)(q, q^{1+p}, -q^{1-2\tau}, -q^{1+p-2\sigma}; q)_n$$

with $D^{0,l}$ *defined in Theorem 5.1.*

Remark 5.4 *The limit case* $q \uparrow 1$ *of the addition formula in Theorem 5.1 and the product formula in Corollary 5.3 to the addition and product formula for the Legendre polynomials, cf.* Askey [1975], *Lecture 4, can be proved using the techniques of* Van Assche and Koornwinder [1991].

5.2 The addition formula for the continuous q-Legendre polynomial In the special case $\sigma = \tau = \mu = 0$ we can obtain the following addition formula for the continuous q-Legendre polynomials as a corollary to Theorem 4.1. The result is a special case of the addition formula for the continuous q-ultraspherical polynomials of Rahman and Verma [1986], (1.24).

Corollary 5.5 *For the continuous q-ultraspherical polynomials defined by*

$$C_n(\cos\psi; \beta|q) = \sum_{k=0}^{n} \frac{(\beta; q)_k (\beta; q)_{n-k}}{(q; q)_k (q; q)_{n-k}} e^{i(n-2k)\psi}$$

we have

$$C_l(\cos\psi;q^2|q^4) = q^l C_l(\cos\phi;q^2|q^4)C_l(\cos\theta;q^2|q^4)$$

$$+\sum_{n=1}^{l}\frac{q^{l-2n}(1+q^{4n})(q^2;q^4)_n}{(-q^2;q^2)_{2n}(q^4;q^4)_n}\begin{bmatrix} l+n \\ 2n\end{bmatrix}_{q^4}^{-1} C_{l-n}(\cos\theta;q^{2+4n}|q^4)C_{l-n}(\cos\phi;q^{2+4n}|q^4)$$

$$\times e^{-in(\theta+\phi)}(q^2e^{2i\theta},q^2e^{2i\phi};q^4)_n\, {}_4\varphi_3\left(\begin{matrix} q^{-4n}, q^{4n}, qe^{i(\phi+\theta)}e^{i\psi}, qe^{i(\phi+\theta)}e^{-i\psi} \\ q^2,\ q^2e^{2i\phi},\ q^2e^{2i\theta}\end{matrix};q^4,q^4\right).$$

Proof We can apply the same reductions as in Koelink [1994a], Section 4 to Theorem 4.1. First observe

$$P(t;\cos\theta,\cos\psi;0;0,0) = \frac{(t^2;q^4)_\infty}{(te^{i\psi+i\theta},te^{i\psi-i\theta},te^{i\theta-i\psi},te^{-i\psi-i\theta};q^4)_\infty}, \qquad |t|<1.$$

The last identity can be observed from $h_n(x;1,1|q^2) = H_n(x|q^4)$, (use e.g., the three-term recurrence relation) and using Rogers's expression for the Poisson kernel of the continuous q-Hermite polynomials, see Askey and Ismail [1983], Bressoud [1980], or by applying the summation formula Gasper and Rahman [1990], (II.18). Now the corollary follows from the following lemma, which has been proved by Mizan Rahman (private communication) using transformation and summation theorems for basic hypergeometric series.

Lemma 5.6 *For $n\in\mathbb{Z}_+$ and the Poisson kernel P defined by* (4.2) *we have*

$$\begin{aligned} &q^{n^2}(qe^{i\theta},q^{-i\theta};q^2)_nP(qe^{i\phi};\cos\psi,\cos\theta;0;0,2n) \\ &\qquad +(-1)^nq^{n^2}(-qe^{i\theta},-q^{-i\theta};q^2)_nP(qe^{i\phi};\cos\psi,\cos\theta;0;0,-2n) \\ &= \frac{2e^{-in\theta}(q^2e^{i\theta};q^4)_n(q^2e^{2i\phi};q^4)_\infty}{(qe^{i\phi+i\psi+i\theta},qe^{i\phi+i\psi-i\theta},qe^{i\phi+i\theta-i\psi},qe^{i\phi-i\psi-i\theta};q^4)_\infty} \\ &\qquad \times {}_4\varphi_3\left(\begin{matrix} q^{-4n}, q^{4n}, qe^{i(\theta+\phi)}e^{i\psi}, qe^{i(\theta+\phi)}e^{-i\psi} \\ q^2,\ q^2e^{2i\phi},\ q^2e^{2i\theta}\end{matrix};q^4,q^4\right).\end{aligned}$$

Since Rahman's proof of Lemma 5.6 is limited to the case $\sigma=\tau=\mu=0$ and the result in Corollary 5.5 is well-known we do not give the proof of Lemma 5.6.

□

Acknowledgements

I thank Mizan Rahman and Sergeĭ Suslov for sending their preprint with Askey and for answering a number of questions. Thanks are also due to the referee for his useful remarks. Most of the work for this paper was done at the Katholieke Universiteit Leuven, and I thank Walter Van Assche and Alfons Van Daele for their hospitality.

References

Askey, R. [1975], Orthogonal Polynomials and Special Functions, CBMS-NSF Regional Conference Series Applied Math., **21**, SIAM, Philadelphia, PA.

Askey, R. and Ismail, M. E. H. [1983], *A generalization of ultraspherical polynomials*, Studies in Pure Mathematics (Erdős, P., ed.), Birkhäuser, Basel, 55–78.

Askey, R. and Ismail, M. E. H. [1984], *Recurrence relations, continued fractions and orthogonal polynomials*, Memoirs Amer. Math. Soc., **49**, no. 300.

Askey, R. A., Rahman, M. and Suslov, S. K. [1994], *On a general q-Fourier transformation with nonsymmetric kernels*, Series 2, no. 21, Carleton University, J. Comp. Appl. Math., to appear.

Askey, R., and Wilson, J. [1985], *Some basic hypergeometric orthogonal polynomials that generalize Jacobi polynomials*, Memoirs Amer. Math. Soc., **54**, no. 319.

Berezanskiĭ, J. M. [1968], Expansions in Eigenfunctions of Selfadjoint Operators, Transl. Math. Monographs 17, Amer. Math. Soc., Providence, RI.

Bressoud, D. M. [1980], *A simple proof of Mehler's formula for q-Hermite polynomials*, Indiana Univ. Math. J., **29**, 577–580.

Chari, V. and Pressley, A. [1994], A Guide to Quantum Groups, Cambridge University Press, Cambridge.

Dijkhuizen, M. S. and Noumi, M. [1995], *A family of quantum projective spaces and related q-hypergeometric orthogonal polynomials*, announcement.

Dombrowski, J. [1990], *Orthogonal polynomials and functional analysis*, in Orthogonal Polynomials: Theory and Practice (Nevai, P., ed.), NATO ASI series C, vol. 294, Kluwer, Dordrecht, 147–161.

Dunkl, C. F. [1977], *An addition theorem for some q-Hahn polynomials*, Monatsh. Math., **85**, 5–37.

Erdélyi, A., Magnus, W., Oberhettinger, F. and Tricomi, F. G. [1953-1955], Higher Transcendental Functions, 3 volumes, McGraw-Hill.

Floreanini, R., Lapointe, L., and Vinet, L. [1994], *A quantum algebra approach to basic multivariable special functions*, J. Phys. A: Math. Gen., **27**, 6781–6797.

Floreanini, R. and Vinet, L. [1992a], *Addition formulas for q-Bessel functions*, J. Math. Phys., **33**, 2984–2988.

Floreanini, R. and Vinet, L. [1992b], *Using quantum algebras in q-special function theory*, Phys. Lett. A, **170**, 21–28.

Floreanini, R. and Vinet, L. [1993a], *On the quantum group and quantum algebra approach to q-special functions*, Lett. Math. Phys., **27**, 179–190.

Floreanini, R. and Vinet, L. [1993b], *An algebraic interpretation of the q-hypergeometric function*, J. Group Theory Phys., **1**, 1–10.

Floreanini, R. and Vinet, L. [1993c], *Quantum algebras and q-special functions*, Ann. Phys., **221**, 53–70.

Floreanini, R. and Vinet, L. [1994], *Generalized q-Bessel functions*, Canad. J. Phys., **72**, 345–354.

Floreanini, R. and Vinet, L. [1995a], *q-Gamma and q-beta functions in quantum algebra representation theory*, J. Comp. Appl. Math., to appear.

Floreanini, R. and Vinet, L. [1995b], *Basic Bessel functions and q-difference equations*, in Proc. 'Symmetries and integrability properties of difference equations', to appear.

Floris, P. G. A. [1994], *Addition formula for q-disk polynomials*, Report W-94-24, University of Leiden, Comp. Math., to appear.

Floris, P. G. A. and Koelink, H. T. [1995], *A commuting q-analogue of the addition formula for disk polynomials*, Report W-95-07, University of Leiden, Constr. Approx., to appear.

Gasper, G. and Rahman, M. [1990], Basic Hypergeometric Series, Encyclopedia Math. Appl. 35, Cambridge University Press, Cambridge.

Groza, V. A. and Kachurik, I. I. [1990], *Addition and multiplication theorems for Krawtchouk, Hahn and Racah q-polynomials*, (in Russian), Dokl. Akad. Nauk Ukrain SSR, Ser. A, **89**, 3–6.

Kalnins, E. G., Manocha, H. L. and Miller, W. [1992], *Models of q-algebra representations: tensor products of special unitary and oscillator algebras*, J. Math. Phys., **33**, 2365–2383.

Kalnins, E. G. and Miller, W. [1994], *Models of q-algebra representations: q-integral transforms and addition theorems*, J. Math. Phys., **35**, 1951–1975.

Kalnins, E. G., Miller, W. and Mukherjee, S. [1993], *Models of q-algebra representations: matrix elements of the q-oscillator algebra*, J. Math. Phys., **34**, 5333–5356.

Kalnins, E. G., Miller, W. and Mukherjee, S. [1994], *Models of q-algebra representations: the group of plane motions*, SIAM J. Math. Anal., **25**, 513–527.

Koelink, H. T. [1993], *A basic analogue of Graf's addition formula and related formulas*, Integral Transforms and Special Functions, **1**, 165–182.

Koelink, H. T. [1994a], *The addition formula for continuous q-Legendre polynomials and associated spherical elements on the* $SU(2)$ *quantum group related to Askey-Wilson polynomials*, SIAM J. Math. Anal., **25**, 197–217.

Koelink, H. T. [1994b], *The quantum group of plane motions and the Hahn-Exton q-Bessel function*, Duke Math. J., **76**, 483–508.

Koelink, H. T. [1995a], *Addition formula for big q-Legendre polynomials from the quantum* $SU(2)$ *group*, Canad. J. Math., **47**, 436–448.

Koelink, H. T. [1995b], *The quantum group of plane motions and basic Bessel functions*, Indag. Math. (N.S.), **6**, 197–211.

Koelink, H. T. [1995c], *Askey-Wilson polynomials and the quantum* $SU(2)$ *group: survey and applications*, Acta Appl. Math., to appear.

Koelink, H. T. [1995d], *Yet another basic analogue of Graf's addition formula*, J. Comp. Appl. Math., to appear.

Koelink, H. T. and Swarttouw, R. F. [1995], *A q-analogue of Graf's addition formula for the Hahn-Exton q-Bessel function*, J. Approx. Theory, **81**, 260–273.

Koelink, H. T. and Verding, J. [1996], *Spectral analysis and the Haar functional on the quantum* $SU(2)$ *group*, Comm. Math. Phys., **177**, 399-415.

Koornwinder, T. H. [1989], *Representations of the twisted $SU(2)$ quantum group and some q-hypergeometric orthogonal polynomials*, Proc. Kon. Ned. Akad. van Wetensch., Ser. A, **92** (Indag. Math. **51**), 97–117.

Koornwinder, T. H. [1990], *Orthogonal polynomials in connection with quantum groups*, inOrthogonal Polynomials: Theory and Practice (Nevai, P., ed.) NATO ASI series C, vol. 294, Kluwer, Dordrecht. 257–292.

Koornwinder, T. H. [1991], *The addition formula for little q-Legendre polynomials and the $SU(2)$ quantum group*, SIAM J. Math. Anal., **22**, 295–301.

Koornwinder, T. H. [1993], *Askey-Wilson polynomials as zonal spherical functions on the $SU(2)$ quantum group*, SIAM J. Math. Anal., **24**, 795–813.

Koornwinder, T. H. and Swarttouw, R. F. [1992], *On q-analogues of the Fourier and Hankel transforms*, Trans. Amer. Math. Soc., **333**, 445–461.

Miller, W. [1968], Lie Theory and Special Functions, Academic Press, New York.

Masuda, T., Mimachi, K., Nakagami, Y., Noumi, M. and Ueno, K. [1991], *Representations of the quantum group $SU_q(2)$ and the little q-Jacobi polynomials*, J. Funct. Anal., **99**, 357–386.

Noumi, M. [1991], *Quantum groups and q-orthogonal polynomials. Towards a realization of Askey-Wilson polynomials on $SU_q(2)$*, in Special Functions (Kashiwara, M., and Miwa, T., eds.), ICM-90 Satellite Conference Proceedings, Springer-Verlag, New York.

Noumi, M. and Mimachi, K. [1990a], *Askey-Wilson polynomials and the quantum group $SU_q(2)$*, Proc. Japan Acad., Ser. A, **66**, 146–149.

Noumi, M. and Mimachi, K. [1990b], *Quantum 2-spheres and big q-Jacobi polynomials*, Comm. Math. Phys., **128**, 521–531.

Noumi, M. and Mimachi, K. [1991], *Rogers's q-ultraspherical polynomials on a quantum 2-sphere*, Duke Math. J., **63**, 65–80.

Noumi, M. and Mimachi, K. [1992a], *Spherical functions on a family of quantum 3-spheres*, Comp. Math., **83**, 19–42.

Noumi, M. and Mimachi, K. [1992b], *Askey-Wilson polynomials as spherical functions on $SU_q(2)$*, in Quantum Groups (Kulish, P. O., ed.), Lecture Notes Math. 1510, Springer-Verlag, New York, 98–103.

Rahman, M. [1988], *An addition theorem and some product formulas for q-Bessel functions*, Canad. J. Math., **40**, 1203–1221.

Rahman, M. [1989], *A simple proof of Koornwinder's addition formula for the little q-Legendre polynomials*, Proc. Amer. Math. Soc., **107**, 373–381.

Rahman, M. and Verma, A. [1986], *Product and addition formulas for the continuous q-ultraspherical polynomials*, SIAM J. Math. Anal., **17**, 1461–1474.

Stanton, D. [1980], *Product formulas for q-Hahn polynomials*, SIAM J. Math. Anal., **11**, 100–107.

Stanton, D. [1981], *Three addition theorems for some q-Krawtchouk polynomials*, Geom. Dedicata, **10**, 403–425.

Stanton, D. [1984], *Orthogonal polynomials and Chevalley groups*, in Special Functions: Group Theoretical Aspects and Applications (Askey, R. A., Koornwinder, T. H., and Schempp, W., eds.), Reidel, Dordrecht, 87–128.

Swarttouw, R. F. [1992], *An addition theorem and some product formulas for the Hahn-Exton q-Bessel functions*, Canad. J. Math., **44**, 867–879.

Vaksman, L. L. and Soibelman, Y. S. [1988], *Algebra of functions on the quantum group* $SU(2)$, Funct. Anal. Appl., **22**, 170–181.

Van Assche, W. and Koornwinder, T. H. [1991], *Asymptotic behaviour for Wall polynomials and the addition formula for little q-Legendre polynomials*, SIAM J. Math. Anal., **22**, 302–311.

Vilenkin, N. J. [1968], Special Functions and the Theory of Group Representations, Transl. Math. Monographs 22, Amer. Math. Soc., Providence, RI.

Vilenkin, N. J. and Klimyk, A. U. [1991 - 1993], Representation of Lie Groups and Special Functions, 3 volumes, Kluwer, Dordrecht.

Watson, G. N. [1944], A Treatise on the Theory of Bessel Functions, 2nd ed., Cambridge University Press, Cambridge.

Woronowicz, S. L. [1987], *Compact matrix pseudogroups*, Comm. Math. Phys., **111**, 613–665.

Fields Institute Communications
Volume **14**, 1997

Special Functions and q-Commuting Variables

Tom H. Koornwinder
Department of Mathematics
University of Amsterdam
Plantage Muidergracht 24
1018 TV Amsterdam, The Netherlands

Abstract. This paper is mostly a survey, with a few new results. The first part deals with functional equations for q-exponentials, q-binomials and q-logarithms in q-commuting variables, and more generally under q-Heisenberg relations. The second part discusses translation invariance of Jackson integrals, q-Fourier transforms and the braided line.

1 Introduction

Identities for special functions often involve several variables, even if the special function itself depends on only one variable. In general these variables are real or complex, so they commute with each other. The theory of quantum groups has been quite successful in producing identities for q-special functions, in particular addition formulas, see, e.g., the survey in Koelink [21, Section 1] and further references given there. Although quantum groups themselves abound of non-commuting variables satisfying certain relations, one usually does not find back a similar type of variables in the resulting q-special function identities. I want to advertize here that special function identities in non-commuting variables should be studied more extensively and systematically. They often provide more elegant formulas than the corresponding identities in commuting variables, and they may be closer to a quantum group theoretical origin and therefore have more canonical properties. Another feature (which may be evaluated in a positive or negative sense) is that such identities are often more algebraic and formal in spirit and further away from Weierstrass type analysis than the identities in commuting variables. The most interesting and challenging cases with non-commuting variables occur when formal infinite series and convergent infinite series mix with each other. One has to be extremely careful there in order to avoid paradoxes, (see Section 9).

The present paper surveys (and extends a little) special function theory involving q-commuting variables x and y (i.e., satisfying the relation $xy = qyx$ with q complex, usually taken between 0 and 1).

1991 *Mathematics Subject Classification.* Primary: 33Dxx, 33D80.

The contents are as follows. In Section 2 we discuss Schützenberger's q-binomial formula. Section 3 deals with various functional equations for q-exponentials, and Section 4 gives some extensions of these results to q-Heisenberg cases. Section 5 describes possible equivalence with formulas involving commuting variables, via the operational interpretation. In Section 6 we discuss the q-logarithm. The next four sections are much inspired by the paper by Kempf & Majid [19]. We discuss translation invariance under a q-commuting translation variable for Jackson integrals over a finite interval (Section 7) and over the interval $(-\infty, \infty)$ (Section 9). In Section 8 we introduce a q-Fourier transform pair in connection with discrete q-Hermite polynomials. While this is in commuting variables, it is related to two types of q-Fourier transforms involving non-commuting variables which have been studied, respectively, by Kempf & Majid [19] and Finkelstein & Marcus [12]. A deeper understanding of many of these results can be obtained by means of Majid's [27] braided quantum groups, in particular the braided line. This is the topic of Section 10. Finally, two further directions are very briefly indicated in Section 11.

Conventions: $\mathbb{Z}_+$ will denote the set of nonnegative integers. From Section 3 on, whenever we work with q, it is supposed that $0 < q < 1$, unless said otherwise. The notation for q-hypergeometric series follows the book by Gasper & Rahman [6].

2 The q-Binomial Formula

Newton's binomial formula says:

$$(x+y)^n = \sum_{k=0}^{n} \binom{n}{k} y^{n-k} x^k, \quad n \in \mathbb{Z}_+. \tag{2.1}$$

Here it is implicitly understood that x and y commute: $xy = yx$. A q-analogue of (2.1) for *q-commuting variables* x, y, i.e., satisfying the relation

$$xy = qyx \tag{2.2}$$

for some $q \in \mathbb{C}$, first appeared in literature in Schützenberger [32], see also Cigler [7, (7)]:

Proposition 2.1 (*q-binomial formula*) Let $q \in \mathbb{C}$. Let $\mathbb{C}_q[x, y]$ be the complex associative algebra with 1 generated by x and y and with relation (2.2). Then the following identity is valid in the algebra $\mathbb{C}_q[x, y]$:

$$(x+y)^n = \sum_{k=0}^{n} \begin{bmatrix} n \\ k \end{bmatrix}_q y^{n-k} x^k, \quad n \in \mathbb{Z}_+. \tag{2.3}$$

Here we used the *q-binomial coefficient*

$$\begin{bmatrix} n \\ k \end{bmatrix}_q := \frac{(q;q)_n}{(q;q)_k\,(q;q)_{n-k}} = \frac{(1-q^n)(1-q^{n-1})\dots(1-q^{n-k+1})}{(1-q)(1-q^2)\dots(1-q^k)}, \tag{2.4}$$

while the *q-shifted factorial* is given by

$$(a;q)_k := (1-a)(1-qa)\dots(1-q^{k-1}a), \quad a \in \mathbb{C},\ k \in \mathbb{Z}_+. \tag{2.5}$$

The recurrence relations below show that the q-binomial coefficient (2.4) is a polynomial in q and therefore remains meaningful for q being a root of unity.

We will give a constructive proof of Proposition 2.1 which goes essentially back to Polya & Alexanderson [29] and which was later written down by Askey [2]. It is straightforward that the monomials $y^l x^k$ $(k, l \in \mathbb{Z}_+)$ form a basis for $\mathbb{C}_q[x,y]$ considered as a linear space and that $(x+y)^n$ will have a unique expansion of the form

$$(x+y)^n = \sum_{k=0}^{n} c_{n,k}\, y^{n-k} x^k \tag{2.6}$$

with the coefficients $c_{n,k}$ also depending on q. It follows immediately from (2.6) that $c_{n,0} = 1 = c_{n,n}$. Also, expansion of $(x+y)^n = (x+y)^{n-1}(x+y)$ and $(x+y)^n = (x+y)(x+y)^{n-1}$, respectively, yields for $n > k > 0$:

$$c_{n,k} = q^k c_{n-1,k} + c_{n-1,k-1}, \qquad c_{n,k} = c_{n-1,k} + q^{n-k} c_{n-1,k-1}. \tag{2.7}$$

Elimination of $c_{n-1,k}$ from these two recurrence equations leaves us with the two-term recurrence

$$c_{n,k} = \frac{1-q^n}{1-q^k}\, c_{n-1,k-1}.$$

Iteration of this last recurrence yields the right-hand side of (2.4). This proves (2.3).

The advantage of this proof is that it is constructive. If one just wants to prove by induction with respect to n that (2.6) holds with $c_{n,k}$ being given by (2.4) then it is sufficient to have only one of the recurrences in (2.7). This is the way in which one usually works in case $q = 1$, where the q-binomial formula (2.3) by continuity becomes the binomial formula (2.1). In that case the two recurrences in (2.7) coincide and it is not possible to get a two-term recurrence formula by elimination.

A second observation, due to Andrews, and written down in Askey [3], is that the q-binomial formula (2.3) is equivalent to an identity in commuting variables. Note that, if the generators x, y of $\mathbb{C}_q[x,y]$ satisfy the q-commutation relations (2.2), then $-yx, y$ also satisfy these relations:

$$(-yx)y = qy(-yx).$$

Hence, we get from (2.3) that

$$(-yx+y)^n = \sum_{k=0}^{n} \begin{bmatrix} n \\ k \end{bmatrix}_q y^{n-k}(-yx)^k. \tag{2.8}$$

The left-hand side of (2.8) equals

$$\begin{aligned} &y(1-x)y(1-x)\ldots y(1-x) \quad (2n \text{ factors}) \\ &= y^n(1-q^{n-1}x)\ldots(1-qx)(1-x) = y^n(x;q)_n. \end{aligned}$$

(Note that the definition (2.5) of q-shifted factorial remains valid for a in any complex associative algebra with 1.) As for the right-hand side of (2.8) note that

$$y^{n-k}(-yx)^k = (-1)^k q^{\frac{1}{2}k(k-1)} y^n x^k$$

and

$$\begin{bmatrix} n \\ k \end{bmatrix}_q = (-1)^k q^{-\frac{1}{2}k(k-1)} q^{nk} \frac{(q^{-n};q)_k}{(q;q)_k}. \tag{2.9}$$

Hence the identity (2.8) can be equivalently written as

$$y^n (x;q)_n = y^n \sum_{k=0}^{n} \frac{(q^{-n};q)_k}{(q;q)_k} (q^n x)^k.$$

Because of the earlier observation about a basis of monomials for $\mathbb{C}_q[x,y]$ we conclude that

$$(x;q)_n = \sum_{k=0}^{n} \frac{(q^{-n};q)_k}{(q;q)_k} (q^n x)^k.$$

This is still an identity in the algebra $\mathbb{C}_q[x,y]$. However, we can map it to an identity in $\mathbb{C}$ by using the algebra homomorphism $\pi: \mathbb{C}_q[x,y] \to \mathbb{C}$ such that, for some $z \in \mathbb{C}$, $\pi(x) = q^n z$ and $\pi(y) = 0$. This yields the formula giving the evaluation of a terminating q-binomial series:

$$(q^{-n}z;q)_n = \sum_{k=0}^{n} \frac{(q^{-n};q)_k}{(q;q)_k} z^k, \quad n \in \mathbb{Z}_+. \tag{2.10}$$

This formula is well-known, see [16, (II.4)]. It was earlier observed by Cigler [7, (8)] that formula (2.10) follows from (2.3). He used an operational interpretation of (2.3). We will discuss such interpretations in a later section. Then we will also see that (2.10) is in fact equivalent to (2.3).

3 Identities for q-Exponentials With q-Commuting Variables

The two *q-exponentials* are defined by

$$e_q(z) := \sum_{n=0}^{\infty} \frac{z^n}{(q;q)_n} = \frac{1}{(z;q)_\infty}, \tag{3.1}$$

$$E_q(z) := \sum_{n=0}^{\infty} \frac{q^{\frac{1}{2}n(n-1)} z^n}{(q;q)_n} = (-z;q)_\infty . \tag{3.2}$$

Here we assume $0 < q < 1$ and $(a;q)_\infty$ is defined as the (convergent) infinite product:

$$(a;q)_\infty := \prod_{j=0}^{\infty} (1 - q^j a).$$

For convergence of the infinite series in (3.1), (3.2) with $z \in \mathbb{C}$ we need $|z| < 1$ in (3.1). However, because of its product representation, e_q has an analytic continuation to $\mathbb{C} \backslash \{q^{-k} \mid k \in \mathbb{Z}\}$. See [16, Section 1.3] for the proofs of the second equalities in (3.1) and (3.2). The two q-exponential series can also be considered as formal power series in the formal variable z. Of course, no convergence condition is needed in that case. From (3.1), (3.2) we have

$$e_q(z)\, E_q(-z) = 1 \tag{3.3}$$

for $|z| < 1$. From the second equalities in (3.1) and (3.2) we see that

$$e_q(qz) = (1-z)\, e_q(z), \quad E_q(z) = (1+z)\, E_q(qz). \tag{3.4}$$

In general, algebraic identities for convergent power series remain valid for the corresponding formal power series. In particular, this applies to (3.3) and (3.4).

Fix $q \in (0,1)$ and let $\mathbb{C}_q[[x,y]]$ be the complex associative algebra with 1 of formal power series

$$\sum_{k,l=0}^{\infty} c_{k,l}\, y^l x^k$$

with arbitrary complex coefficients $c_{k,l}$ and where x, y satisfy relation (2.2), i.e., $xy = qyx$. The following Proposition generalizing the classical functional equation $e^x e^y = e^{x+y}$ for commuting variables x, y, was given first by Schützenberger [32], see also Cigler [7, (10)].

Proposition 3.1 In the algebra $\mathbb{C}_q[[x,y]]$ we have the identities

$$e_q(x+y) = e_q(y)e_q(x), \tag{3.5}$$
$$E_q(x+y) = E_q(x)\,E_q(y). \tag{3.6}$$

Proof By means of the q-binomial formula (2.3) we can write the left-hand side of (3.5) as an element of $\mathbb{C}_q[[x,y]]$, and next rewrite it as the right-hand side:

$$\begin{aligned} e_q(x+y) &= \sum_{n=0}^{\infty}\sum_{k=0}^{n} \frac{1}{(q;q)_n} \begin{bmatrix} n \\ k \end{bmatrix}_q y^{n-k}x^k = \sum_{n=0}^{\infty}\sum_{k=0}^{n} \frac{1}{(q;q)_{n-k}(q;q)_k}\, y^{n-k}x^k \\ &= \sum_{k,l=0}^{\infty} \frac{1}{(q;q)_l(q;q)_k}\, y^l x^k = e_q(y)e_q(x). \end{aligned}$$

This settles (3.5). For the proof of (3.6) note that, in view of (3.3), $E_q(x+y)$ is a left and right inverse to $e_q(-x-y)$, and $E_q(x)E_q(y)$ is a left and right inverse to $e_q(-y)e_q(-x)$. Now use (3.5). □

The reader is warned that the apparent symmetry in x and y in the left hand side of (3.5) does not allow us to conclude that $e_q(x+y) = e_q(x)e_q(y)$, since the relation $xy = qyx$ is not symmetric in x and y. The next Proposition gives a formula for $e_q(x)e_q(y)$ in the algebra $\mathbb{C}_q[[x,y]]$, i.e., for the right-hand side of (3.5) with the order of the two factors interchanged. It is a special case of a more general result given first in operational form by Rogers [30], which we will discuss in the next section. See also Gelfand & Fairlie [17, (46)], Floreanini & Vinet [13, formula (23d)], Faddeev & Volkov [11, p.314, formula (2)], Faddeev & Kashaev [10, Section 2], A. N. Kirillov [20, Section 2.5, Lemma 9], and McDermott & Solomon [28, (10)].

Proposition 3.2 In the algebra $\mathbb{C}_q[[x,y]]$ we have the identities

$$\begin{aligned} e_q(x)\,e_q(y) &= e_q(y-yx)\,e_q(x) && (3.7)\\ &= e_q(x+y-yx) && (3.8)\\ &= e_q(y)\,e_q(-yx)\,e_q(x) && (3.9)\\ &= e_q(y)\,e_q(x-yx), && (3.10)\\ E_q(y)\,E_q(x) &= E_q(x+y+yx), && (3.11)\\ &= E_q(x)\,E_q(yx)\,E_q(y). && (3.12) \end{aligned}$$

Proof Because $e_q(x)$ is invertible as a formal power series (cf. (3.3)), formula (3.7) can equivalently be stated as

$$e_q(x)\,e_q(y)\,e_q(x)^{-1} = e_q(y - yx). \tag{3.13}$$

For any two formal power series $f(z)$ and $g(z)$, with $f(z)$ being invertible and $g(z) = \sum_{k=0}^{\infty} c_k z^k$, we have

$$f(x)g(y)f(x)^{-1} = \sum_{k=0}^{\infty} c_k\, f(x)\, y^k\, f(x)^{-1} = \sum_{k=0}^{\infty} c_k\, \big(f(x) y f(x)^{-1}\big)^k = g\big(f(x)\, y\, f(x)^{-1}\big)$$

as identities in $\mathbb{C}_q[[x,y]]$. In particular,

$$e_q(x)\,e_q(y)\,e_q(x)^{-1} = e_q\big(e_q(x)\,y\,e_q(x)^{-1}\big). \tag{3.14}$$

Now, since $xy = qyx$ and by (3.4) we have

$$e_q(x)\,y\,e_q(x)^{-1} = y\,e_q(qx)\,e_q(x)^{-1} = y\,(1-x)\,e_q(x)\,e_q(x)^{-1} = y\,(1-x). \tag{3.15}$$

Together with (3.14) this settles (3.13) and hence (3.7).

Next it follows from Proposition 3.1 that the right-hand side of (3.7) equals (3.8), since $x(y - yx) = q(y - yx)x$. The equalities (3.9) and (3.10) also follow by application of Proposition 3.1. Finally, (3.11) and (3.12) follow from (3.8) and (3.9) in a similar way as we obtained (3.6). □

Note that the equalities (3.7)–(3.11) reduce to the classical identities $e^x e^y = e^y e^x = e^{x+y}$ if we replace x, y by $(1-q)x, (1-q)y$ and let $q \uparrow 1$ on using

$$\lim_{q\uparrow 1} e_q((1-q)z) = e^z = \lim_{q\uparrow 1} E_q((1-q)z), \tag{3.16}$$

cf. [16, (1.3.17)].

As a corollary to both Proposition 3.1 and Proposition 3.2 we have a functional equation for the *q-Gaussians*

$$g_q(x) := e_{q^2}(-x^2), \quad G_q(x) := E_{q^2}(-x^2) \tag{3.17}$$

in q-commuting variables. First note that that, for $z \in \mathbb{C}$ with $|z| < 1$, we have

$$e_{q^2}(-z^2) = \frac{1}{(-z^2;q^2)_\infty} = \frac{1}{(iz;q)_\infty(-iz;q)_\infty} = e_q(iz)\,e_q(-iz).$$

Thus the equality

$$e_{q^2}(-z^2) = e_q(iz)\,e_q(-iz) \tag{3.18}$$

is also valid for arbitrary real z or as an identity for formal power series.

Corollary 3.3 In the algebra $\mathbb{C}_q[[x,y]]$ we have the identities

$$e_{q^2}(-(x+y)^2) = e_{q^2}(-y^2)\, e_q(-yx)\, e_{q^2}(-x^2), \tag{3.19}$$
$$E_{q^2}(-(x+y)^2) = E_{q^2}(-x^2)\, E_q(-yx)\, E_{q^2}(-y^2). \tag{3.20}$$

Proof We first apply (3.18), next Proposition 3.1 (twice), next Proposition 3.2 and finally (3.18) again (twice):

$$\begin{aligned} e_{q^2}(-(x+y)^2) &= e_q(i(x+y))\, e_q(-i(x+y)) = e_q(iy)\, e_q(ix)\, e_q(-iy)\, e_q(-ix) \\ &= e_q(iy)\, e_q(-iy)\, e_q(-yx)\, e_q(ix)\, e_q(-ix) = e_{q^2}(-y^2)\, e_q(-yx)\, e_{q^2}(-x^2). \end{aligned}$$

This yields (3.19). Then formula (3.20) follows by taking the inverse on both sides and replacing x, y by ix, iy, respectively. □

Let us next discuss generalizations of the previous results in this section for the case of non-terminating q-binomial series

$${}_1\phi_0(a;\,;q,z) := \sum_{k=0}^{\infty} \frac{(a;q)_k}{(q;q)_k}\, z^k = \frac{(az;q)_\infty}{(z;q)_\infty}, \quad |z|<1,\ a\in\mathbb{C}, \tag{3.21}$$

see [16, (II.3)]. Formula (3.21) can be rewritten as

$${}_1\phi_0(a;\,;q,z) = E_q(-az)\, e_q(z), \tag{3.22}$$

and this remains valid as an identity for formal power series. The termwise limit for $q \uparrow 1$ of ${}_1\phi_0(q^a;\,;q,z)$ is

$${}_1F_0(a;\,;z) := \sum_{k=0}^{\infty} \frac{(a)_k}{k!}\, z^k = (1-z)^{-a}.$$

The functional equation $(1-x)^{-a}(1-y)^{-a} = (1-x-y+xy)^{-a}$ in commuting variables x, y can equivalently be written as

$${}_1F_0(a;\,;x)\, {}_1F_0(a;\,;y) = {}_1F_0(a;\,;x+y-yx). \tag{3.23}$$

We now give two q-analogues of (3.23), valid in the algebra $\mathbb{C}_q[[x,y]]$.

Proposition 3.4 In the algebra $\mathbb{C}_q[[x,y]]$ (so $xy = qyx$) we have for $a \in \mathbb{C}$ the identities

$${}_1\phi_0(a;\,;q,x)\, {}_1\phi_0(a;\,;q,y) = {}_1\phi_0(a;\,;q,x+y-yx), \tag{3.24}$$
$${}_1\phi_0(a;\,;q,y)\, {}_1\phi_0(a;\,;q,x) = {}_1\phi_0(a;\,;q,x+y-ayx). \tag{3.25}$$

Proof In view of (3.22) the identities (3.24), (3.25) can be equivalently written as

$$\begin{aligned} e_q(x)\, E_q(-ax)\, E_q(-ay)\, e_q(y) &= E_q(-a(x+y-yx))\, e_q(x+y-yx), \\ E_q(-ay)\, e_q(y)\, e_q(x)\, E_q(-ax) &= e_q(x+y-ayx)\, E_q(-a(x+y-ayx)). \end{aligned}$$

In view of (3.6), (3.8), (3.5) and (3.11) these identities are in their turn equivalent to

$$\begin{aligned} e_q(x)\, E_q(-a(x+y))\, e_q(y) &= E_q(-a(x+y-yx))\, e_q(x)\, e_q(y), \\ E_q(-ay)\, e_q(x+y)\, E_q(-ax) &= e_q(x+y-ayx)\, E_q(-ay)\, E_q(-ax). \end{aligned}$$

Once more, these identities can be rewritten into equivalent forms:

$$E_q(-a(x+y)) = (e_q(x))^{-1} E_q(-a(x+y-yx))\, e_q(x),$$
$$e_q(x+y) = e_q(ay)\, e_q(x+y-ayx)\, (e_q(ay))^{-1}.$$

By a similar argument as in the proof of Proposition 3.2 these two identities will follow if we can show that

$$x+y = (e_q(x))^{-1} (x+y-yx)\, e_q(x), \tag{3.26}$$
$$x+y = e_q(ay)\, (x+y-ayx)\, (e_q(ay))^{-1}. \tag{3.27}$$

As for (3.26), its right-hand side can be rewritten as

$$x + (e_q(x))^{-1} y(1-x)\, e_q(x) = x + e_q(x)^{-1} y\, e_q(qx) = x + e_q(x)^{-1} e_q(x)\, y = x+y,$$

where we used (3.4). Formula (3.27) can be proved by a similar argument. □

Formula (3.25) was given by Faddeev & Volkov [11, p.315, multiplication rule for $s(\lambda, w)$], see also Kirillov [20, Section 2.5, Exercise 3]. The terminating cases of (3.24), (3.25) (i.e., $a = q^{-n}$, see Kirillov [20, Section 2.5, Lemma 10]) are:

Corollary 3.5 In the algebra $\mathbb{C}_q[x,y]$ we have for $n \in \mathbb{Z}_+$ the identities

$$(x;q)_n\, (y;q)_n = (x+y-q^n yx;q)_n, \quad (y;q)_n\, (x;q)_n = (x+y-yx;q)_n. \tag{3.28}$$

4 Identities for q-Exponentials With q-Heisenberg Relations

Proposition 3.2 can be generalized to an identity in the algebra $\mathbb{C}_{qHeis}[[x,y,c]]$ of formal power series

$$\sum_{k,l,m=0}^{\infty} a_{k,l,m}\, c^m y^l x^k$$

with arbitrary complex coefficients $a_{k,l,m}$ and where x, y, c satisfy the *q-Heisenberg relations*

$$xy - qyx = (1-q)c, \quad xc = cx, \quad yc = cy. \tag{4.1}$$

Here $q \in (0,1)$ is fixed as before. Note that c is a central element of $\mathbb{C}_{qHeis}[[x,y,c]]$. Note that, by adding the relation $c = 0$, we can map the algebra $\mathbb{C}_{qHeis}[[x,y,c]]$ homomorphically onto the algebra $\mathbb{C}_q[[x,y]]$. It is also interesting to remark that the algebra $\mathbb{C}_{qHeis}[[x,y,c]]$ is isomorphic to the algebra of formal power series in x, y, z with relations

$$xy - yx = (1-q)z, \quad xz - qzx = 0, \quad zy - qyz = 0. \tag{4.2}$$

Just let z and c be related by

$$z = c - yx. \tag{4.3}$$

We need the following identity in $\mathbb{C}_{qHeis}[[x,y,c]]$, which is proved by induction with respect to n:

$$x^n y = q^n y x^n + (1-q^n) c x^{n-1}. \tag{4.4}$$

The generalization below of Proposition 3.2 was first given by Rogers [30] (in operational form, see Bowman [6] for a modern treatment). The same result was found later, independently, by Gelfand & Fairlie [17, (46)], McDermott & Solomon [28, (10)] and Kashaev [18, (4.19)].

Proposition 4.1 In the algebra $\mathbb{C}_{qHeis}[[x,y,c]]$ (i.e., with relations (4.1)) we have the identities

$$e_q(x)\,e_q(y) = e_q(y - yx + c)\,e_q(x) \tag{4.5}$$
$$= e_q(y)\,e_q(-yx + c)\,e_q(x) \tag{4.6}$$
$$= e_q(y)\,e_q(x - yx + c). \tag{4.7}$$

Proof As in the previous proof we have equality (3.14). By (4.4) we see that

$$e_q(x)\,y = \sum_{n=0}^{\infty} \frac{x^n y}{(q;q)_n} = \sum_{n=0}^{\infty} \frac{q^n y x^n}{(q;q)_n} + \sum_{n=1}^{\infty} \frac{(1-q^n)cx^{n-1}}{(q;q)_n} = y\,e_q(qx) + c\,e_q(x).$$

Hence, by (3.4),

$$e_q(x)\,y = (y - yx + c)\,e_q(x). \tag{4.8}$$

Now (4.5) follows from (3.14) and (4.8). The other two identities follow by application of Proposition 3.1. □

Remark 4.2 Because of (4.2) and (4.3), formula (4.6) can be rewritten as

$$e_q(x)\,e_q(y) = e_q(y)\,e_q\big((1-q)^{-1}[x,y]\big)\,e_q(x). \tag{4.9}$$

After rescaling we have, still as an identity in $\mathbb{C}_{qHeis}[[x,y,c]]$:

$$e_q\big((1-q)x\big)\,e_q\big((1-q)y\big) = e_q\big((1-q)y\big)\,e_q\big((1-q)[x,y]\big)\,e_q\big((1-q)x\big). \tag{4.10}$$

Replace x, y, c in (4.1) by $X, Y, (1-q)^{-1}C$, respectively, and let next $q \uparrow 1$ in (4.1) and (4.10). Then we obtain the identity

$$e^X e^Y = e^Y e^C e^X \quad \text{or equivalently} \quad e^X e^Y = e^{Y+C} e^X \tag{4.11}$$

in the algebra of formal power series in X, Y, C satisfying the *Heisenberg relations*

$$[X,Y] = C, \quad [X,C] = 0, \quad [Y,C] = 0. \tag{4.12}$$

Note that the second identity in (4.11) immediately follows from

$$e^X Y e^{-X} = \exp(\operatorname{ad} X)\,Y = Y + [X,Y] = Y + C,$$

where $(\operatorname{ad} U)\,V := [U,V] = UV - VU$. A slightly deeper identity in the Heisenberg algebra is obtained by applying the *Baker-Campbell-Hausdorff formula* (see for instance Varadarajan [35, Section 2.15]) to the case of relations (4.12). This yields

$$e^Y e^X = e^{X+Y-\frac{1}{2}C} \quad \text{or equivalently} \quad e^{X+Y} = e^Y e^{\frac{1}{2}C} e^X. \tag{4.13}$$

Formula (4.13) has often been observed in literature. A q-analogue of (4.13) in a q-Heisenberg algebra (but not precisely the algebra $\mathbb{C}_{qHeis}[[x,y,c]]$ with relations (4.1)) was found by Gelfand & Fairlie [17, (49)]. In fact, the result follows easily from (3.5):

Proposition 4.3 In the algebra of formal power series in x, w, z under relations

$$xw - qwx = (1-q)z^2, \quad xz = qzx, \quad zw = qwz \tag{4.14}$$

we have the identity

$$e_q(x+w) = e_q(w)\, e_{q^2}(z^2)\, e_q(x). \tag{4.15}$$

Proof It follows from relations (4.14) that $(x-z)(w+z) = q(w+z)(x-z)$. Hence, by repeated application of (3.5):

$$e_q(x+w) = e_q(z+w)\, e_q(x-z) = e_q(w)\, e_q(z)\, e_q(-z)\, e_q(x).$$

Now the result follows by application of (3.18). □

Remark 4.4 The algebra of formal power series in x, w, v under relations

$$xw - qwx = (1-q)v, \quad xv = q^2 vx, \quad vw = q^2 wv \tag{4.16}$$

is isomorphically embedded into the algebra of formal power series in x, w, z under relations (4.14) by adding the relations $v = z^2$ to relations (4.16). This is seen by observing that the algebra of polynomials in x, w, v under relations (4.16) has a basis of elements $w^k x^l v^m$ $(k, l, m \in \mathbb{Z}_+)$ and that the algebra of polynomials in x, w, z under relations (4.14) has a basis of elements $w^k x^l z^m$ $(k, l, m \in \mathbb{Z}_+)$ (use Bergman's [4] diamond lemma). Thus it follows from Proposition 4.3 that, in the algebra of formal power series in x, w, v under relations (4.16), we have the identity

$$e_q(x+w) = e_q(w)\, e_{q^2}(v)\, e_q(x). \tag{4.17}$$

Remark 4.5 In relations (4.16) replace x, w, v by $(1-q)X, (1-q)Y, (1-q)C$ and let $q \uparrow 1$. Then X, Y, C will satisfy the Heisenberg relations (4.12) and identity (4.17) becomes the second identity in (4.13).

We may also rewrite (4.15) as an identity in $\mathbb{C}_{qHeis}[[x, y, c]]$. Just put $w := y[x, y]$ in (4.2). Then x, w, z satisfy relations (4.14). Thus (4.15) becomes in terms of x, y, z:

$$e_q(x + y[x,y]) = e_q(y[x,y])\, e_{q^2}\big((1-q)^{-1}[x,y]^2\big)\, e_q(x). \tag{4.18}$$

After rescaling we have, still under relations (4.2):

$$e_q\big((1-q)(x+y[x,y])\big) = e_q\big((1-q)y[x,y]\big)\, e_{q^2}\big((1-q)[x,y]^2\big)\, e_q\big((1-q)x\big). \tag{4.19}$$

Now make substitutions of x, y, z into X, Y, C as we earlier did for (4.10). Next let $q \uparrow 1$ in (4.1) and (4.19). Then we obtain the identity

$$e^{X+YC} = e^{YC}\, e^{\frac{1}{2}C^2}\, e^X$$

under Heisenberg relations (4.12). This identity is equivalent to (4.13).

Remark 4.6 It seems somewhat arbitrary that we stipulate relations (4.14) in order to obtain identity (4.15). In fact, they arise from a much more general Ansatz. First rewrite (4.17) equivalently as

$$E_q(-w)\, e_q(x+w)\, E_q(-x) = e_{q^2}(v).$$

Now we ask more generally under which minimal set of relations for x and w we have that

$$E_q(-w)\, e_q(x+w)\, E_q(-x) = f(v) \tag{4.20}$$

for some formal power series f in one variable and some element v which is homogeneous of degree 2 in x and w, i.e., a linear combination of x^2, w^2, xw, wx. Surprisingly, the answer is that $v = (1-q)^{-1}(xw - qwx)$, $f = e_{q^2}$ and that the relations are

$$x(xw - qwx) = q^2(xw - qwx)x, \quad (xw - qwx)y = q^2 y(xw - qwx), \tag{4.21}$$

so we recover (4.17) under relations (4.16) as the unique solution of our problem. In fact, expansion of the left-hand side of (4.20) up to quadratic terms yields $1 + (xw - qwx)/(q;q)_2$. It follows from our Ansatz that all terms of homogeneous odd degree in the expansion of the left-hand side of (4.20) must vanish. The vanishing of the third degree terms precisely yields the relations (4.21).

5 Equivalence Between Identities in the Non-Commuting and the Commuting Case

In the previous sections we saw many examples of identities in non-commuting variables. It is sometimes possible to rewrite these identities in terms of commuting variables, usually in various different ways. The idea is always the following. Suppose our identity in non-commuting variables lives in a certain algebra $\mathcal{A}$. Let π be a representation of the algebra $\mathcal{A}$ on a suitable linear space $\mathcal{F}$ of functions (for instance the space of polynomials or formal power series in one complex variable). Suppose that there is a subset $\{f_m\}$ of $\mathcal{F}$ such that, for all $a \in \mathcal{A}$, we have the implication: $\pi(a)\, f_m = 0$ for all $m \implies a = 0$. An identity $a = b$ in $\mathcal{A}$ is then equivalent to the collection of identities $\pi(a)\, f_m = \pi(b)\, f_m$ for all m.

As an example, fix $q \in (0,1)$, let x, y be subject to the relation $xy = qyx$, and let $\mathcal{A}$ be the algebra $\mathbb{C}_q[x,y]$ of polynomials in x, y (cf. Section 2) or the algebra $\mathbb{C}_q[[x,y]]$ of formal power series in x, y (cf. Section 3). Let $\mathbb{C}_q[x,y]$ resp. $\mathbb{C}_q[[x,y]]$ act on the space $\mathcal{F}$ of polynomials resp. formal power series in one complex variable z by an algebra representation π such that

$$(\pi(x)f)(z) := qz\, f(qz), \quad (\pi(y)f)(z) := z\, f(z). \tag{5.1}$$

Then $\pi(x)\,\pi(y) = q\,\pi(y)\,\pi(x)$, so π preserves the relation $xy = qyx$.

By induction with respect to k we see that

$$(\pi(x^k)f)(z) = q^{\frac{1}{2}\,k(k+1)}\, z^k\, f(q^k z) \quad (k \in \mathbb{Z}_+),$$

so

$$(\pi(y^l x^k)f)(z) = z^l\, (\pi(x^k)f)(z) = q^{\frac{1}{2}\,k(k+1)}\, z^{k+l}\, f(q^k z) \quad (k, l \in \mathbb{Z}_+).$$

Thus, by a little abuse of notation,

$$\pi(y^l x^k)\, z^m = q^{\frac{1}{2}k(k+1)+km}\, z^{k+l+m} \quad (k, l, m \in \mathbb{Z}_+). \tag{5.2}$$

Now we let a formal power series in x, y act on a formal power series in z in a σ-additive way:

$$\begin{aligned} \pi\left(\sum_{k,l} c_{k,l}\, y^l x^k\right) \sum_m a_m z^m &= \sum_{k,l,m} c_{k,l}\, a_m\, q^{\frac{1}{2}k(k+1)+km}\, z^{k+l+m} \\ &= \sum_{n=0}^{\infty} z^n \sum_{k+l+m=n} c_{k,l}\, a_m\, q^{\frac{1}{2}k(k+1)+km}. \end{aligned}$$

We see that the result is again a well-defined formal power series in z.

The reader is warned that a σ-additive extension of a representation as we gave above, is not always possible. Then one has to work with representations by unbounded operators on a Hilbert space, and identities satisfied by formal series may no longer hold in the representation, see for instance Woronowicz [36]. (I thank S. L. Woronowicz for pointing this out to me.)

Next we show that, if $a = \sum_{k,l} c_{k,l}\, y^l x^k$ and $\pi(a)\, z^m = 0$ for all $m \in \mathbb{Z}_+$, then $a = 0$. Indeed, if

$$\sum_{k,l} c_{k,l}\, \pi(y^l x^k)\, z^m = 0 \quad \text{for all } m \in \mathbb{Z}_+$$

then

$$\sum_{k,l} c_{k,l}\, q^{\frac{1}{2}k(k+1)+km}\, z^{k+l} = 0 \quad \text{for all } m \in \mathbb{Z}_+.$$

Hence for all $n \in \mathbb{Z}_+$ we have

$$\sum_{k=0}^{n} c_{k,n-k}\, q^{\frac{1}{2}k(k+1)}\, (q^m)^k = 0 \quad \text{for all } m \in \mathbb{Z}_+.$$

This shows that the $c_{k,n-k}$ are 0.

Now we will see how we can rewrite the identities (2.3) and (3.5) involving noncommuting variables into equivalent commutative form by means of the representation π of $\mathbb{C}_q[[x,y]]$ (or $\mathbb{C}_q[x,y]$). For fixed $n \in \mathbb{Z}_+$ the q-binomial formula (2.3) is equivalent to the set of identities

$$\pi\big((x+y)^n\big)\, z^m = \sum_{k=0}^{n} \begin{bmatrix} n \\ k \end{bmatrix}_q \pi(y^{n-k} x^k)\, z^m \quad (m \in \mathbb{Z}_+). \tag{5.3}$$

By induction with respect to n we see that

$$\pi\big((x+y)^n\big)\, z^m = (-q^{m+1}; q)_n\, z^{m+n}. \tag{5.4}$$

By also using (5.2) and (2.9) we see that (5.3) is equivalent to

$$(-q^{m+1}; q)_n\, z^{m+n} = \sum_{k=0}^{n} \frac{(q^{-n}; q)_k}{(q; q)_k} (-q^{m+n+1})^k\, z^{m+n} \quad (m \in \mathbb{Z}_+).$$

We can divide both sides of this last identity by z^{m+n}. Thus we have shown that (2.3) is equivalent to the terminating q-binomial sum (2.10) considered for all $z = -q^{n+m+1}$ ($m \in \mathbb{Z}_+$), which in its turn is equivalent to (2.10) considered for all $z \in \mathbb{C}$.

Let us next handle identity (3.5) in this way. Observe first that $(\pi(y)f)(z) = z f(z)$ implies that

$$\big(\pi(g(y))\, f\big)(z) = g(z)\, f(z) \tag{5.5}$$

for any two formal power series f and g in one variable. Identity (3.5) is equivalent to the set of identities

$$\pi\left(e_q(x+y)\right)\, z^m = \pi\big(e_q(y)\, e_q(x)\big)\, z^m \quad (m \in \mathbb{Z}_+).$$

Now use (5.5) and expand $e_q(x+y)$ and $e_q(x)$. We obtain that (3.5) is equivalent to

$$\sum_{n=0}^{\infty} \frac{\pi\left((x+y)^n\right)\, z^m}{(q;q)_n} = e_q(z) \sum_{k=0}^{\infty} \frac{\pi(x^k) z^m}{(q;q)_k} \quad (m \in \mathbb{Z}_+).$$

By (5.4) and (5.2) this can be rewritten as

$$\sum_{n=0}^{\infty} \frac{(-q^{m+1};q)_n\, z^{m+n}}{(q;q)_n} = e_q(z) \sum_{k=0}^{\infty} \frac{q^{\frac{1}{2}k(k+1)+km}\, z^{k+m}}{(q;q)_k} \qquad (m \in \mathbb{Z}_+).$$

On dividing both sides by z^m and on using the first equality in (3.2) we see that (3.5) is equivalent to the series of identities

$${}_1\phi_0(-q^{m+1};;q,z) = e_q(z)\, E_q(q^{m+1}z) \quad (m \in \mathbb{Z}_+),$$

i.e., to special cases of the evaluation of non-terminating q-hypergeometric series, cf. (3.22).

I invite the reader to experiment with several other representations π of algebras $\mathbb{C}_q[x,y]$ or $\mathbb{C}_q[[x,y]]$ on the space of polynomials or formal power series in z, for instance

$$\big(\pi(x)f\big)(z) := f(qz), \quad \big(\pi(y)f\big)(z) := z\, f(z), \tag{5.6}$$
$$\big(\pi(x)f\big)(z) := (D_q f)(z), \quad \big(\pi(y)f\big)(z) := f(qz), \tag{5.7}$$
$$\big(\pi(x)f\big)(z) := f(z+\log q), \quad \big(\pi(y)f\big)(z) := e^z\, f(z), \tag{5.8}$$
$$\big(\pi(x)f\big)(z) := e^z\, f(z), \quad \big(\pi(y)f\big)(z) := f(z-\log q). \tag{5.9}$$

In (5.7) we used the notation for the *q-derivative*

$$(D_q f)(z) := \frac{f(z)-f(qz)}{(1-q)z}\,. \tag{5.10}$$

For instance, consider the q-binomial formula (2.3) in representation (5.6), with both sides acting on $e_q(z)$, in order to arrive at a special case of the q-Chu-Vandermonde sum [16, (II.6)] (one upper parameter zero). Also consider identity (3.9) in representation (5.8) or identity (3.12) in representation (5.9), with both sides acting on functions $e^{i\mu z}$, in order to arrive at an identity in commuting variables which is equivalent to the summation formula (3.21) of non-terminating q-binomial series. Here the infinite sum results from the action of the left-hand sides of (3.9) or (3.12) on $e^{i\mu z}$. Faddeev & Kashaev [10, Section 2] point out an alternative for the action of these left-hand sides. They observe, in representation (5.9), that the action of $g_1(y)g_2(x)$ (g_1 and g_2 suitable functions) on $e^{i\mu z}$ can be written as multiplication by a certain double integral involving g_1 and g_2 (the symbol of the product of the operators $\pi(g_1(y))$ and $\pi(g_2(y))$). This argument looks quite formal and the convergence of the resulting double integral is not clear.

A wealth of further results in the spirit of this section is contained in Cigler [7], [8]. He also gives applications to continuous q-Hermite polynomials and to q-Laguerre polynomials and he develops a q-analogue of Rota's umbral calculus [31].

A well-known representation π of the algebra $\mathbb{C}_{qHeis}[[x,y,c]]$ (see (4.1), (4.2), (4.3)) on the space of formal power series is given by $\pi(x) := (1-q)\,c\,D_q$, $(\pi(y)f)(t) := t\,f(t)$. See Rogers [30] for many results using this representation, or a modern treatment by Bowman [6].

6 The q-Logarithm

Euler's dilogarithm is defined by

$$Li_2(z) := \sum_{n=1}^{\infty} \frac{z^n}{n^2}, \quad |z| < 1.$$

A. N. Kirillov [20, (2.52)] defines the following q-analogue ($0 < q < 1$ as before):

$$Li_2(z;q) := \sum_{n=1}^{\infty} \frac{z^n}{n(1-q^n)}, \quad |z| < 1. \tag{6.1}$$

Formally we have the termwise limit

$$\lim_{q\uparrow 1}(1-q)\, Li_2(z;q) = Li_2(z). \tag{6.2}$$

Kirillov [20, Section 2.5, Lemma 8] observes the following remarkable formula:

$$Li_2(z;q) = \log\bigl(e_q(z)\bigr), \quad |z| < 1. \tag{6.3}$$

For the proof, note that

$$Li_2(qz;q) = Li_2(z;q) + \log(1-z)$$

(substitute the corresponding power series). Hence

$$\exp\bigl(Li_2(z;q)\bigr) = \frac{\exp\bigl(Li_2(qz;q)\bigr)}{1-z} = \ldots = \frac{\exp\bigl(Li_2(q^k z;q)\bigr)}{(z;q)_k}.$$

On taking limits for $k \to \infty$, the right-hand side tends to

$$\frac{\exp\bigl(Li_2(0;q)\bigr)}{(z;q)_\infty} = \exp(0)\, e_q(z) = e_q(z).$$

A more precise formulation of (6.2) going back to Ramanujan (see Berndt [5, Ch. 27, Entry 6]) gives an asymptotic series for $Li_2(z;q)$ (or $\log(e_q(z))$ in rising powers of $-\log q$ as $q \uparrow 1$. Kirillov [20, Section 2.5, Corollary 10] and Ueno & Nishizawa [34] derive this asymptotic series by using Euler-Maclaurin's summation formula.

Faddeev & Kashaev [10, Section 2], see also Kirillov [20, Theorem I], indicate that Rogers' five-term identity for Euler's dilogarithm can be obtained as a limit case as $q \uparrow 1$ of (3.9) or (3.12). This uses (6.3) and (6.2). Faddeev and Kashaev also use the representation (5.9) of the left-hand side of (3.12) by means of a double integral (see the end of Section 5). Their arguments are quite formal.

Next we consider a q-analogue of

$$-\log(1-z) = \sum_{n=1}^{\infty} \frac{z^n}{n}, \quad |z| < 1.$$

We define and notate it as

$$\log_q(z) := \sum_{n=1}^{\infty} \frac{z^n}{1-q^n} \tag{6.4}$$

and we consider it either as a convergent power series for $|z| < 1$ or as a formal power series. Note that we have formally the termwise limit

$$\lim_{q\uparrow 1} (1-q)\log_q(z) = -\log(1-z). \tag{6.5}$$

It follows from (6.1) and (6.4) that

$$\log_q(z) = z\, Li_2'(z;q).$$

Hence, by (6.3):

$$\log_q(z) = \frac{z\, e_q'(z)}{e_q(z)}. \tag{6.6}$$

Another interesting formula is

$$\log_q(z) = -\frac{\partial}{\partial a}\, {}_1\phi_0(a;\,;q,z)\Big|_{a=1}. \tag{6.7}$$

It is the q-analogue of

$$-\log(1-z) = \frac{\partial}{\partial a}(1-z)^{-a}\Big|_{a=0}.$$

For the proof of (6.7) note that

$$\frac{\partial}{\partial a}\left(\frac{(az;q)_\infty}{(z;q)_\infty}\right) = \frac{\partial}{\partial a}\left(\frac{e_q(z)}{e_q(az)}\right) = -\frac{z\, e_q(z)\, e_q'(az)}{e_q(az)^2}.$$

Hence

$$\frac{\partial}{\partial a}\left(\frac{(az;q)_\infty}{(z;q)_\infty}\right)\Bigg|_{a=1} = -\frac{z\, e_q'(z)}{e_q(z)}.$$

Now apply (3.21) and (6.6).

Note also the q-derivative (see (5.10)) of $\log_q$:

$$(1-q)\,(D_q \log_q)(z) = \sum_{n=0}^{\infty} z^n = \frac{1}{1-z}. \tag{6.8}$$

The formulas (6.3), (6.6), (6.7) and (6.8) are of hybrid nature because they use classical objects (the logarithm, the classical derivative and the function $z \mapsto (1-z)^{-1}$, respectively) in a context of q-special functions.

It is a natural question to look for the inverse function to $\log_q$, analogous to the function $x \mapsto 1 - e^{-x}$ being inverse to the function $y \mapsto -\log(1-y)$. However, the inverse function to $\log_q$ does not seem to have a nice explicit expression. (I thank C. Krattenthaler for checking this by use of Maple.) Some alternative way to find a q-analogue is as follows. By the chain rule $f(g(x)) = x$ implies $f'(g(x))g'(x) = 1$. Now the following q-analogue holds:

$$(D_q f)(g(x))\,(D_q g)(x) = 1, \quad g(x) := 1 - e_q(-(1-q)x), \quad f(y) := (1-q)\log_q(y). \tag{6.9}$$

Note however that $(D_q g)(f(y))\,(D_q f)(y) \neq 1$ for f and g as in (6.9).

The following Proposition gives a q-analogue of the classical functional equation

$$\log(xy) = \log x + \log y, \text{ or equivalently } -\log\big((1-x)(1-y)\big) = -\log(1-x) - \log(1-y),$$

for commuting variables x, y. This Proposition was independently found by A. N. Kirillov [20, Section 2.5, Exercise 11] and by the author.

Proposition 6.1 In the algebra $\mathbb{C}_q[[x,y]]$ we have the identity

$$\log_q(x+y-yx) = \log_q(x) + \log_q(y). \tag{6.10}$$

Proof Since $(x-yx)y = qy(x-yx)$ we get by the q-binomial formula (2.3) that

$$\begin{aligned}(x+y-yx)^n &= \sum_{k=0}^{n} \begin{bmatrix} n \\ k \end{bmatrix}_q y^{n-k}(x-yx)^k \\ &= \sum_{k=0}^{n} y^{n-k}\,(y;q)_k\,x^k.\end{aligned}$$

Hence

$$\begin{aligned}\log_q(x+y-yx) &= \sum_{n=1}^{\infty}\left(\frac{1}{1-q^n}\,y^n + \sum_{k=1}^{n}\frac{1}{1-q^n}\begin{bmatrix} n \\ k \end{bmatrix}_q y^{n-k}\,(y;q)_k\,x^k\right) \\ &= \log_q(y) + \sum_{k=1}^{\infty}\left(\sum_{n=k}^{\infty}\frac{1}{1-q^n}\begin{bmatrix} n \\ k \end{bmatrix}_q y^{n-k}\right)(y;q)_k\,x^k.\end{aligned}$$

So we are done if we can show that

$$\left(\sum_{n=k}^{\infty}\frac{1-q^k}{1-q^n}\begin{bmatrix} n \\ k \end{bmatrix}_q y^{n-k}\right)(y;q)_k = 1. \tag{6.11}$$

It is sufficient to prove (6.11) for complex y with $|y|<1$. Then

$$\begin{aligned}\sum_{n=k}^{\infty}\frac{1-q^k}{1-q^n}\begin{bmatrix} n \\ k \end{bmatrix}_q y^{n-k} &= \sum_{m=0}^{\infty}\frac{1-q^k}{1-q^{m+k}}\begin{bmatrix} m+k \\ k \end{bmatrix}_q y^m \\ = \sum_{m=0}^{\infty}\frac{1-q^k}{1-q^{m+k}}\,\frac{(q^{k+1};q)_m}{(q;q)_m}\,y^m &= \sum_{m=0}^{\infty}\frac{(q^k;q)_m}{(q;q)_m}\,y^m = \frac{(q^k y;q)_\infty}{(y;q)_\infty} = \frac{1}{(y;q)_k},\end{aligned}$$

where we used in the forelast identity the evaluation (3.21) of a non-terminating q-binomial series. □

A somewhat formal, but very quick proof of Proposition 6.1 is obtained from (3.24). Just differentiate both sides of (3.24) with respect to a, put $a=1$ and use (6.7).

Yet another proof is obtained from (3.5), which we use in the form $e_q(t(x+y)) = e_q(ty)\,e_q(tx)$ with $t\in\mathbb{R}$. Differentiate both sides with respect to t, put $t=1$ and use (6.6) in order to obtain

$$e_q(x+y)\,\log_q(x+y) = e_q(y)\,(\log_q(y)+\log_q(x))\,e_q(x).$$

Next, replace $e_q(x+y)$ by $e_q(y)\,e_q(x)$ and pull the $e_q(x)$ factor through $\log_q(x+y)$, on using (3.15).

7 Jackson Integral in q-Commuting Variables

This section reviews results from Kempf & Majid [19, Section 1]. We start with a q-analogue of Taylor series for q-commuting variables.

Fix $q \in (0,1)$. The *q-derivative* $D_q f$ of a function f on $\mathbb{R}$ was already defined in (5.10). If $f(x) = x^n$ then $(D_q f)(x) = \dfrac{1-q^n}{1-q}\, x^{n-1}$, so we can let D_q and its iterates D_q^k act on polynomials or formal power series in x. We have

$$D_q^k x^n = \frac{(q^n;q^{-1})_k}{(1-q)^k}\, x^{n-k}.$$

Hence, in the algebra $\mathbb{C}_q[x,y]$ we can rewrite the q-binomial formula (2.3) as

$$(x+y)^n = \sum_{k=0}^{n} \frac{(1-q)^k}{(q;q)_k}\, y^k\, D_q^k x^n. \tag{7.1}$$

Now let $f(x) := \sum_{n=0}^{\infty} c_n x^n$ be a formal power series. Then, in the algebra $\mathbb{C}_q[[x,y]]$ of formal power series in x, y under relation $xy = qyx$, we have

$$\begin{aligned} f(x+y) &= \sum_{n=0}^{\infty} c_n (x+y)^n = \sum_{n=0}^{\infty}\sum_{k=0}^{n} c_n \frac{(1-q)^k}{(q;q)_k}\, y^k\, D_q^k x^n \\ &= \sum_{k=0}^{\infty} \frac{(1-q)^k}{(q;q)_k}\, y^k \sum_{n=k}^{\infty} c_n D_q^k x^n = \sum_{k=0}^{\infty} \frac{(1-q)^k}{(q;q)_k}\, y^k\, D_q^k\, f(x). \end{aligned}$$

Thus we have proved:

Proposition 7.1 Let f be a formal power series in one variable. Let $xy = qyx$. Then, in the algebra $\mathbb{C}_q[[x,y]]$ we have for each $m \in \mathbb{Z}_+$ the identity

$$\begin{aligned} f(x+y) &= \sum_{k=0}^{\infty} \frac{1}{(q;q)_k}\, y^k\, ((1-q)D_q)^k\, f(x) && (7.2) \\ &= \sum_{k=0}^{m-1} \frac{1}{(q;q)_k}\, y^k\, ((1-q)D_q)^k\, f(x) + y^m\, g_m(x,y) && (7.3) \end{aligned}$$

for a suitable element $g_m(x,y)$ of $\mathbb{C}_q[[x,y]]$.

If we write $\mathcal{O}(y^m)$ instead of $y^m\, g_m(x,y)$ then (7.3) implies in particular, for $m = 2$, that

$$f(x+y) = f(x) + y\, D_q\, f(x) + \mathcal{O}(y^2) \quad (xy = qyx), \tag{7.4}$$

as a q-analogue in q-commuting variables of the classical formula

$$f(x+y) = f(x) + yf'(x) + \mathcal{O}(y^2) \quad (xy = yx).$$

Fix $q \in (0,1)$. For a function f on $\mathbb{R}$, the *Jackson integral* is defined by

$$\int_0^x f(t)\, d_q t := (1-q) \sum_{k=0}^{\infty} f(q^k x)\, q^k x, \tag{7.5}$$

where $x \in \mathbb{R}$, provided the sum on the right-hand side converges absolutely, for instance if f is bounded near zero. The Jackson integral avant la lettre of $f(t) := t^n$ was already computed by Fermat:

$$\int_0^x t^n \, d_q t = \frac{1-q}{1-q^{n+1}} \, x^{n+1}. \tag{7.6}$$

Thus, if

$$f(x) := \sum_{n=0}^{\infty} c_n x^n \tag{7.7}$$

is a formal power series in a formal (not necessarily real or complex) variable x then we obtain its Jackson integral

$$\int_0^x f(t) \, d_q t = \sum_{n=0}^{\infty} c_n \, \frac{1-q}{1-q^{n+1}} \, x^{n+1} \tag{7.8}$$

also as a formal power series.

Now let $xy = qyx$. Then we obtain in the algebra $\mathbb{C}_q[x,y]$ by (7.6) and a two-fold application of the q-binomial formula (2.3):

$$\begin{aligned}
\int_0^{x+y} t^n \, d_q t &= \frac{1-q}{1-q^{n+1}} \, (x+y)^{n+1} \\
&= \frac{1-q}{1-q^{n+1}} \, y^{n+1} + \frac{1-q}{1-q^{n+1}} \sum_{l=0}^{n} \begin{bmatrix} n+1 \\ l \end{bmatrix}_q y^l x^{n+1-l} \\
&= \frac{1-q}{1-q^{n+1}} \, y^{n+1} + (1-q) \sum_{l=0}^{n} \begin{bmatrix} n \\ l \end{bmatrix}_q \frac{1}{1-q^{n-l+1}} \, y^l x^{n-l+1} \\
&= \frac{1-q}{1-q^{n+1}} \, y^{n+1} + (1-q) \sum_{l=0}^{n} \sum_{k=0}^{\infty} \begin{bmatrix} n \\ l \end{bmatrix}_q y^l (q^k x)^{n-l+1} \\
&= \frac{1-q}{1-q^{n+1}} \, y^{n+1} + (1-q) \sum_{k=0}^{\infty} (q^k x + y)^n \, q^k x \\
&= \int_0^y t^n \, d_q t + \int_0^x (t+y)^n \, d_q t,
\end{aligned}$$

where we use the definition

$$\int_0^x f(t+y) \, d_q t := (1-q) \sum_{k=0}^{\infty} f(q^k x + y) \, q^k x.$$

Thus we have shown that

$$\int_0^{x+y} t^n \, d_q t = \int_0^y t^n \, d_q t + \int_0^x (t+y)^n \, d_q t \quad (xy = qyx)$$

and therefore also for formal power series (7.7):

$$\int_0^{x+y} f(t) \, d_q t = \int_0^y f(t) \, d_q t + \int_0^x f(t+y) \, d_q t \quad (xy = qyx). \tag{7.9}$$

Now recall the definition

$$\int_a^b f(t)\,d_qt := \int_0^a f(t)\,d_qt - \int_0^b f(t)\,d_qt, \tag{7.10}$$

where the two Jackson integrals on the right-hand side are defined by (7.5). Definition (7.10) can also be used for formal variables a, b. Thus by (7.9) and (7.10) we have the following Proposition:

Proposition 7.2 Let f be a formal power series in one variable. Let $xy = qyx$. Then, in the algebra $\mathbb{C}_q[[x,y]]$ we have

$$\int_y^{x+y} f(t)\,d_qt = \int_0^x f(t+y)\,d_qt. \tag{7.11}$$

Thus the translation invariance of the Riemann integral, which seemed to be destroyed when q-deforming it to the Jackson integral, can be preserved when we work with q-commuting variables.

By way of example we give a fourth proof of the functional equation (6.10) for the q-logarithm in q-commuting variables. First observe, by straightforward application of (7.8) and (6.4), that

$$\log_q(x) = \int_0^x (1-t)^{-1}\,d_qt. \tag{7.12}$$

Now let us work in $\mathbb{C}_q[[x,y]]$. Then

$$\log_q(x+y-xy;q) - \log(y;q) = \int_0^{x+y-yx} (1-t)^{-1}\,d_qt - \int_0^y (1-t)^{-1}\,d_qt$$
$$= \int_0^{x-yx} (1-(t+y))^{-1}\,d_qt = (1-q)\sum_{k=0}^{\infty}\left(1-q^k(x-yx)-y\right)^{-1} q^k(x-yx)$$
$$= (1-q)\sum_{k=0}^{\infty}\left((1-y)(1-q^kx)\right)^{-1}(1-y)q^kx = (1-q)\sum_{k=0}^{\infty}(1-q^kx)^{-1}q^kx$$
$$= \int_0^x (1-t)^{-1}\,d_qt = \log_q(x),$$

where we applied (7.11) in the second equality. Thus we have given a new proof of (6.10).

8 q-Hermite Polynomials and a q-Fourier Transform Pair

The *discrete q-Hermite I polynomials* (see Koekoek & Swarttouw [22, Section 3.28] and references given there) are given by

$$\begin{aligned} h_n(x;q) :=& x^n\, {}_2\phi_0(q^{-n}, q^{-n+1};;q^2, q^{2n-1}x^{-2}) \\ =& (q;q)_n \sum_{k=0}^{[n/2]} \frac{(-1)^k\, q^{k(k-1)}\, x^{n-2k}}{(q^2;q^2)_k\,(q;q)_{n-2k}}. \end{aligned} \tag{8.1}$$

They are orthogonal polynomials satisfying the orthogonality relations

$$\int_{-1}^1 h_m(x;q)\,h_n(x;q)\,E_{q^2}(-q^2x^2)\,d_qx = b_q\, q^{\frac{1}{2}n(n-1)}\,(q;q)_n\,\delta_{m,n}, \tag{8.2}$$

where

$$b_q := (1-q)\,(q,-q,-1;q)_\infty. \tag{8.3}$$

There is the following generating function:

$$\frac{(t^2;q^2)_\infty}{(xt;q)_\infty} = E_{q^2}(-t^2)\,e_q(xt) = \sum_{n=0}^{\infty} \frac{h_n(x;q)}{(q;q)_n}\,t^n \quad (|xt|<1). \tag{8.4}$$

Several useful formulas can be derived from this generating function. First we can expand a monomial:

$$x^n = (q;q)_n \sum_{k=0}^{[\frac12 n]} \frac{h_{n-2k}(x;q)}{(q^2;q^2)_k\,(q;q)_{n-2k}}. \tag{8.5}$$

For the proof, multiply both sides of (8.4) with $e_{q^2}(t^2)$ and next compare coefficients of t.

Next we have

$$\sum_{k=0}^{m} \frac{(q^{-m};q)_k}{(q;q)_k}\,q^k\,h_k(x;q)\,x^{m-k} = \begin{cases} (-1)^n\,q^{-n^2}\,(q;q^2)_n & \text{if } m=2n,\\ 0 & \text{if } m=2n+1. \end{cases} \tag{8.6}$$

For the proof, multiply both sides of (8.4) with $E_q(-xt)$ and next compare coefficients of t.

From the generating function (8.4) together with the orthogonality relations (8.2) we obtain

$$\int_{-1}^{1} e_q(-ixt)\,h_n(x;q)\,E_{q^2}(-q^2x^2)\,d_qx = b_q\,q^{\frac12 n(n-1)}\,i^{-n}\,t^n\,e_{q^2}(-t^2). \tag{8.7}$$

For the proof, replace t by $-it$ in (8.4), multiply both sides with $h_n(x;q)\,E_{q^2}(-q^2x^2)$ $e_{q^2}(-t^2)$ and q-integrate both sides over $[-1,1]$.

The *discrete q-Hermite II polynomials* (see Koekoek & Swarttouw [22, Section 3.29] and references given there) are given by

$$\widetilde{h}_n(x;q) := x^n\,{}_2\phi_1(q^{-n},q^{-n+1};0;q^2,-q^2x^{-2}) \tag{8.8}$$

$$= (q;q)_n \sum_{k=0}^{[\frac12 n]} \frac{(-1)^k\,q^{-2nk}\,q^{k(2k+1)}\,x^{n-2k}}{(q^2;q^2)_k\,(q;q)_{n-2k}}. \tag{8.9}$$

(There is a slight error in [22, (3.29.1)], which is corrected in formula (8.8) above.) They are related to the discrete q-Hermite I polynomials by

$$h_n(ix;q^{-1}) = i^n\,\widetilde{h}_n(x;q). \tag{8.10}$$

They are orthogonal polynomials satisfying orthogonality relations given by a Jackson integral over $(-\infty,\infty)$. Let us use the notation

$$\int_0^{\gamma.\infty} f(t)\,d_qt := (1-q) \sum_{k=-\infty}^{\infty} f(q^k\gamma)\,q^k\gamma \tag{8.11}$$

for the Jackson integral over $(0,\infty)$ of a function defined on $\{q^k\gamma \mid k\in\mathbb{Z}\}$ for some $\gamma\in(0,\infty)$, where we suppose that the sum on the right-hand side of (8.11) absolutely converges. So the definition is dependent on γ but invariant under the transform $\gamma \mapsto q\gamma$. For f defined on $\{\pm q^k\gamma \mid k \in \mathbb{Z}\}$ we can also define the Jackson integral of f over $(-\infty, \infty)$, again depending on γ and invariant under $\gamma \mapsto q\gamma$:

$$\int_{-\gamma.\infty}^{\gamma.\infty} f(t)\, d_qt := \left(\int_0^{\gamma.\infty} f(t)\, d_qt - \int_0^{-\gamma.\infty} f(t)\, d_qt \right) \tag{8.12}$$

$$=(1-q) \sum_{k=-\infty}^{\infty} \big(f(q^k\gamma) + f(-q^k\gamma)\big) q^k\gamma. \tag{8.13}$$

We again suppose that the infinite sums are absolutely convergent.

The orthogonality relations for the discrete q-Hermite II polynomials are:

$$\int_{-\gamma.\infty}^{\gamma.\infty} \widetilde{h}_m(x;q)\, \widetilde{h}_n(x;q)\, e_{q^2}(-x^2)\, d_qx = c_q(\gamma)\, q^{-n^2}\, (q;q)_n\, \delta_{m,n}, \tag{8.14}$$

where

$$c_q(\gamma) := \frac{2(1-q)\,(q^2, -q\gamma^2, -q\gamma^{-2}; q^2)_\infty\, \gamma}{(-\gamma^2, -q^2/\gamma^2, q; q^2)_\infty}. \tag{8.15}$$

Note that the orthogonality measure is not uniquely determined. There is the generating function

$$\frac{(-xt;q)_\infty}{(-t^2;q^2)_\infty} = e_{q^2}(-t^2)\, E_q(xt) = \sum_{n=0}^{\infty} \frac{q^{\frac{1}{2}n(n-1)}}{(q;q)_n}\, \widetilde{h}_n(x;q)\, t^n. \tag{8.16}$$

Now we can obtain formulas analogous to (8.5)–(8.7), either by using (8.16) just as (8.4) was used in the discrete q-Hermite I case, or by using (8.10):

$$x^n = (q;q)_n \sum_{k=0}^{[\frac{1}{2}n]} \frac{q^{-2nk+3k^2}}{(q;q)_{n-2k}\, (q^2;q^2)_k}\, \widetilde{h}_{n-2k}(x;q), \tag{8.17}$$

$$\sum_{k=0}^{m} \frac{(q^{-m};q)_k}{(q;q)_k}\, q^{mk}\, \widetilde{h}_k(x;q)\, x^{m-k} = \begin{cases} (-1)^n\, (q;q^2)_n & \text{if } m = 2n, \\ 0 & \text{if } m = 2n+1, \end{cases} \tag{8.18}$$

$$\int_{-\gamma.\infty}^{\gamma.\infty} E_q(iqxt)\, \widetilde{h}_n(x;q)\, e_{q^2}(-x^2)\, d_qx = c_q(\gamma)\, q^{-\frac{1}{2}n(n-1)}\, i^n\, t^n\, E_{q^2}(-q^2t^2). \tag{8.19}$$

Now we can combine formulas for cases I and II of the discrete q-Hermite polynomials. It follows from (8.5), (8.7) and (8.9) that

$$\int_{-1}^{1} e_q(-ixt)\, x^n\, E_{q^2}(-q^2x^2)\, d_qx = b_q\, q^{\frac{1}{2}n(n-1)}\, i^{-n}\, \widetilde{h}_n(t;q)\, e_{q^2}(-t^2) \tag{8.20}$$

and it follows from (8.17), (8.19) and (8.1) that

$$\int_{-\gamma.\infty}^{\gamma.\infty} E_q(iqxt)\, x^n\, e_{q^2}(-x^2)\, d_qx = c_q(\gamma)\, q^{-\frac{1}{2}n(n-1)}\, i^n\, h_n(t;q)\, E_{q^2}(-q^2t^2). \tag{8.21}$$

By comparing (8.7) with (8.21) or (8.20) with (8.19), we arrive at the following pair of q-Fourier transforms, which are inverse to each other when acting on suitable functions:

Theorem 8.1 Let

$$(\mathcal{F}_q f)(y) := \frac{1}{b_q} \int_{-1}^{1} e_q(-ixy)\, f(x)\, d_q x, \tag{8.22}$$

$$(\widetilde{\mathcal{F}}_{q,\gamma} g)(x) := \frac{1}{c_q(\gamma)} \int_{-\gamma.\infty}^{\gamma.\infty} E_q(iqxy)\, g(y)\, d_q y. \tag{8.23}$$

Then the following two statements are equivalent:
(a) $f(x) =$ polynomial $(x) \times E_{q^2}(-q^2x^2)$ and $g = \mathcal{F}_q f$;
(b) $g(y) =$ polynomial $(y) \times e_{q^2}(-y^2)$ and $f = \widetilde{\mathcal{F}}_{q,\gamma} g$.
In particular,

$$f(x) = h_n(x;q)\, E_{q^2}(-q^2x^2) \Longleftrightarrow g(y) = q^{\frac{1}{2}n(n-1)}\, i^{-n}\, y^n\, e_{q^2}(-y^2); \tag{8.24}$$

$$f(x) = x^n\, E_{q^2}(-q^2x^2) \Longleftrightarrow g(y) = q^{\frac{1}{2}n(n-1)}\, i^{-n}\, \widetilde{h}_n(y;q)\, e_{q^2}(-y^2). \tag{8.25}$$

By q-integration by parts and by using (3.4) we can see how $\mathcal{F}_q$ and $\widetilde{\mathcal{F}}_{q,\gamma}$ send a q-derivative operator to a multiplication operator and a multiplication operator to a q-derivative operator. For this purpose we also need a variant D_q^+, called *forward q-derivative*, of the (*backward*) q-derivative $D_q = D_q^-$ as defined in (5.10):

$$(D_q^- f)(x) := \frac{f(x) - f(qx)}{(1-q)x}, \quad (D_q^+ f)(x) := \frac{f(q^{-1}x) - f(x)}{(1-q)x}. \tag{8.26}$$

Proposition 8.2 If f is continuous at 0 and $f(q^{-1}) = 0 = f(-q^{-1})$ then

$$(1-q)\,(\mathcal{F}_q(D_q^+ f))(y) = iy\,(\mathcal{F}_q f)(y).$$

If g is continuous at 0 and $\lim_{n\to\infty} E_q(\pm ix\gamma q^{-n+1})\, g(\pm\gamma q^{-n}) = 0$ then

$$(1-q)\,(\widetilde{\mathcal{F}}_{q,\gamma}(D_q^- g))(x) = -ix\,(\widetilde{\mathcal{F}}_{q,\gamma} g)(x).$$

It follows from (8.24), (8.25) and Proposition 8.2 that (with $D_q^{\pm}$ acting on functions of x):

$$(1-q)\, D_q^+\big(h_n(x;q)\, E_{q^2}(-q^2x^2)\big) = -q^{-n}\, h_{n+1}(x;q)\, E_{q^2}(-q^2x^2), \tag{8.27}$$

$$(1-q)\, D_q^-\big(\widetilde{h}_n(x;q)\, e_{q^2}(-x^2)\big) = -q^n\, \widetilde{h}_{n+1}(x;q)\, e_{q^2}(-x^2). \tag{8.28}$$

These two formulas can also be proved independently. Thus, if (8.27) and (8.28) are given then the general case of (8.24) and (8.25) follows from the special case $n = 0$ by means of Proposition 8.2.

Note that iteration of (8.27), (8.28) yields Rodrigues type formulas

$$h_n(x;q) = (-1)^n\, q^{\frac{1}{2}n(n-1)}\, e_{q^2}(q^2x^2)\, (1-q)^n\, (D_q^+)^n E_{q^2}(-q^2x^2), \tag{8.29}$$

$$\widetilde{h}_n(x;q) = (-1)^n\, q^{-\frac{1}{2}n(n-1)}\, E_{q^2}(x^2)\, (1-q)^n\, (D_q^-)^n\, e_{q^2}(-x^2). \tag{8.30}$$

All formulas, given in this section, involving discrete q-Hermite polynomials until now are analogues of formulas for classical Hermite polynomials: just take limits as $q \uparrow 1$ (after possibly some rescaling). However, there are some other nice formulas for classical Hermite polynomials for which the q-analogue can be better given with q-commuting variables. For instance, in the algebra $\mathbb{C}_q[x,y]$ $(xy = qyx)$ we have:

$$h_n(x+y;q) = \sum_{k=0}^{n} \begin{bmatrix} n \\ k \end{bmatrix}_q y^{n-k}\, h_k(x;q). \tag{8.31}$$

For the proof, let t be scalar and use the generating function (8.4) and the functional equation (3.5):

$$\sum_{n=0}^{\infty} \frac{h_n(x+y;q)}{(q;q)_n}\, t^n = e_q((x+y)t)\, E_{q^2}(-t^2) = e_q(yt)\, e_q(xt)\, E_{q^2}(-t^2)$$
$$= \sum_{l=0}^{\infty} \frac{y^l}{(q;q)_l}\, t^l \sum_{k=0}^{\infty} \frac{h_k(x;q)}{(q;q)_k}\, t^k = \sum_{n=0}^{\infty} \left(\sum_{k=0}^{n} \begin{bmatrix} n \\ k \end{bmatrix}_q y^{n-k}\, h_k(x;q) \right) \frac{t^n}{(q;q)_n}.$$

Another example of a formula for Hermite polynomials without q-analogue in commuting variables is the formula expanding $H_n(\lambda x)$ in terms of a series of $H_m(x)$:

$$H_n(\lambda x) = n! \sum_{k=0}^{[\frac{1}{2}n]} \frac{(-1)^k\, (1-\lambda^2)^k\, \lambda^{n-2k}}{(n-2k)!\, k!}\, H_{n-2k}(x). \tag{8.32}$$

This formula can be proved by use of the generating function

$$e^{2xt-t^2} = \sum_{n=0}^{\infty} \frac{H_n(x)\, t^n}{n!}.$$

Just write

$$e^{2\lambda xt - t^2} = e^{2\lambda xt - \lambda^2 t^2}\, e^{-(1-\lambda^2)t^2},$$

expand all factors in terms of powers of t and compare coefficients of t^n on both sides. Note that we used that $e^{-t^2} = e^{-\lambda^2 t^2}\, e^{-(1-\lambda^2)t^2}$, which strongly suggests to use q-commuting variables for a q-analogue.

Suppose now that λ, μ satisfy the relation $\lambda\mu = q^{\frac{1}{2}}\mu\lambda$ and let x and t be scalar. Then $\lambda^2\mu^2 = q^2\mu^2\lambda^2$, so, by (3.6):

$$e_q(\lambda xt)\, E_{q^2}(-(\lambda^2+\mu^2)t^2) = \big(e_q(\lambda xt)\, E_{q^2}(-\lambda^2 t^2)\big) E_{q^2}(-\mu^2 t^2).$$

Now expand factors in terms of powers of t by using the generating function (8.4) and compare coefficients of t^n on both sides. Then we obtain a q-analogue of (8.32):

$$\sum_{k=0}^{[\frac{1}{2}n]} \frac{(-1)^k\, q^{k(k-1)}\, \lambda^{n-2k}\, (\lambda^2+\mu^2)^k\, x^{n-2k}}{(q;q)_{n-2k}\, (q^2;q^2)_k} = \sum_{k=0}^{[\frac{1}{2}n]} \frac{(-1)^k\, q^{k(k-1)}\, \lambda^{n-2k}\, \mu^{2k}\, h_{n-2k}(x;q)}{(q;q)_{n-2k}\, (q^2;q^2)_k}. \tag{8.33}$$

The left-hand side of (8.33), after multiplication by $(q;q)_n$, can be considered as $h_n(x;q)$ being kind of rescaled by means of λ and μ. Formula (8.33) can be used in order to arrive at a q-analogue of the Fourier transform sending $H_n(x)\, e^{-\frac{1}{2}x^2}$ to a constant multiple of this, or more generally $H_n(x)\, e^{-ax^2}$ to a constant multiple of $H_n((4a(1-a))^{-\frac{1}{2}}x)\, e^{-x^2/(4a)}$ (see [9, (1.10.8), (2.10.10)]). Just combine (8.21) with (8.33). I do not give details, but the reader should compare with very related results in Finkelstein & Marcus [12], where the quantum group $SU_q(2)$ is also brought into the game.

The q-Fourier transform $\widetilde{\mathcal{F}}_{q,\gamma}$ (see (8.23)) also occurs in Kempf & Majid [19, Section VIB]. However, the Fourier kernel is written there as $e_q(iyx)$ with x and y

q-commuting. Their discussion is tied up very much with the notion of translation invariant Jackson integral over $(-\infty,\infty)$ and with $\mathbb{C}_q[x]$ considered as a braided Hopf algebra, see the next two sections for a few more details.

As far as I know, a purely analytic approach to the q-Fourier transform pair in Theorem 8.1 has not appeared earlier in literature. Of course, there are many further questions, e.g., extension of the transforms to a bigger class of functions and continuity properties of the transforms.

9 Translation Invariance of Jackson Integral Over $(-\infty,\infty)$

The results of this section are essentially due to Kempf & Majid [19, Section IV], but the approach is different.

In (8.12), (8.13) we gave the definition of a Jackson integral over $(-\infty,\infty)$. We cannot extend this definition to the case $\int_{-x.\infty}^{x.\infty} f(t)\,d_qt$ where f is a formal power series and x is a formal variable (compare with (7.8) for the Jackson integral from 0 to x), since the Jackson integral of $f(t) := t^n$ over $(-\infty,\infty)$ is not well-defined.

Still, in view of the classical formula

$$\int_{-\infty}^{\infty} f(t)\,dt = \int_{-\infty}^{\infty} f(t+y)\,dt \quad (y\in\mathbb{R}),$$

valid for absolutely convergent integrals, we would like to find an extension of the Jackson integral translation invariance (7.11) on finite intervals for q-commuting variables to the case of an infinite interval. So we would like to have that

$$\int_{-x.\infty}^{x.\infty} f(t)\,d_qt = \int_{-x.\infty}^{x.\infty} f(t+y)\,d_qt \quad (xy=qyx) \tag{9.1}$$

for suitable formal power series f for which both sides of (9.1) have meaning.

Let me first give a completely formal proof of (9.1), as a limit case of (7.11). For $r\in\mathbb{Z}_+$ it follows from (7.11) that

$$\int_{-q^{-r}(x+q^ry)}^{q^{-r}(x+q^ry)} f(t)\,d_qt = \int_{-q^{-r}x}^{q^{-r}x} f(t+y)\,d_qt \quad (xy=qyx).$$

Now let $r\to\infty$. Then equality (9.1) is obtained as a formal limit case.

Evidently this argument gives only heuristic evidence for the validity of (9.1). Let me give next a still formal, but more satisfactory proof of (9.1) for suitable functions f. By formal substitution of (8.13) in the right-hand side of (9.1) and by (7.2) we have

$$\begin{aligned}\int_{-x.\infty}^{x.\infty} f(t+y)\,d_qt &= (1-q)\sum_{k=-\infty}^{\infty}\left(f(q^kx+y)+f(-q^kx+y)\right)q^kx\\ &= \sum_{m=0}^{\infty} y^m\,(1-q)\sum_{k=-\infty}^{\infty}\left(f_m(q^kx)+f_m(-q^kx)\right)q^kx\\ &= \sum_{m=0}^{\infty} y^m\int_{-x.\infty}^{x.\infty} f_m(t)\,d_qt,\end{aligned} \tag{9.2}$$

where

$$f_m(z) = \frac{1}{(q;q)_m}\,((1-q)D_q)^m\,f(z). \tag{9.3}$$

If f is a function on $\mathbb{R}$ then so is f_m. So the Jackson integrals $\int_{-x.\infty}^{x.\infty} f_m(t)\,d_qt$ will have concrete meaning for $x \in \mathbb{R}$ and if the sums defining the Jackson integral are convergent. However, x must q-commute with y, so x cannot be in $\mathbb{R}$. We can circumvent this dilemma by passing to a suitable representation of the relation $xy = qyx$. Let us take a slight extension of the representation (5.6), now using the dot notation instead of π:

$$x.g(z) := \gamma\, g(qz), \quad y.g(z) := z\, g(z). \tag{9.4}$$

Here $\gamma \in \mathbb{R}\backslash\{0\}$ is fixed and $g(z)$ is a formal power series. Thus

$$x.z^k = \gamma q^k z^k \quad (k \in \mathbb{Z}_+),$$

which we will formally extend to

$$f(x).z^k = f(\gamma q^k)\, z^k \quad (k \in \mathbb{Z}_+)$$

if f is a function on $\mathbb{R}$. Hence

$$\left(\int_{-x.\infty}^{x.\infty} f(t)\,d_qt\right).z^k = \left(\int_{-\gamma q^k.\infty}^{\gamma q^k.\infty} f(t)\,d_qt\right) z^k = \left(\int_{-\gamma.\infty}^{\gamma.\infty} f(t)\,d_qt\right) z^k$$

provided that the sums defining the Jackson integral

$$I_f(\gamma) := \int_{-\gamma.\infty}^{\gamma.\infty} f(t)\,d_qt$$

converge absolutely. Note that $I_f(\gamma) = I_f(q\gamma)$. For such f we conclude that, for any formal power series g:

$$\left(\int_{-x.\infty}^{x.\infty} f(t)\,d_qt\right).g(z) = I_f(\gamma)\,g(z). \tag{9.5}$$

Now take up (9.2) again in the representation (9.4). We find

$$\left(\int_{-x.\infty}^{x.\infty} f(t+y)\,d_qt\right).g(z) = \sum_{m=0}^{\infty} I_{f_m}(\gamma)\,y^m.g(z) = \sum_{m=0}^{\infty} I_{f_m}(\gamma)\,z^m\,g(z),$$

while

$$\left(\int_{-x.\infty}^{x.\infty} f(t)\,d_qt\right).g(z) = I_f(\gamma)\,g(z).$$

So formula (9.1) in the representation (9.4) is equivalent with the vanishing of $I_{f_m}(\gamma)$ $(m = 1, 2, \ldots)$. It is easy to find a class of functions f for which these numbers vanish. Note that each f_m $(m \in \mathbb{Z}_+)$ is a q-derivative of another function.

Lemma 9.1 Let $\gamma \in \mathbb{R}\backslash\{0\}$ and let f be a function on $\{\pm\gamma q^k \mid k \in \mathbb{Z}\}$ such that $\lim_{k\to\infty} f(q^k\gamma) = \lim_{k\to\infty} f(-q^k\gamma)$ and $\lim_{k\to\infty} f(\pm q^{-k}\gamma) = 0$. Then $\int_{-\gamma.\infty}^{\gamma.\infty} (D_qf)(t)\,d_qt = 0$.

Proof It follows by summation by parts that

$$\int_{-\gamma.\infty}^{\gamma.\infty} (D_qf)(t)\,d_qt = \lim_{m,n\to\infty} \big(f(q^{-m}\gamma) - f(q^n\gamma) - f(-q^{-m}\gamma) + f(-q^n\gamma)\big). \quad \square$$

Proposition 9.2 Let $\gamma \in \mathbb{R}\backslash\{0\}$ and let f be a function on $\{\pm\gamma q^k \mid k \in \mathbb{Z}\} \cup \{0\}$ such that, for all $m \in \mathbb{Z}_+$, $D_q^m f$ is continuous at 0 and

$$|(D_q^m f)(\pm q^{-k}\gamma)| = \mathcal{O}(q^{(1+\varepsilon)k}) \quad \text{as } k \to \infty \tag{9.6}$$

for certain $\varepsilon > 0$. Then f satisfies the translation invariance (9.1) in the representation (9.4). If, moreover, the estimate (9.6) is satisfied for all $\varepsilon > 0$ (so f and all its q-derivatives are *rapidly decreasing* on the domain of definition), then f multiplied with any polynomial also satisfies the translation invariance (9.1) in the representation (9.4).

Proof Because of (9.3) and the estimate on $D_q^m f$, the Jackson integrals defining the $I_{f_m}(\gamma)$ converge absolutely. For each $m > 0$, f_m is a q-derivative, so $I_{f_m}(\gamma) = 0$ by Lemma 9.1 and the estimate for $D_q^{m-1} f$. For the proof of the last statement use the q-Leibniz rule $(D_q(fg))(x) = f(x)\,(D_q g)(x) + (D_q f)(x)\,g(qx)$. □

Note that the continuity at 0 in the Proposition is satisfied if f is the restriction of a function which is C^∞ on a neighbourhood of 0.

As an example we consider Jackson integrals involving the q-Gaussian $g_q(x) = e_{q^2}(-x^2)$ (cf. (3.17)). This function satisfies the conditions of the Proposition (including the rapid decreasing property). In fact, it follows by induction with respect to m that $(D_q^m g_q)(x) = p_m(x)\,e_{q^2}(-x^2)$ with p_m a polynomial of degree m (more concretely a discrete q-Hermite II polynomial, see (8.30)). Now observe that

$$|e_{q^2}(-x^2)| = \prod_{k=0}^{\infty} |1 + q^{2k}x^2|^{-1} \le (1 + q^{2n}x^2)^{-n} = \mathcal{O}(|x|^{-2n}) \quad \text{as } x \to \pm\infty$$

for all $n \in \mathbb{Z}_+$. Thus we know that (9.1) in the representation (9.4) is valid for $f(x) := x^m\,e_{q^2}(-x^2)$. Let us see the implications of this result for q-special functions, by which we can make contact with the results of Section 8.

It follows from (8.1) that

$$h_m(0;q) = \begin{cases} (-1)^n\,q^{n(n-1)}\,(q;q^2)_n & \text{if } m = 2n, \\ 0 & \text{if } m = 2n+1. \end{cases} \tag{9.7}$$

Hence the case $t = 0$ of (8.21) yields

$$\int_{-\gamma.\infty}^{\gamma.\infty} t^m\,e_{q^2}(-t^2)\,d_q t = \begin{cases} c_q(\gamma)\,q^{-n^2}(q;q^2)_n & \text{if } m = 2n, \\ 0 & \text{if } m = 2n+1 \end{cases} \tag{9.8}$$

where $c_q(\gamma)$ is given by (8.15). In fact, the case $m = 2n$ of (9.8) is a consequence of Ramanujan's ${}_1\psi_1$ summation formula, see [16, (II.29)].

Consider (9.1) with $f(x) := x^m\,e_{q^2}(-x^2)$. The left-hand side of (9.1), when acting as an operator on a formal power series $g(z)$ in the representation (9.4), can be evaluated as

$$\left(\int_{-x.\infty}^{x.\infty} t^m\,e_{q^2}(-t^2)\,d_q t\right).g(z) = \begin{cases} c_q(\gamma)\,q^{-n^2}(q;q^2)_n\,g(z) & \text{if } m = 2n, \\ 0 & \text{if } m = 2n+1. \end{cases} \tag{9.9}$$

We expand the right-hand side of (9.1), still formally, as

$$\begin{aligned}&\int_{-x.\infty}^{x.\infty} e_{q^2}(-(t+y)^2)\,(t+y)^m\,d_qt\\&=\sum_{k=0}^{m}\begin{bmatrix}m\\k\end{bmatrix}_q e_{q^2}(-y^2)\int_{-x.\infty}^{x.\infty} e_q(-yt)\,e_{q^2}(-t^2)\,y^{m-k}\,t^k\,d_qt\\&=\sum_{k=0}^{m} q^{-(k+1)(m-k)}\begin{bmatrix}m\\k\end{bmatrix}_q e_{q^2}(-y^2)\Bigl(\int_{-x.\infty}^{x.\infty} e_q(-yt)\,e_{q^2}(-t^2)\,t^k\,d_qt\Bigr)y^{m-k} \qquad (9.10)\end{aligned}$$

where we used (2.3) and (3.19) for the first equality, while the second equality follows from (8.13) and the q-commutation of x and y. Now we give meaning to the right-hand side of (9.10) as an operator acting on a formal power series $g(z)$ in the representation (9.4). Consider first:

$$\begin{aligned}&\Bigl(\int_{-x.\infty}^{x.\infty} e_q(-yt)\,e_{q^2}(-t^2)\,t^k\,d_qt\Bigr).g(z)\\&=\sum_{l=0}^{\infty}\frac{(-1)^l q^{\frac12 l(l-1)}y^l}{(q;q)_l}\Bigl(\int_{-x.\infty}^{x.\infty} e_{q^2}(-t^2)\,t^{k+l}\,d_qt\Bigr).g(z)\\&=\sum_{l=0}^{\infty}\frac{(-1)^l q^{\frac12 l(l-1)}z^l}{(q;q)_l}\Bigl(\int_{-\gamma.\infty}^{\gamma.\infty} e_{q^2}(-t^2)\,t^{k+l}\,d_qt\Bigr)g(z)\\&=\Bigl(\int_{-\gamma.\infty}^{\gamma.\infty} E_q(-zt)\,e_{q^2}(-t^2)\,t^k\,d_qt\Bigr)g(z)\\&=c_q(\gamma)\,q^{-\frac12 k(k-1)}\,i^k\,h_k(iq^{-1}z;q)\,E_{q^2}(z^2)\,g(z). \qquad (9.11)\end{aligned}$$

Here we used (9.5) and (8.21). After substitution of this result in (9.10) we obtain that the right-hand side of (9.10) acting on $g(z)$ becomes:

$$c_q(\gamma)\sum_{k=0}^{m}\frac{(q^{-m};q)_k}{(q;q)_k}\,q^k\,i^{-m}\,h_k(iq^{-1}z;q)\,(iq^{-1}z)^{m-k}\,g(z). \qquad (9.12)$$

We know, at least formally, that (9.9) and (9.12) must be equal to each other. But this equality can equivalently be written as (8.6), which we had already proved in an elementary way.

Let us next consider whether (9.1) holds when f equals the second q-Gaussian (cf. (3.17))

$$G_q(x) := E_{q^2}(-x^2) = (x^2;q^2)_\infty.$$

Note that $G_q(\pm q^{-m}) = 0$ for $m \in \mathbb{Z}_+$. So $G_q(\pm\gamma q^{-m}) = 0$ for m a sufficiently large integer if γ is an integer power of q. However, if γ is not an integer power of q then $|G_q(\gamma q^{-m})|$ increases faster than C^m for any $C > 1$ as $m \to \infty$. Indeed, take $n \in \mathbb{Z}$ such that $\gamma^2 q^{-2n-2} - 1 \ge C$. Then, for $m \ge n$:

$$|G_q(\gamma q^{-m})| \ge (\gamma^2 q^{-2n-2} - 1)^{m-n}\,|G(\gamma q^{-n})|.$$

So the Jackson integral $\int_{-\gamma.\infty}^{\gamma.\infty} G_q(t)\,t^m\,d_qt$ (the analogue of (9.8)) only converges absolutely for $\pm\gamma$ being an integer power of q and then it turns down to computing the Jackson integral over $[-q,q]$.

It follows from (8.9) that

$$\widetilde{h}_m(0;q) = \begin{cases} (q;q^2)_n\,(-1)^n\,q^{n-2n^2} & \text{if } m=2n, \\ 0 & \text{if } m=2n+1. \end{cases} \tag{9.13}$$

Hence the case $t=0$ of (8.20) yields

$$\int_{-q}^{q} E_{q^2}(-t^2)\,t^m\,d_qt = \begin{cases} b_q\,q^{2n+1}\,(q;q^2)_n & \text{if } m=2n, \\ 0 & \text{if } m=2n+1, \end{cases} \tag{9.14}$$

where b_q is given by (8.3). Alternatively, formula (9.14) can also be obtained by a completely elementary computation.

Since G_q is a C^∞-function and since it vanishes on the set $\{\pm q^{-m} \mid m \in \mathbb{Z}_+\}$, it clearly satisfies all conditions of Proposition 9.2 for $\gamma = 1$. So the function $f(x) := x^m\,E_{q^2}(-x^2)$ will satisfy (9.1) in the representation (9.4) for $\gamma = 1$, i.e., in the representation (5.6). In order to see the implications of this, we can imitate what we did for the other q-Gaussian g_q. The left-hand side of (9.1), when acting as an operator on a formal power series $g(z)$ in the representation (5.6), can be evaluated as

$$\left(\int_{-x.\infty}^{x.\infty} t^m\,E_{q^2}(-t^2)\,d_qt\right).\,g(z) = \begin{cases} b_q\,q^{2n+1}(q;q^2)_n\,g(z) & \text{if } m=2n, \\ 0 & \text{if } m=2n+1 \end{cases} \tag{9.15}$$

by using (9.14). We expand the right-hand side of (9.1), still formally, as

$$\begin{aligned} &\int_{-x.\infty}^{x.\infty} (t+y)^m\,E_{q^2}(-(t+y)^2)\,d_qt \\ &= \sum_{k=0}^{m} \begin{bmatrix} m \\ k \end{bmatrix}_q y^{m-k} \int_{-x.\infty}^{x.\infty} t^k\,E_{q^2}(-t^2)\,E_q(-yt)\,E_{q^2}(-y^2)\,d_qt \\ &= \sum_{k=0}^{m} \begin{bmatrix} m \\ k \end{bmatrix}_q y^{m-k} \left(\int_{-x.\infty}^{x.\infty} t^k\,E_{q^2}(-t^2)\,E_q(-yt)\,d_qt\right) E_{q^2}(-q^{-2}y^2). \end{aligned} \tag{9.16}$$

Here we used (2.3) and (3.20). Now we give meaning to the right-hand side of (9.16) as an operator acting on a formal power series $g(z)$ in the representation (5.6). First derive, analogous to the proof of (9.11) but now using (8.20), that

$$\begin{aligned} \left(\int_{-x.\infty}^{x.\infty} t^k\,E_{q^2}(-t^2)\,E_q(-yt)\,d_qt\right).\,g(z) &= \left(\int_{-q}^{q} t^k\,E_{q^2}(-t^2)\,e_{q^2}(-q^{-2}zt)\,d_qt\right) g(z) \\ &= q b_q\,i^{-k}\,q^{\frac{1}{2}k(k+1)}\,e_{q^2}(q^{-2}z^2)\,\widetilde{h}_k(-iq^{-1}z;q)\,g(z). \end{aligned}$$

After substitution of this result in (9.16) we obtain that the right-hand side of (9.16) acting on $g(z)$ becomes

$$q b_q \sum_{k=0}^{m} \frac{(q^{-m};q)_k}{(q;q)_k}\,q^{(m+1)k}\,i^k\,z^{m-k}\,\widetilde{h}_k(-iq^{-1}z;q)\,g(z). \tag{9.17}$$

The right-hand side of (9.15) must be equal to (9.17). But this equality can equivalently be written as (8.17), which we had already proved in an elementary way.

Thus we have seen in this section that the translation invariance (9.1) for the case that $f(t) = t^m g_q(t)$ or $t^m G_q(t)$ turns down to the identities (8.6) and (8.17) for discrete q-Hermite polynomials.

10 Braided Hopf Algebras

In this section we introduce braided Hopf algebras and show the relevance of this structure for the results of Section 8. First we recall the notion of an ordinary Hopf algebra (see for instance Abe [1], Sweedler [33], Koornwinder [25, Section 1]).

We will work over the field of complex numbers, so a *linear space* will mean a complex linear space. If V and W are linear spaces then the *tensor product* $V \otimes W$ will be the linear space which is the algebraic tensor product of V and W, so $V \otimes W$ will be spanned by the elements $v \otimes w$ $(v, w \in V)$. The tensor products $V \otimes \mathbb{C}$ and $\mathbb{C} \otimes V$ will be naturally identified with V.

By an *algebra* we will mean a complex associative algebra with identity element 1. The field of complex numbers is an algebra in an evident way. If $\mathcal{A}$ is an algebra then the mapping $m: \mathcal{A} \otimes \mathcal{A} \to \mathcal{A}$ will denote the linear extension of the bilinear mapping $(a_1, a_2) \mapsto a_1 a_2 : \mathcal{A} \times \mathcal{A} \to \mathcal{A}$. If $\mathcal{A}$ and $\mathcal{B}$ are algebras then an *algebra homomorphism* $\phi: \mathcal{A} \to \mathcal{B}$ is a linear mapping satisfying $\phi(a_1 a_2) = \phi(a_1)\phi(a_2)$ and $\phi(1) = 1$.

If V is a linear space then the *flip operator* is the linear operator $\sigma: V \otimes V \to V \otimes V$ such that $\sigma(v \otimes w) = w \otimes v$. If $\mathcal{A}$ is an algebra then we give an algebra structure to $\mathcal{A} \otimes \mathcal{A}$ by putting $(a \otimes b)(c \otimes d) := ac \otimes bd$ and by extending this to a bilinear mapping of $(\mathcal{A}\otimes\mathcal{A}) \times (\mathcal{A}\otimes\mathcal{A})$ to $\mathcal{A}\otimes\mathcal{A}$. Let $m_{\mathcal{A}\otimes\mathcal{A}}: \mathcal{A}\otimes\mathcal{A}\otimes\mathcal{A}\otimes\mathcal{A} \to \mathcal{A}\otimes\mathcal{A}$ be the linear operator m corresponding to this algebra structure. Thus $m_{\mathcal{A}\otimes\mathcal{A}}$ acting on $a \otimes b \otimes c \otimes d$ can be written as the composition of two operators:

$$a \otimes b \otimes c \otimes d \longrightarrow a \otimes c \otimes b \otimes d \longrightarrow ac \otimes bd,$$

so

$$m_{\mathcal{A}\otimes\mathcal{A}} = (m_{\mathcal{A}} \otimes m_{\mathcal{A}}) \circ (\mathrm{id} \otimes \sigma \otimes \mathrm{id}). \tag{10.1}$$

Definition 10.1 A *Hopf algebra* is an algebra $\mathcal{A}$ equipped with three additional operators $\Delta: \mathcal{A} \to \mathcal{A} \otimes \mathcal{A}$ (*comultiplication*), $\varepsilon: \mathcal{A} \to \mathbb{C}$ (*counit*) and $S: \mathcal{A} \to \mathcal{A}$ (*antipode*), where Δ and ε are algebra homomorphisms and S is a linear mapping, and where the following additional properties are satisfied:

$$(\Delta \otimes \mathrm{id}) \circ \Delta = (\mathrm{id} \otimes \Delta) \circ \Delta \quad (\textit{coassociativity}), \tag{10.2}$$

$$(\varepsilon \otimes \mathrm{id}) \circ \Delta = \mathrm{id} = (\mathrm{id} \otimes \varepsilon) \circ \Delta, \tag{10.3}$$

$$\big(m \circ (S \otimes \mathrm{id}) \circ \Delta\big)(a) = \varepsilon(a)\,1 = \big(m \circ (\mathrm{id} \otimes S) \circ \Delta\big)(a) \quad (a \in \mathcal{A}). \tag{10.4}$$

It can be shown as a consequence of this definition that the antipode is anti-multiplicative and anti-comultiplicative:

$$\begin{aligned} &S(ab) = S(b)S(a), \quad S(1) = 1, \\ &(S \otimes S) \circ \sigma \circ \Delta = \Delta \circ S, \quad \varepsilon \circ S = \varepsilon. \end{aligned} \tag{10.5}$$

In the definition of Hopf algebra the flip operator σ entered in the specification (10.1) of the algebra structure of $\mathcal{A} \otimes \mathcal{A}$, and this algebra structure is needed since $\Delta: \mathcal{A} \to \mathcal{A} \otimes \mathcal{A}$ is required to be an algebra homomorphism. In a *braided Hopf algebra* the role of σ is taken over by some other bijective linear mapping $\Psi: \mathcal{A} \otimes \mathcal{A} \to \mathcal{A} \otimes \mathcal{A}$, the so-called *braiding*. The multiplication in $\mathcal{A} \otimes \mathcal{A}$ is now defined by

$$(a \otimes b)(c \otimes d) := (m \otimes m)(\mathrm{id} \otimes \Psi \otimes \mathrm{id})(a \otimes b \otimes c \otimes d). \tag{10.6}$$

The braiding Ψ has to satisfy some further axioms such that multiplication in $\mathcal{A}\otimes\mathcal{A}$ is associative and comultiplication in $\mathcal{A}\otimes\mathcal{A}$ is coassociative, which I will not give here. The definition of Hopf algebra is now precisely as in Definition 10.1, but with modified multiplication rule in $\mathcal{A}\otimes\mathcal{A}$. Braided Hopf algebras were introduced by Majid, see [27] and references given there. I refer to his papers for further details.

By way of example, consider the *braided line* $\mathcal{A} := \mathbb{C}_q[x]$ (see Koornwinder [23, Section 6.8], Majid [26]). As an algebra it is just the algebra $\mathbb{C}[x]$ of polynomials in one variable x, generated by the element x, and with basis $1, x, x^2, \ldots$. Now let $q \in (0,1)$ and introduce the braiding Ψ by specifying it on a basis of $\mathcal{A}\otimes\mathcal{A}$:

$$\Psi(x^k \otimes x^l) := q^{kl}\, x^l \otimes x^k \quad (k,l \in \mathbb{Z}_+).$$

So for multiplication in $\mathcal{A}\otimes\mathcal{A}$ we will have:

$$\begin{aligned}(x^k \otimes 1)(1 \otimes x^l) &= x^k \otimes x^l,\\ (1 \otimes x^k)(x^l \otimes 1) &= q^{kl}\, x^l \otimes x^k,\\ (x^{k_1} \otimes x^{k_2})\,(x^{l_1} \otimes x^{l_2}) &= q^{k_2 l_1}\, x^{k_1+l_1} \otimes x^{k_2+l_2}.\end{aligned}$$

Hence

$$\begin{aligned}\big((x^{k_1} \otimes x^{k_2})\,(x^{l_1} \otimes x^{l_2})\big)\,(x^{m_1} \otimes x^{m_2}) &= q^{k_2 l_1 + k_2 m_1 + l_2 m_1}\, x^{k_1+l_1+m_1} \otimes x^{k_2+l_2+m_2}\\ &= (x^{k_1} \otimes x^{k_2})\,\big((x^{l_1} \otimes x^{l_2})\,(x^{m_1} \otimes x^{m_2})\big),\end{aligned}$$

which proves the associativity.

Note that $\mathcal{A}\otimes\mathcal{A}$ can be considered as the algebra with generators $1 \otimes x$ and $x \otimes 1$ and with relation $(1 \otimes x)(x \otimes 1) = q(x \otimes 1)(1 \otimes x)$, so it is isomorphic with the algebra $\mathbb{C}_q[x,y]$ under the isomorphisms $x^l \otimes x^k \mapsto y^l x^k$.

More generally we can make the n-fold tensor product $\otimes^n \mathcal{A}$ into an algebra by the rule

$$(x^{k_1} \otimes \cdots \otimes x^{k_n})(x^{l_1} \otimes \cdots \otimes x^{l_n}) = q^{\sum_{i>j} k_i l_j}\, x^{k_1+l_1} \otimes \cdots \otimes x^{k_n+l_n}.$$

A simple computation shows that the multiplication is associative. Futhermore, the linear subspace of $\otimes^n \mathcal{A}$ spanned by the elements

$$1 \otimes \cdots \otimes 1 \otimes x^{k_1} \otimes 1 \otimes \cdots \otimes 1 \otimes x^{k_2} \otimes 1 \otimes \cdots\cdots \otimes 1 \otimes x^{k_r} \otimes 1 \otimes \cdots \otimes 1$$

(non-zero powers of x only allowed at positions $i_1, i_2, \ldots, i_r$) is a subalgebra of $\otimes^n \mathcal{A}$ isomorphic to $\otimes^r \mathcal{A}$.

Since the comultiplication $\Delta\colon \mathcal{A} \to \mathcal{A}\otimes\mathcal{A}$ has to be an algebra homomorphism, it is sufficient to define it on the generator x of $\mathcal{A}$:

$$\Delta(x) := x \otimes 1 + 1 \otimes x.$$

Then

$$\big((\Delta \otimes \mathrm{id}) \circ \Delta\big)(x) = x \otimes 1 \otimes 1 + 1 \otimes x \otimes 1 + 1 \otimes 1 \otimes x = \big((\mathrm{id} \otimes \Delta) \circ \Delta\big)(x),$$

so the coassociativity, being valid on the generator $x \in \mathcal{A}$, will be valid in general. If f is a polynomial in one variable then

$$\Delta(f(x)) = f(x \otimes 1 + 1 \otimes x). \tag{10.7}$$

For the comultiplication applied to a general basis element $x^n \in \mathcal{A}$ we find

$$\Delta(x^n) = \sum_{k=0}^{n} \begin{bmatrix} n \\ k \end{bmatrix}_q x^{n-k} \otimes x^k. \tag{10.8}$$

Indeed, by rewriting the two sides of (10.8) we have to prove that

$$(x \otimes 1 + 1 \otimes x)^n = \sum_{k=0}^{n} \begin{bmatrix} n \\ k \end{bmatrix}_q (x \otimes 1)^{n-k} (1 \otimes x)^k,$$

and this is true by the q-binomial formula (2.3) since $(1 \otimes x)(x \otimes 1) = q(x \otimes 1)(1 \otimes x)$.

In order to find the counit $\varepsilon: \mathcal{A} \to \mathbb{C}$ we apply the first identity of (10.3) to x^n and we use (10.8):

$$x^n = (\varepsilon \otimes \mathrm{id})(\Delta(x^n)) = \sum_{k=0}^{n} \begin{bmatrix} n \\ k \end{bmatrix}_q \varepsilon(x^{n-k})\, x^k,$$

which yields

$$\varepsilon(x^n) = \begin{cases} 0 & \text{if } n > 0, \\ 1 & \text{if } n = 0. \end{cases} \tag{10.9}$$

Finally we look for the existence of an antipode $S: \mathcal{A} \to \mathcal{A}$. It turns out that $S(x^n)$ is uniquely found by letting one of the identitities of (10.4) (say the second one) act on the basis elements x^n of $\mathcal{A}$. We obtain

$$S(x^n) := (-1)^n\, q^{\frac{1}{2}n(n-1)}\, x^n. \tag{10.10}$$

Indeed, the second identity of (10.4) yields

$$\sum_{k=0}^{n} \begin{bmatrix} n \\ k \end{bmatrix}_q x^{n-k}\, S(x^k) = \big(m \circ (\mathrm{id} \otimes S) \circ \Delta\big)(x^n) = \varepsilon(x^n)\, 1 = \delta_{n,0}\, 1, \tag{10.11}$$

so $S(x^n)$ is obtained by recurrence with respect to n. Now the left-hand side of (10.11) with $S(x^k)$ given by (10.10) becomes

$$\sum_{k=0}^{n} \begin{bmatrix} n \\ k \end{bmatrix}_q (-1)^k\, q^{\frac{1}{2}k(k-1)}\, x^n = \sum_{k=0}^{n} \frac{(q^{-n};q)_k}{(q;q)_k}\, q^{nk}\, x^n = {}_1\phi_0(q^{-n};\,;q,q^n)\, x^n = (1;q)_n\, x^n,$$

which equals the right-hand side of (10.11). The proof that the first identity of (10.4) acting on x^n also holds, amounts to the same computation as we just gave.

Observe that

$$S(x^{m+n}) = q^{mn}\, S(x^m)\, S(x^n),$$

$$\Delta(S(x^n)) = \sum_{k=0}^{n} \begin{bmatrix} n \\ k \end{bmatrix}_q q^{k(n-k)}\, S(x^{n-k}) \otimes S(x^k).$$

Hence

$$S \circ m = m \circ (S \otimes S) \circ \Psi, \quad \Delta \circ S = (S \otimes S) \circ \Psi \circ \Delta. \tag{10.12}$$

This can be considered as an analogue of (10.5) for the braided case, where the flip σ is replaced by the braiding Ψ.

We can reformulate formula (8.31) by using the above comultiplication:

$$\Delta(h_n(x;q)) = \sum_{k=0}^{n} \begin{bmatrix} n \\ k \end{bmatrix}_q x^{n-k} \otimes h_k(x;q). \tag{10.13}$$

When we apply $m \circ (S \otimes \mathrm{id}) \circ \Delta$ to both sides of (10.13) then we obtain

$$h_n(0;q) = \sum_{k=0}^{n} \begin{bmatrix} n \\ k \end{bmatrix}_q (-1)^{n-k} q^{\frac{1}{2}(n-k)(n-k-1)} x^{n-k} h_k(x;q).$$

In view of (9.7) this is just (8.6).

We can extend the above braided Hopf algebra structure of $\mathbb{C}_q[x]$ to the algebra $\mathbb{C}_q[[x]]$ of formal power series in x. Then Δ, ε and S are well-defined in a termwise way on $\mathbb{C}_q[[x]]$, by using (10.8), (10.9) and (10.10). For $\Delta(f(x))$, f being a formal power series, we can still use (10.7). In particular, for Δ, ε and S acting on $e_q(x)$ we obtain

$$\Delta(e_q(x)) = e_q(x) \otimes e_q(x), \quad \varepsilon(e_q(x)) = 1, \quad S(e_q(x)) = E_q(-x). \tag{10.14}$$

The first identity, which was already observed in Koornwinder [23, Section 6.8], follows from (and is equivalent to) (3.5) since $\Delta(e_q(x)) = e_q(x \otimes 1 + 1 \otimes x)$ and $(1 \otimes x)(x \otimes 1) = q(x \otimes 1)(1 \otimes x)$. Now it follows from (10.4) and (10.14) that

$$1 = \varepsilon(e_q(x))1 = \big(m \circ (S \otimes \mathrm{id}) \circ \Delta\big)(e_q(x)) = S(e_q(x))\, e_q(x) = E_q(-x)\, e_q(x),$$

i.e., we have reobtained (3.3).

Formula (9.1) (the translation invariance of the Jackson integral over $(-x.\infty, x.\infty)$) can also be rephrased in terms of the above comultiplication. We will work in a very formal way. Let $\int$ be the linear operator defined by

$$\textstyle\int(f(x)) := \displaystyle\int_{-x.\infty}^{x.\infty} f(t)\, d_q t.$$

If y is an element with the property that $xy = q^k yx$ for some integer k, then y commutes with $\int(f(x))$, so $\int(f(x))$ may be considered as a scalar. Now we can rewrite (9.1) as

$$\big((\mathrm{id} \otimes \textstyle\int) \circ \Delta\big)(f(x)) = \textstyle\int(f(x))\, 1. \tag{10.15}$$

Indeed, the left-hand side can formally be written as

$$\int_{-x.\infty}^{x.\infty} f(x \otimes 1 + 1 \otimes t)\, d_q t$$

and $(1 \otimes t)(x \otimes 1) = q(x \otimes 1)(1 \otimes t)$ if $t = q^k x$ for some integer k.

Next we will consider the q-Fourier transform $\mathcal{F}_q$ defined by (8.22) from the point of view of this comultiplication. We fix $q \in (0,1)$ and we rewrite (8.22) very formally as

$$\mathcal{F}_y(f(x)) := \int_{-x.\infty}^{x.\infty} e_q(-ity)\, f(t)\, d_q t \quad (y \in \mathbb{R}).$$

Then the linear operator $\mathcal{F}_y$, like $\int$, will map to a space of scalars. It follows from (10.15) and (3.5) that

$$\begin{aligned}\mathcal{F}_y(f(x))\,1 &= (\mathrm{id}\otimes\textstyle\int)\big(\Delta(e_q(-ixy))\,\Delta(f(x))\big)\\ &= (\mathrm{id}\otimes\textstyle\int)\Big(\big(e_q(-ixy)\otimes e_q(-ixy)\big)\,\Delta(f(x))\Big)\\ &= (\mathrm{id}\otimes\mathcal{F}_y)(\Delta(f(x))\,e_q(-ixy)).\end{aligned}$$

Now multiply the left-hand side and the right-hand side both by $E_q(ixy)$. Then we obtain

$$(\mathrm{id}\otimes\mathcal{F}_y)(\Delta(f(x))) = \mathcal{F}_y(f(x))\,E_q(ixy). \tag{10.16}$$

This formula is a q-analogue of the well-known property of the Fourier transform that

$$\int_{-\infty}^{\infty} e^{-ity}\,f(x+t)\,dt = e^{ixy}\int_{-\infty}^{\infty} e^{-ity}\,f(t)\,dt.$$

Formula (10.16) can also be written as

$$(\mathrm{id}\otimes\textstyle\int)\big((1\otimes e_q(-ixy))\,\Delta(f(x))\big) = (\mathrm{id}\otimes\textstyle\int)\Big(\big((S\otimes\mathrm{id})\circ\Delta\big)(e_q(-ixy))\,(1\otimes f(x))\Big).$$

More generally we have

$$(\mathrm{id}\otimes\textstyle\int)\big((1\otimes g(x))\,\Delta(f(x))\big) = (\mathrm{id}\otimes\textstyle\int)\Big(\big((S\otimes\mathrm{id})\circ\Delta\big)(g(x))\,(1\otimes f(x))\Big), \tag{10.17}$$

which is a q-analogue of

$$\int_{-\infty}^{\infty} g(t)\,f(x+t)\,dt = \int_{-\infty}^{\infty} g(t-x)\,f(t)\,dt.$$

Formula (10.17) is given by Kempf & Majid in [19, (139)] with diagrammatic proof in their Figure 2(b). See this paper also for many further results about Jackson integral and q-Fourier transform in connection with the braided line.

11 Further results

In this final section, I very briefly mention three further results.

1. Special function identities involving non-commuting variables satisfying relations which may be more complicated than q-commutation occur very naturally as addition formulas obtained from quantum groups. A prototype of this is the addition formula for little q-Legendre polynomials, in the context of $SU_q(2)$. See Koornwinder [24], where the formula with non-commuting variables is next equivalently rewritten in commuting variables. In the context of $U_q(n)$ a similar but more complicated addition formula in non-commuting variables was obtained by Floris [14] for q-disk polynomials. Next Floris & Koelink [15] found an equivalent form in commuting variables of this addition formula. See Koelink [21] for other examples involving $SU_q(2)$ of rewriting addition formulas from non-commutative form into commutative form.
2. Some q-hypergeometric series which are not summable when parameters and argument commute, may suddenly become summable when these variables do not commute but satisfy certain relations. G. Andrews (private communication) showed me a surprising example of this involving a ${}_{m+1}\phi_m$. Many further results in this direction, often in operational form, can be found in Bowman [6].

3. The following was communicated to me by A. Yu. Volkov. The second part of formula (3.28) nicely generalizes to an identity in x, y, c with q-Heisenberg relations (4.1):

$$(y;q)_n\,(x;q)_n = \prod_{k=0}^{n-1}(1-q^k(x+y-yx+c)+q^{2k}c).$$

The proof is by induction with respect to n. On letting $n\to\infty$ and taking the inverse on both sides, we obtain an addition to Proposition 4.1.

Acknowledgements

This paper was written while the author was a guest at the Institute Mittag-Leffler, Djursholm, Sweden during October, November 1995. I am grateful to the director of the Institute and to the organizers of the special year "Analysis on Lie Groups" for hospitality. I also thank G. Andrews, R. Askey, D. Bowman, A. N. Kirillov, C. Krattenthaler, A. Yu. Volkov and S. L. Woronowicz for helpful remarks.

References

[1] Abe, E. [1980] *Hopf algebras*, Cambridge University Press.

[2] Askey, R. [1994] *Abstract of "Extensions of Hermite polynomials and other orthogonal polynomials"*, SIAM Newsletter "Orthogonal Polynomials and Special Functions" **5**, no.1, 14.

[3] Askey, R. [1996] *A brief introduction to the world of q*, in Symmetries and integrability of difference equations, (D. Levi and P. Winternitz, eds.), CRM Proceedings and Lecture Notes 9, Amer. Math. Soc., Providence.

[4] Bergman, G. M. [1978] *The diamond lemma for ring theory*, Advances in Math. **29**, 178–218.

[5] Berndt, B. C. [1994] *Ramanujan's notebooks, Part IV*, Springer.

[6] Bowman, D. *q-Difference operators and symmetric expansions*, in preparation.

[7] Cigler, J. [1979] *Operatormethoden für q-Identitäten*, Monatsh. Math. **88**, 87–105.

[8] Cigler, J. [1981] *Operatormethoden für q-Identitäten II: q-Laguerre-Polynome*, Monatsh. Math. **91**, 105–117.

[9] Erdélyi, A. [1954] *Tables of integral transforms, Vol. I*, McGraw-Hill.

[10] Faddeev, L. and Kashaev, R. M. [1994] *Quantum dilogarithm*, Modern Phys. Lett. **A 9**, 427–434.

[11] Faddeev, L. and Volkov, A. Yu. [1993] *Abelian current algebra and the Virasoro algebra on the lattice*, Phys. Lett. **B 315**, 311–318.

[12] Finkelstein, R. and Marcus, E. [1995] *Transformation theory of the q-oscillator*, J. Math. Phys. **36**, 2652–2672.

[13] Floreanini, R. and Vinet, L. [1993] Automorphisms of the q-oscillator algebra and basic orthogonal polynomials, Phys. Lett. **A 180**, 393–401.

[14] Floris, P. G. A. *Addition formula for q-disk polynomials*, Compositio Math., to appear.

[15] Floris, P. G. A. and Koelink, H. T. *A commuting q-analogue of the addition formula for disk polynomials*, Constr. Math., to appear.

[16] Gasper, G. and Rahman, M. [1990] *Basic hypergeometric series*, Encyclopedia of Mathematics and its Applications 35, Cambridge University Press.

[17] Gelfand, I. M. and Fairlie, D. B. [1991] *The algebra of Weyl symmetrised polynomials and its quantum extension*, Comm. Math. Phys. **136**, 487–499.

[18] Kashaev, R. M. [1995] *Heisenberg double and pentagon relation*, preprint, q-alg/9503005, to appear in Algebraic Analysis.

[19] Kempf, A. and Majid, S. [1994] *Algebraic q-integration and Fourier theory on quantum and braided spaces*, J. Math. Phys. **35**, 6802–6837.

[20] Kirillov, A. N. [1995] *Dilogarithm identities*, Progress Theor. Phys. Supplement No. 118, 61–142.

[21] Koelink, H. T. *Addition formulas for q-special functions*, Proceedings in Special Functions, q-Series and Related Topics, Toronto, June 1995, The Fields Institute Commumications Series (to appear).

[22] Koekoek, R. and Swarttouw, R. F. [1994] *The Askey scheme of hypergeometric orthogonal polynomials and its q-analogue*, Report 94-05, Technical University Delft, Faculty of Technical Mathematics and Informatics.

[23] Koornwinder, T. H. [1990] *Orthogonal polynomials in connection with quantum groups*, in Orthogonal polynomials: theory and practice (P. Nevai, ed.), NATO ASI Series C 294, Kluwer, 257–292.

[24] Koornwinder, T. H. [1991] *The addition formula for little q-Legendre polynomials and the $SU(2)$ quantum group*, SIAM J. Math. Anal. **22**, 295–301.

[25] Koornwinder, T. H. *Compact quantum groups and q-special functions*, in Representations of Lie groups and quantum groups (V. Baldoni and M. A. Picardello, eds.), Pitman Research Notes in Mathematics Series 311, Longman Scientific & Technical, 46–128.

[26] Majid, S. [1992] *C-statistical quantum groups and Weyl algebras*, J. Math. Phys. **33**, 3431–3444.

[27] Majid, S. [1994] *Algebras and Hopf algebras in braided categories*, in Advances in Hopf algebras, Marcel Dekker, 55–105.

[28] McDermott, R. J. and Solomon, A. I. [1994] *An analogue of the unitary displacement operator for the q-oscillator*, J. Phys. **A 27**, 2037–2043.

[29] Polya, G. and Alexanderson, G. L. [1970] *Gaussian binomial coefficients*, Elem. Math. **26**, 102–108.

[30] Rogers, L. J. [1893] *On the expansion of some infinite products*, Proc. London Math. Soc. **24**, 337–352.

[31] Rota, G.-C., Kahaner, D. and Odlyzko, A. [1973] *Finite operator calculus*, J. Math. Anal. Appl. **42**, 684–760.

[32] Schützenberger, M. P. [1953] *Une interprétation de certaines solutions de l'équation fonctionnelle: $F(x+y) = F(x)F(y)$*, C. R. Acad. Sci. Paris **236**, 352–353.

[33] Sweedler, M. E. [1969] *Hopf algebras*, Benjamin.

[34] Ueno, K. and Nishizawa, M. [1995] *Quantum groups and zeta functions*, in Quantum groups: formalism and applications (J. Lukierski et al., eds.), PWN, Warsaw.

[35] Varadarajan, V. S. [1974] *Lie groups, Lie algebras and their representations*, Prentice-Hall.

[36] Woronowicz, S. L. [1992] *Operator equalities related to quantum $E(2)$ group*, Comm. Math. Phys. **144**, 417–428.

Fields Institute Communications
Volume **14**, 1997

Multivariable Askey-Wilson Polynomials and Quantum Complex Grassmannians

Masatoshi Noumi and Mathijs S. Dijkhuizen
Department of Mathematics, Kobe University
Rokko, Kobe 657, Japan

Tetsuya Sugitani
Department of Mathematical Sciences, University of Tokyo
Komaba, Meguro-Ku, Tokyo 153, Japan

Abstract. We present a one-parameter family of constant solutions of the reflection equation and define a family of quantum complex Grassmannians endowed with a transitive action of the quantum unitary group. By computing the radial part of a suitable Casimir operator, we identify the zonal spherical functions (i.e., infinitesimally bi-invariant matrix coefficients of finite-dimensional irreducible representations) as multivariable Askey-Wilson polynomials containing two continuous and two discrete parameters.

0 Introduction

Every classical compact symmetric space M can be realized as an orbit of square matrices, the action of the symmetry group G being given by $X \mapsto TXT^t$ or $X \mapsto TXT^*$ ($X \in M$, $T \in G$). This approach has proved to be fruitful in the study of quantum symmetric spaces as well. The proper way to quantize the space of square matrices endowed with either of the above-mentioned group actions would be to make use of commutation relations given by one of two types of so-called *reflection equations.* These equations pop up at various places in mathematics, notably in the theory of quantum integrable systems (cf. Kulish [1991]). The quantized symmetric space M_q is then realized as the quantum orbit containing a certain "classical" or $\mathbb{C}$-valued point in the quantum space of square matrices (i.e., a $\mathbb{C}$-valued evaluation homomorphism on the quantized algebra of functions). These "classical" points are naturally in 1-1 correspondence with constant solutions J of the reflection equation.

There is another, slightly different, but closely related, way to explain the role of the reflection equation in the theory of quantum symmetric spaces. Given any square matrix J with complex coefficients, there is a straightforward way to define

1991 *Mathematics Subject Classification.* Primary: 33D80, 33D55; Secondary: 17B37.

The second author (M.S. D.) acknowledges financial support by the Japan Society for the Promotion of Science (JSPS) and the Netherlands Organization for Scientific Research (NWO).

a two-sided co-ideal $\mathfrak{k}_J$ in the quantized universal enveloping algebra using the L-operators introduced in Reshetikhin-Faddeev-Takhtajan [1990]. The reflection equation now more or less guarantees the existence of sufficiently many $\mathfrak{k}_J$-fixed vectors. This fact is obviously central to the study of zonal spherical functions. If the matrix J satisfies the reflection equation, the invariant ring corresponding to the coideal $\mathfrak{k}_J$ defines a quantum homogeneous G_q-space that is G_q-isomorphic to the quantum orbit passing through the "classical" point associated with J.

In Noumi [1993], the first author used the ideas sketched above to construct and analyse quantum analogues of the symmetric spaces $GL(n)/SO(n)$ and $GL(2n)/Sp(n)$. The corresponding zonal spherical functions were shown to be expressible in terms of Macdonald's symmetric polynomials corresponding to root system A_{n-1} (cf. Macdonald [1988]). As a follow-up, the first and third author extended this approach to all classical compact symmetric spaces (Noumi-Sugitani [1995]) . In all these cases, the zonal spherical functions could be expressed as Macdonald's polynomials (Macdonald [1988]) or Koornwinder's multivariable Askey-Wilson polynomials (Koornwinder [1992]), depending on the restricted root system of the symmetric space.

The quantum symmetric spaces treated in Noumi [1993], Noumi-Sugitani [1995], however, do not exhaust all known examples of quantum analogues of classical compact symmetric spaces. In the early stages of quantum group theory, Podleś had already defined a continuously parametrized (parameter different from q!) family of mutually non-isomorphic $SU_q(2)$-homogeneous spaces. Each of these spaces can be regarded as a quantum analogue of the classical 2-sphere. The zonal spherical functions on these quantum spheres were analysed by Mimachi and the first author (Noumi-Mimachi [1990]), and by Koornwinder [1993]. They can be expressed as a certain subfamily of Askey-Wilson polynomials (Askey-Wilson [1985]) or limit cases thereof.

Recently, the first and second authors generalized these results to quantum complex projective spaces of arbitrary dimension (Dijkhuizen-Noumi [1995]) . They studied a family of quantum projective spaces which were introduced in Korogodsky-Vaksman [1992] and depend on a continuous parameter. The zonal spherical functions were expressed as Askey-Wilson polynomials depending on two continuous and one discrete parameter or limit cases thereof. See Dijkhuizen [1994] for a more extensive discussion of the relation between these parametrized families of quantum symmetric spaces and the examples treated in Noumi [1993], Noumi-Sugitani [1995].

In this paper, we announce some results which constitute a generalization of Dijkhuizen-Noumi [1995]. We present a one-parameter family of constant solutions to one type of reflection equation, and use them to define a family of quantum homogeneous $U_q(n)$-spaces that can be regarded as a quantum analogue of a complex Grassmannian of arbitrary rank. It turns out that the zonal spherical functions on these quantum Grassmannian spaces are expressed as a subfamily of Koornwinder's multivariable Askey-Wilson polynomials depending on two continuous and two discrete parameters. This last result is proved by computing the radial part of a suitable Casimir operator in the quantized universal enveloping algebra, a method which was first used by Koornwinder (Koornwinder [1993]) in the $SU_q(2)$-case, and subsequently generalized by the first author (Noumi [1993]) to higher rank quantum symmetric spaces.

This paper does not contain any proofs. Full proofs and more details will be given elsewhere.

1 Recall on multivariable Askey-Wilson polynomials

We recall the definition of multivariable Askey-Wilson polynomials (Koornwinder [1992]). Our notation is slightly different from Koornwinder [1992].

Let $P_\Sigma := \bigoplus_{1\le k\le l} \mathbb{Z}\varepsilon_k$ denote the weight lattice of the root system BC_l, and P_Σ^+ the cone of dominant weights in P_Σ. A dominant weight $\lambda = \sum_{k=1}^l \lambda_k\varepsilon_k \in P_\Sigma$ is then characterized by the condition $\lambda_1 \ge \ldots \ge \lambda_l \ge 0$. Let $\le$ denote the usual dominance ordering on weights. Recall that $\lambda \le \mu$ if and only if $\sum_{i=1}^k \lambda_i \le \sum_{i=1}^k \mu_i$ for all $1 \le k \le l$.

We will use the notation $x_k := e^{\varepsilon_k}$ $(1 \le k \le \ell)$ to refer to the formal exponentials. Let $\mathbb{C}[x^{\pm 1}] := \mathbb{C}[x_1^{\pm 1}, \ldots, x_l^{\pm 1}]$ denote the algebra of Laurent polynomials in the variables x_k $(1 \le k \le l)$. $\mathbb{C}[x^{\pm 1}]$ can be viewed as the group algebra of the weight lattice P_Σ. The Weyl group $W := \mathbb{Z}_2^l \rtimes \mathfrak{S}_l$ of BC_l acts on $\mathbb{C}[x^{\pm 1}]$ in a natural way by permutations and sign changes of the x_k. Let $\mathbb{C}[x^{\pm 1}]^W \subset \mathbb{C}[x^{\pm 1}]$ denote the subalgebra of W-invariant elements.

Let $0 < q < 1$. Write $T_{q,x_k}\colon \mathbb{C}[x^{\pm 1}] \to \mathbb{C}[x^{\pm 1}]$ for the q-shift operator sending x_k to qx_k. Koornwinder's q-difference operator $D_{\varepsilon_1}\colon \mathbb{C}[x^{\pm 1}] \to \mathbb{C}[x^{\pm 1}]$ is defined by

$$D_{\varepsilon_1} := \sum_{k=1}^{l} \left(\Phi_k^+(x)T_{q,x_k} + \Phi_k^-(x)T_{q,x_k}^{-1}\right) - \Phi^0(x), \tag{1.1}$$

where

$$\begin{aligned}
\Phi_k^+(x) &:= \frac{(1-ax_k)(1-bx_k)(1-cx_k)(1-dx_k)}{(1-x_k^2)(1-qx_k^2)} \prod_{i\neq k} \frac{(tx_k - x_i)(tx_kx_i - 1)}{(x_k - x_i)(x_kx_i - 1)}, \\
\Phi_k^-(x) &:= \frac{(x_k-a)(x_k-b)(x_k-c)(x_k-d)}{(x_k^2-1)(x_k^2-q)} \prod_{i\neq k} \frac{(x_k - tx_i)(x_kx_i - t)}{(x_k - x_i)(x_kx_i - 1)}, \\
\Phi^0(x) &:= \sum_{k=1}^{l} \Phi_k^+(x) + \Phi_k^-(x).
\end{aligned} \tag{1.2}$$

Here we assume that the parameters a, b, c, d, t are complex numbers satisfying

$$0 < t < 1, \quad -q \le abcd < 1. \tag{1.3}$$

The operator D_{ε_1} maps $\mathbb{C}[x^{\pm 1}]^W$ into itself.

It can be shown that there is a unique family $(P_\lambda)_{\lambda\in P_\Sigma^+}$ in $\mathbb{C}[x^{\pm 1}]^W$ such that

$$\begin{aligned}
&\text{(i)} \quad P_\lambda = m_\lambda + \sum_{\mu<\lambda} c_{\lambda\mu} m_\mu, \\
&\text{(ii)} \quad D_{\varepsilon_1} P_\lambda = c_{\lambda\lambda} P_\lambda.
\end{aligned} \tag{1.4}$$

Here $m_\lambda := \sum_{\mu\in W\lambda} x^\mu$ is the orbit sum corresponding to $\lambda \in P_\Sigma^+$. One has

$$c_{\lambda\lambda} = \sum_{k=1}^{l} \left(q^{-1}abcdt^{2l-k-1}(q^{\lambda_k} - 1) + t^{k-1}(q^{-\lambda_k} - 1)\right) \quad (\lambda \in P_\Sigma^+). \tag{1.5}$$

Note that $c_{\lambda\lambda} \neq c_{\mu\mu}$ for $\lambda > \mu$, provided (1.3) is satisfied (cf. Stokman-Koornwinder [1995], Propostion 4.6). The elements

$$P_\lambda = P_\lambda(x; a, b, c, d; q, t) \tag{1.6}$$

will be called *multivariable Askey-Wilson polynomials* (associated with the root system BC_l). For certain values of the parameters one reobtains Macdonald's polynomials (Macdonald [1988]) corresponding to the pairs (BC_l, B_l) or (BC_l, C_l).

Assume now for simplicity that $0 < t < 1$ and that a, b, c, d are real numbers with $|a|, |b|, |c|, |d| < 1$. Let $T_\Sigma^{\mathbb{C}} := (\mathbb{C}^*)^l$ denote the complex torus of dimension l. We denote its natural compact real form by T_Σ. Elements of $\mathbb{C}[x^{\pm 1}]$ can be naturally viewed as polynomial functions on $T_\Sigma^{\mathbb{C}}$.

Recall the notation of *q-shifted factorials*:

$$(a;q)_n := \prod_{k=0}^{n-1}(1-aq^k),\ (a_1,\dots,a_s;q)_n := \prod_{j=1}^{s}(a_j;q)_n,\ (a;q)_\infty := \lim_{n\to\infty}(a;q)_n. \tag{1.7}$$

Under the above-mentioned conditions on a, b, c, d, t, the infinite product

$$\Delta^+ := \prod_{i=1}^{l} \frac{(x_i^2;q)_\infty}{(ax_i, bx_i, cx_i, dx_i;q)_\infty} \cdot \prod_{1\le i<j\le l} \frac{(x_i/x_j;q)_\infty}{(tx_i/x_j;q)_\infty} \frac{(x_ix_j;q)_\infty}{(tx_ix_j;q)_\infty} \tag{1.8}$$

defines a continuous function on T_Σ. If we put

$$\Delta(x) := \Delta^+(x)\Delta^+(x^{-1}), \tag{1.9}$$

then Δ is a positive continuous function on T_Σ. The Askey-Wilson polynomials $P_\lambda(x; a, b, c, d; q, t)$ are mutually orthogonal with respect to the weight function Δ:

$$\int_{T_\Sigma} P_\lambda(x) P_\mu(x^{-1})\Delta(x)\,\mathrm{d}x = 0, \quad \lambda \neq \mu. \tag{1.10}$$

Here $\mathrm{d}x$ denotes the Haar measure on the real torus T_Σ.

For more general values of the parameters, an explicit expression of the orthogonality measure was recently written down by Stokman [1996].

2 Quantum complex Grassmannians

Let us fix some notation for the quantum unitary group (cf. Jimbo [1986], Reshetikhin-Faddeev-Takhtajan [1990], Noumi [1993]). More details can be found in Noumi [1993]. Let $0 < q < 1$ and $n \geq 2$. Let $\mathcal{A}_q = \mathcal{A}_q(U(n))$ denote the algebra of functions on the *quantum unitary group* $U_q(n)$. The matrix elements of the vector representation V with basis $(v_i)_{1\le i\le n}$ are written $t_{ij} \in \mathcal{A}_q$ $(1 \le i, j \le n)$. They satisfy the commutation relations

$$RT_1T_2 = T_2T_1R, \tag{2.1}$$

where $T := (t_{ij})_{1\le i,j\le n}$ is an $n \times n$ matrix with coefficients in $\mathcal{A}_q$, $T_1 := T \otimes \mathrm{id}$ and $T_2 := \mathrm{id} \otimes T$ are Kronecker matrix products, and $R \in \mathrm{End}(V \otimes V)$ is the invertible $n^2 \times n^2$ matrix defined by

$$R := \sum_{ij} q^{\delta_{ij}} e_{ii} \otimes e_{jj} + (q - q^{-1}) \sum_{i>j} e_{ij} \otimes e_{ji}. \tag{2.2}$$

The $e_{ij} \in \mathrm{End}(V)$ denote the standard matrix units with respect to the basis (v_i). We put

$$R^+ := PRP, \qquad R^- := R^{-1}, \tag{2.3}$$

where $P \in \mathrm{End}(V \otimes V)$ is the usual permutation operator. The Hopf $*$-algebra $\mathcal{A}_q$ is spanned by the coefficients of its finite-dimensional unitary corepresentations (cf. Woronowicz [1987], Noumi-Yamada-Mimachi [1993], Dijkhuizen-Koornwinder [1994]).

Let P denote the free $\mathbb{Z}$-module of rank n with canonical basis $(\varepsilon_i)_{1\leq i\leq n}$. The *quantized universal enveloping algebra* $\mathcal{U}_q = \mathcal{U}_q(\mathfrak{gl}(n))$ is the algebra generated by the symbols q^h $(h \in P^*)$ and e_i, f_i $(1 \leq i \leq n-1)$ subject to the well-known quantized Weyl-Serre relations. The algebra $\mathcal{U}_q$ is also generated by the so-called *L-operators* L^+_{ij}, $L^-_{ij} \in \mathcal{U}_q$ (cf. Jimbo [1986], Reshetikhin-Faddeev-Takhtajan [1990], Noumi [1993]). The matrices $L^\pm := \sum_{ij} e_{ij} \otimes L^\pm_{ij}$ with coefficients in $\mathcal{U}_q$ satisfy the following relations:

$$R^+ L_1^\epsilon L_2^\epsilon = L_2^\epsilon L_1^\epsilon R^+ \ (\epsilon = \pm), \quad R^+ L_1^+ L_2^- = L_2^- L_1^+ R^+, \tag{2.4}$$

where $L_1^\pm := L^\pm \otimes \mathrm{id}$ and $L_2^\pm := \mathrm{id} \otimes L^\pm$ are Kronecker matrix products. The Hopf $*$-algebra structure on $\mathcal{U}_q$ is determined by:

$$\Delta(L^\pm_{ij}) = \sum_k L^\pm_{ik} \otimes L^\pm_{kj}, \quad \varepsilon(L^\pm_{ij}) = \delta_{ij}, \quad (L^\pm_{ij})^* = S(L^\mp_{ji}), \tag{2.5}$$

for $1 \leq i,j \leq n$. We put $\tau = * \circ S: \mathcal{U}_q \to \mathcal{U}_q$. Then

$$\tau(L^\pm_{ij}) = L^\mp_{ji} \quad (1 \leq i,j \leq n). \tag{2.6}$$

Recall that a left $\mathcal{U}_q$-module W is called *P-weighted* if it has a vector space basis consisting of weight vectors with weights in P. The cone $P^+ \subset P$ of *dominant weights* consists by definition of all weights $\lambda = \sum_k \lambda_k \varepsilon_k \in P$ such that $\lambda_1 \geq \ldots \geq \lambda_n$. There is a 1-1 correspondence $\lambda \longleftrightarrow V(\lambda)$ between dominant weights and irreducible P-weighted finite-dimensional left $\mathcal{U}_q$-modules such that $\lambda \in P^+$ is the highest weight of $V(\lambda)$ (cf. Rosso [1988]). Recall that $\lambda \in P^+$ is called a highest weight of a left $\mathcal{U}_q$-module W if there exists a non-zero vector $w \in W$ such that $q^h \cdot w = q^{\langle h,\lambda\rangle} v$ and $e_i \cdot w = 0$ for all $1 \leq i \leq n-1$.

Given a left $\mathcal{U}_q$-module W, we define a right $\mathcal{U}_q$-module structure on the vector space conjugate to W by putting

$$w \cdot u := u^* \cdot w \quad (w \in W, u \in \mathcal{U}_q). \tag{2.7}$$

The corresponding right $\mathcal{U}_q$-module will be denoted by W°.

The Hopf $*$-algebra pairing $\langle\cdot\,,\,\cdot\rangle$ between $\mathcal{U}_q$ and $\mathcal{A}_q$ is determined by

$$\langle L_1^\pm, T_2 \rangle = R^\pm, \quad \langle L^\pm, \mathrm{det}_q \rangle = q^{\pm 1}\,\mathrm{id}\,. \tag{2.8}$$

This pairing is nondegenerate in the sense that the canonical mapping $\mathcal{A}_q \to \mathrm{Hom}_{\mathbb{C}}(\mathcal{U}_q, \mathbb{C})$ is injective. Under this duality, the finite-dimensional corepresentations of $\mathcal{A}_q$ are in 1-1 correspondence with the finite-dimensional P-weighted representations of $\mathcal{U}_q$.

Using the pairing $\langle\cdot\,,\,\cdot\rangle$ between $\mathcal{U}_q$ and $\mathcal{A}_q$, one can identify $\mathcal{A}_q$ with a subspace of the algebraic linear dual of $\mathcal{U}_q$. One defines a $\mathcal{U}_q$-bimodule structure on $\mathcal{A}_q$ by putting:

$$u \cdot a := (\mathrm{id} \otimes u) \circ \Delta(a), \quad a \cdot u := (u \otimes \mathrm{id}) \circ \Delta(a) \quad (u \in \mathcal{U}_q,\, a \in \mathcal{A}_q). \tag{2.9}$$

The action of the L-operators is given by:

$$L_1^\varepsilon \cdot T_2 = T_2 R^\varepsilon, \quad T_2 \cdot L_1^\varepsilon = R^\varepsilon T_2 \quad (\varepsilon = \pm). \tag{2.10}$$

The multiplication $\mathcal{A}_q \otimes \mathcal{A}_q \to \mathcal{A}_q$ and the unit mapping $\mathbb{C} \to \mathcal{A}_q$ are $\mathcal{U}_q$-bimodule homomorphisms. In other words, $\mathcal{A}_q$ is an *algebra with two-sided $\mathcal{U}_q$-symmetry*. One has the following decomposition of $\mathcal{A}_q$ into irreducible $\mathcal{U}_q$-bimodules:

$$\mathcal{A}_q = \bigoplus_{\lambda \in P^+} V(\lambda) \otimes V(\lambda)^\circ. \tag{2.11}$$

Here the subspace $W(\lambda) := V(\lambda) \otimes V(\lambda)^\circ \subset \mathcal{A}_q$ is spanned by the matrix coefficients of the (co-)representation $V(\lambda)$. The decomposition (2.11) can also be characterized as the simultaneous eigenspace decomposition of $\mathcal{A}_q$ with respect to the natural action of the center $\mathcal{Z}\mathcal{U}_q \subset \mathcal{U}_q$. Let $h\colon \mathcal{A}_q \to \mathbb{C}$ denote the Haar functional on $\mathcal{A}_q$ (cf. Woronowicz [1987], Dijkhuizen-Koornwinder [1994]). Then $\langle a, b\rangle := h(b^* a)$ defines a positive definite inner product on $\mathcal{A}_q$ with respect to which the subspaces $W(\lambda) \subset \mathcal{A}_q$ are mutually orthogonal (Schur orthogonality).

We now proceed to define a family of quantum Grassmannians. Let X be an $n \times n$ matrix with coefficients in any ring. Consider the following *reflection equation*:

$$R_{12} X_1 R_{12}^{-1} X_2 = X_2 R_{21}^{-1} X_1 R_{21}. \tag{2.12}$$

Here $X_1 := X \otimes \mathrm{id}$ and $X_2 := \mathrm{id} \otimes X$ are Kronecker matrix products, $R_{12} = R$, $R_{21} = PRP$ (P being the permutation operator). Define the algebra $\mathcal{C}_q = \mathcal{C}_q(n)$ of functions on the quantum space of *q-Hermitian matrices* as the algebra generated by the symbols x_{ij} $(1 \le i, j \le n)$ subject to the relations given by the reflection equation (2.12). There is a unique $*$-operation on $\mathcal{C}_q$ such that $x_{ij}^* = x_{ji}$. In shorthand notation, we write $X^* = X$.

Proposition 2.1 *There is a unique $*$-algebra homomorphism $\delta\colon \mathcal{C}_q \to \mathcal{A}_q \otimes \mathcal{C}_q$ such that*

$$\delta(x_{ij}) = \sum_{r,s} t_{ir} t_{js}^* \otimes x_{rs} \quad \text{or} \quad \delta(X) = TXT^*. \tag{2.13}$$

The mapping δ is a comodule mapping, hence defines an action of the quantum unitary group $U_q(n)$ on the quantum space of q-Hermitian matrices.

In the terminology of Dijkhuizen-Koornwinder [1994], a *classical point* in the quantum space of q-Hermitian matrices is a $*$-algebra homomorphism $\tilde{\varepsilon}\colon \mathcal{C}_q \to \mathbb{C}$. By the definition of the algebra $\mathcal{C}_q$, such mappings $\tilde{\varepsilon}$ are in 1-1 correspondence with $n \times n$ matrices J with complex coefficients satisfying the reflection equation (2.12) and such that $J^* = J$. The correspondence is given by $\tilde{\varepsilon}(X) = J$.

In the remainder of this paper, we fix an integer $1 \le l \le \left[\frac{n}{2}\right]$. Let σ be a real parameter. Define an $n \times n$ matrix J^σ by putting

$$J^\sigma := \sum_{1 \le k \le l} q^\sigma (q^{-\sigma} - q^\sigma) e_{kk} + \sum_{l < k < l'} e_{kk} - \sum_{k \le l \text{ or } k \ge l'} q^\sigma e_{kk'}, \tag{2.14}$$

where $k' := n+1-k$ $(1 \le k \le n)$.

Proposition 2.2 *The matrix J^σ satisfies the reflection equation (2.12) for any value of the parameter $\sigma \in \mathbb{R}$.*

Define a mapping $\Psi^\sigma : \mathcal{C}_q \to \mathcal{A}_q$ by putting $\Psi^\sigma := (\mathrm{id} \otimes \tilde{\varepsilon}^\sigma) \circ \delta$. Then Ψ^σ is a $*$-algebra homomorphism intertwining the natural (left) coactions of $\mathcal{A}_q$ on $\mathcal{C}_q$ and itself. We write $\mathcal{B}_q^\sigma := \Psi^\sigma(\mathcal{C}_q) \subset \mathcal{A}_q$ for the image of Ψ^σ. $\mathcal{B}_q^\sigma$ is a $*$-subalgebra and left coideal in $\mathcal{A}_q$. It can be viewed as the algebra of functions on the $U_q(n)$-orbit of the classical point defined by J^σ in the quantum space of q-Hermitian matrices. The action of $U_q(n)$ on this quantum orbit is transitive by construction (cf. Dijkhuizen-Koornwinder [1994]). If $q = 1$ then the algebra B_q^σ $(\sigma \in \mathbb{R})$ is $U(n)$-isomorphic with the algebra of functions on the complex Grassmannian of rank l endowed with its natural $U(n)$-action.

The next obvious step is to look for a stabilizer "subgroup". Define an $n \times n$ matrix M^σ with coefficients in $\mathcal{U}_q$ by putting

$$M^\sigma := L^+ J^\sigma - J^\sigma L^- \in \mathrm{End}(V) \otimes \mathcal{U}_q. \tag{2.15}$$

The subspace $\mathfrak{k}^\sigma \subset \mathcal{U}_q$ is by definition spanned by the coefficients of the matrix M^σ.

Lemma 2.3 *The subspace $\mathfrak{k}^\sigma \subset \mathcal{U}_q$ is a τ-invariant two-sided coideal.*

An element $a \in \mathcal{A}_q$ is called left *$\mathfrak{k}^\sigma$-invariant* if $\mathfrak{k}^\sigma \cdot a = 0$.

Proposition 2.4 *The subalgebra $\mathcal{B}_q^\sigma \subset \mathcal{A}_q$ consists precisely of all left $\mathfrak{k}^\sigma$-invariant elements in $\mathcal{A}_q$.*

In other words, the two-sided coideal $\mathfrak{k}^\sigma$ can be regarded as an infinitesimal stabilizer of the quantum orbit defined by the matrix J^σ. It can be shown that, in the limit $q \to 1$, the coideal $\mathfrak{k}^\sigma \subset \mathcal{U}_q$ tends to a Lie subalgebra of $\mathfrak{gl}(n, \mathbb{C})$ which is conjugate to the Lie algebra $\mathfrak{k}$ of the subgroup $U(l) \times U(n-l) \subset U(n)$.

Let W be a finite-dimensional left $\mathcal{U}_q$-module. A vector $w \in W$ is called $\mathfrak{k}^\sigma$-fixed if $\mathfrak{k}^\sigma \cdot w = 0$. The subspace of $\mathfrak{k}^\sigma$-fixed vectors in W is written $W_{\mathfrak{k}^\sigma}$. We have the following key lemma:

Lemma 2.5 *Let V denote the vector representation of $\mathcal{U}_q$, V^* its contragredient with dual basis (v_i^*). The element $w_{J^\sigma} := \sum_{i,j} J_{ij}^\sigma v_i \otimes v_j^* \in V \otimes V^*$ is a $\mathfrak{k}^\sigma$-fixed vector.*

Given any complex matrix J, one can define a matrix M_J with coefficients in $\mathcal{U}_q$ as in (2.15), and a corresponding two-sided coideal $\mathfrak{k}_J \subset \mathcal{U}_q$. In order that the element $w_J := \sum_{i,j} J_{ij} v_i \otimes v_j^* \in V \otimes V^*$ (V vector representation) be a $\mathfrak{k}_J$-fixed vector it is necessary and sufficient for the matrix J to satisfy the reflection equation (2.12).

Let us identify elements $\lambda \in P^+$ with sequences $(\lambda_1, \ldots, \lambda_n)$ of integers such that $\lambda_1 \ge \ldots \ge \lambda_n$.

Theorem 2.6 *For any $\lambda \in P^+$, the subspace $V(\lambda)_{\mathfrak{k}^\sigma}$ is at most one-dimensional. There are non-zero $\mathfrak{k}^\sigma$-fixed vectors in $V(\lambda)$ if and only if $\lambda \in P^+$ is of the form*

$$\lambda = (\mu_1, \ldots, \mu_l, 0, \ldots, 0, -\mu_l, \ldots, -\mu_1), \quad \mu_1 \geq \ldots \mu_l \geq 0. \tag{2.16}$$

A dominant weight $\lambda \in P^+$ and the corresponding representation $V(\lambda)$ are called *spherical* if λ is of the form (2.16). The set of spherical dominant weights will be denoted by $P_{\mathfrak{k}}^+$.

Corollary 2.7 *One has the following irreducible decomposition of the right $\mathcal{U}_q$-module (or left $\mathcal{A}_q$-comodule) $\mathcal{B}_q^\sigma$:*

$$\mathcal{B}_q^\sigma = \bigoplus_{\lambda \in P_{\mathfrak{k}}^+} V(\lambda)^\circ. \tag{2.17}$$

3 Zonal spherical functions

Let σ, τ be real parameters. Define $\mathcal{H}^{(\sigma,\tau)} \subset \mathcal{A}_q$ as the subspace of left $\mathfrak{k}^\sigma$-invariant and right $\mathfrak{k}^\tau$-invariant elements. It is a $*$-subalgebra of $\mathcal{A}_q$.

Proposition 3.1 *If we put $\mathcal{H}^{(\sigma,\tau)}(\lambda) := \mathcal{H}^{(\sigma,\tau)} \cap W(\lambda)$ ($\lambda \in P^+$) then we have the decomposition*

$$\mathcal{H}^{(\sigma,\tau)} = \bigoplus_{\lambda \in P_{\mathfrak{k}}^+} \mathcal{H}^{(\sigma,\tau)}(\lambda). \tag{3.1}$$

Each of the subspaces $\mathcal{H}^{(\sigma,\tau)}(\lambda)$ ($\lambda \in P_{\mathfrak{k}}^+$) is one-dimensional.

Any non-zero element of $\mathcal{H}^{(\sigma,\tau)}(\lambda)$ ($\lambda \in P_{\mathfrak{k}}^+$) is called a *zonal spherical function* corresponding to the spherical weight λ.

Let $\mathcal{A}(\mathbb{T}) := \mathbb{C}[z_1^{\pm 1}, \ldots, z_n^{\pm 1}]$ be the algebra of Laurent polynomials in the variables z_i ($1 \leq i \leq n$). There is a unique Hopf $*$-algebra structure on $\mathcal{A}(\mathbb{T})$ such that

$$\Delta(z_i) = z_i \otimes z_i, \quad \varepsilon(z_i) = 1, \quad z_i^* = z_i^{-1} \quad (1 \leq i \leq n). \tag{3.2}$$

There is a unique surjective Hopf $*$-algebra morphism

$$|_{\mathbb{T}} \colon \mathcal{A}_q \longrightarrow \mathcal{A}(\mathbb{T})$$

mapping $t_{ij} \in \mathcal{A}_q$ onto $\delta_{ij} z_i \in \mathcal{A}(\mathbb{T})$. The torus $\mathbb{T}$ can be viewed as a subgroup of the quantum unitary group $U_q(n)$. The mapping $|_{\mathbb{T}}$ then is the corresponding restriction of functions. Let us introduce new variables x_i ($1 \leq i \leq l$) in the algebra $\mathcal{A}(\mathbb{T})$ by putting

$$x_1 := z_1 z_n^{-1}, \quad x_2 := z_2 z_{n-1}^{-1}, \quad \ldots, \quad x_l := z_l z_{n+1-l}^{-1}. \tag{3.3}$$

Note that the Weyl group W of the root system BC_l acts naturally on the subalgebra of $\mathcal{A}(\mathbb{T})$ generated by the $x_i^{\pm 1}$. The subalgebra of W-invariant elements is written $\mathbb{C}[x^{\pm 1}]^W = \mathbb{C}[x_1^{\pm 1}, \ldots, x_l^{\pm 1}]^W$.

Theorem 3.2 *The mapping* $_{|\mathbb{T}}: \mathcal{A}_q \longrightarrow \mathcal{A}(\mathbb{T})$ *maps* $\mathcal{H}^{(\sigma,\tau)}$ *injectively onto its image* $\mathcal{H}^{(\sigma,\tau)}_{|\mathbb{T}} \subset \mathcal{A}(\mathbb{T})$. *One has*

$$\mathcal{H}^{(\sigma,\tau)}_{|\mathbb{T}} = \mathbb{C}[x_1^{\pm 1}, \ldots, x_l^{\pm 1}]^W. \tag{3.4}$$

In particular, the algebra $\mathcal{H}^{(\sigma,\tau)}$ *is commutative.*

Consider the Casimir operator (cf. Reshetikhin-Faddeev-Takhtajan [1990])

$$C := \sum_{ij} q^{2(n-i)} L^+_{ij} S(L^-_{ji}) \in \mathcal{U}_q. \tag{3.5}$$

C is central in $\mathcal{U}_q$ and acts as a scalar on each subspace $W(\lambda)$ ($\lambda \in P^+$). The corresponding eigenvalue is

$$\chi_\lambda(C) = \sum_{k=1}^{n} q^{2(\lambda_k + n - k)}. \tag{3.6}$$

Since C is central in $\mathcal{U}_q$, the left action of C on $\mathcal{A}_q$ preserves the subalgebra $\mathcal{H}^{(\sigma,\tau)}$, hence acts as a scalar on each of the subspaces $\mathcal{H}^{(\sigma,\tau)}(\lambda)$ ($\lambda \in P^+_\Sigma$). Namely, each zonal spherical function $\varphi \in \mathcal{H}^{(\sigma,\tau)}(\lambda)$ satisfies the equation $C \cdot \varphi = \chi_\lambda(C)\, \varphi$.

There is a uniquely determined linear operator $D: \mathcal{H}^{(\sigma,\tau)}_{|\mathbb{T}} \to \mathcal{H}^{(\sigma,\tau)}_{|\mathbb{T}}$ (called *the radial part* of the Casimir operator C) such that on $\mathcal{H}^{(\sigma,\tau)}$ we have

$$_{|\mathbb{T}} \circ C = D \circ {}_{|\mathbb{T}}, \tag{3.7}$$

where the symbol C denotes the left action of the element $C \in \mathcal{U}_q$ on $\mathcal{H}^{(\sigma,\tau)} \subset \mathcal{A}_q$. It is possible to compute an explicit expression for the radial part D. Recall that with any spherical weight $\lambda = (\mu_1, \ldots, \mu_l, 0, \ldots, 0, -\mu_l, \ldots, -\mu_1)$ in the weight lattice P one can naturally associate a dominant weight $\mu = (\mu_1, \ldots, \mu_l)$ in the weight lattice P_Σ.

Theorem 3.3 *The radial part*

$$D - \chi_\lambda(C)\,\mathrm{id} : \mathbb{C}[x^{\pm 1}]^W \to \mathbb{C}[x^{\pm 1}]^W \tag{3.8}$$

is a constant multiple of Koornwinder's q-difference operator $D_{\varepsilon_1} - c_{\mu\mu}\,\mathrm{id}$, *defined as in (1.1) with base* q^2 *and parameters*

$$a = -q^{\sigma+\tau+1},\ b = -q^{-\sigma-\tau+1},\ c = q^{\sigma-\tau+1},\ d = q^{-\sigma+\tau+2(n-2l)+1},\ t = q^2. \tag{3.9}$$

Note that the parameters (3.9) satisfy condition (1.3).

For any $\lambda \in P^+_{\mathfrak{k}}$, let us fix a non-zero element $\varphi(\lambda) \in \mathcal{H}^{(\sigma,\tau)}(\lambda)$ (zonal spherical function). Using Theorem (3.3) and the definition of the Askey-Wilson polynomials P_μ we can prove:

Theorem 3.4 *The restriction $\varphi(\lambda)_{|\mathbb{T}}$ of the zonal spherical function $\varphi(\lambda)$ ($\lambda \in P_{\mathfrak{k}}^{+}$) to the toral subgroup $\mathbb{T} \subset U_q(n)$ is equal to the Askey-Wilson polynomial*

$$P_\mu(x; -q^{\sigma+\tau+1}, -q^{-\sigma-\tau+1}, q^{\sigma-\tau+1}, q^{-\sigma+\tau+2(n-2l)+1}; q^2, q^2) \qquad (3.10)$$

up to a scalar multiple.

Remark 3.5 As stated in Section 2, the subspaces $W(\lambda)$ are mutually orthogonal with respect to the inner product on $\mathcal{A}_q$ defined in terms of the Haar functional. This means that the restricted zonal spherical functions $\varphi(\lambda)_{|\mathbb{T}}$ are mutually orthogonal with respect to the inner product on the algebra $\mathcal{H}_{|\mathbb{T}}^{(\sigma,\tau)}$ induced by the restriction of the Haar functional.

Remark 3.6 In the limit $\sigma \to \pm\infty$, the coideal $\mathfrak{k}^\sigma$ tends to a two-sided coideal in $\mathcal{U}_q$ which can be viewed as a direct q-analogue of the Lie subalgebra $\mathfrak{gl}(l)\oplus\mathfrak{gl}(n-l) \subset \mathfrak{gl}(n)$. The corresponding quantum Grassmannian has the usual Plancherel decomposition and can be regarded as the quotient of the quantum unitary group $U_q(n)$ by the quantum subgroup $U_q(l) \times U_q(n-l)$. The limit transition of our zonal spherical functions to this case is expected to be consistent with the limit transition (cf. Stokman-Koornwinder [1995]) of general multivariable Askey-Wilson polynomials to multivariable big q-Jacobi or little q-Jacobi polynomials (cf. Stokman [1995]).

References

Askey, R. and Wilson, J. [1985] *Some basic hypergeometric polynomials that generalize Jacobi polynomials,* Mem. Amer. Math. Soc. **54**, no. 319.

Dijkhuizen, M. S. [1994] *Some remarks on the construction of quantum symmetric spaces,* Proceedings of a conference on Representation Theory of Lie groups, Lie algebras, and their Quantum Analogues, Twente (NL), Dec. 1994 Acta Appl. Math. (to appear).

Dijkhuizen, M. S. and Koornwinder, T. H. [1994] *Quantum homogeneous spaces, duality and quantum 2-spheres,* Geom. Dedicata **52**, 291-315.

Dijkhuizen, M. S. and Koornwinder, T. H. [1994] *CQG algebras: a direct algebraic approach to compact quantum groups,* Lett. Math. Phys. **32**, 315-330.

Dijkhuizen, M. S. and Noumi, M. [1995] *A family of quantum projective spaces and related q-hypergeometric orthogonal polynomials,* preprint.

Drinfel'd, V. G. [1986] *Quantum groups,* in: Proceedings ICM Berkeley (1986), ed. A.M. Gleason, Amer. Math. Soc., Providence, RI, 798-820.

Jimbo, M. [1986] *A q-analogue of $U(\mathfrak{gl}(n))$ Hecke algebra and the Yang-Baxter equation,* Lett. Math. Phys. **11**, 247-252.

Koornwinder, T. H. [1993] *Askey-Wilson polynomials as zonal spherical functions on the $SU(2)$ quantum group,* SIAM J. Math. Anal. **24**, no. 3, 795-813.

Koornwinder, T. H. [1992] *Askey-Wilson polynomials for root systems of type BC,* in: "Hypergeometric functions on domains of positivity, Jack polynomials, and applications", (Richards, D. S. P., ed.) Contemp. Math. 138, Amer. Math. Soc., Providence, RI, pp. 189-204.

Korogodsky, L. I. and Vaksman L. L. [1992] *Quantum G-spaces and Heisenberg algebra,* in: "Quantum Groups", (Kulish, P. P., ed.) Lecture Notes in Math. 1510, Springer-Verlag, pp. 56-66.

Kulish, P. P. [1991] *Quantum groups and quantum algebras as symmetries of dynamical systems*, preprint YITP/K-959.

Macdonald, I. G. [1988] *Orthogonal polynomials associated with root systems*, preprint.

Noumi, M. [1993] *Macdonald's symmetric polynomials as zonal spherical functions on some quantum homogeneous spaces,* Adv. Math. (to appear).

Noumi, M. and Mimachi, K. [1990] *Quantum 2-spheres and big q-Jacobi polynomials,* Comm. Math. Phys. **128**, 521-531.

Noumi, M. and Sugitani, T. [1995] *Quantum symmetric spaces and related q-orthogonal polynomials*, in: "Group Theoretical Methods in Physics", Proceedings XX ICGTMP, Toyonaka (Japan), 1994, (A. Arima et al., ed.), World Scientific, Singapore, pp. 28-40.

Noumi, M., Yamada, H. and Mimachi, K. [1993] *Finite dimensional representations of the quantum group $GL_q(n;\mathbb{C})$ and the zonal spherical functions on $U_q(n-1)\backslash U_q(n)$,* Japanese J. Math. **19**, 31-80.

Podleś, P. [1987] *Quantum spheres,* Lett. Math. Phys. **14**, 193-202.

Reshetikhin, N., Faddeev, L. D. and Takhtajan, L. A. [1990] *Quantization of Lie groups and Lie algebras,* Leningrad Math. J. **1**, 193-225.

Rosso, M. [1988] *Finite-dimensional representations of the quantum analog of a complex simple Lie algebra,* Comm. Math. Phys. **117**, 581-593.

Stokman, J. V. [1995] *Multivariable big and little q-Jacobi polynomials,* SIAM J. Math. Appl. (to appear).

Stokman, J. V. *Multivariable BC type Askey-Wilson polynomials with partly discrete orthogonality measure*, preprint University of Amsterdam.

Stokman, J. V. and Koornwinder, T. H. [1995] *Limit transitions for BC type multivariable orthogonal polynomials,* Canadian J. Math. (to appear).

Woronowicz, S. L. [1987] *Compact matrix pseudogroups,* Comm. Math. Phys. **111**, 613-665.

Fields Institute Communications
Volume **14**, 1997

A Mathematica q-Analogue of Zeilberger's Algorithm Based on an Algebraically Motivated Approach to q-Hypergeometric Telescoping

Peter Paule
Research Institute for Symbolic Computation
J. Kepler University Linz
A–4040 Linz, Austria

Axel Riese
Research Institute for Symbolic Computation
J. Kepler University Linz
A–4040 Linz, Austria

Abstract. *Mathematica* implementations, available by email request, of q-analogues of Gosper's and Zeilberger's algorithm are described. Nontrivial examples are given in order to illustrate the usage of these packages. The algorithms are based on a new approach to q-hypergeometric telescoping in which a new algebraic concept, q-greatest factorial factorization (qGFF), plays a fundamental role.

1 Introduction

Based on Gosper's [1978] algorithm for indefinite hypergeometric summation, Zeilberger's algorithm for proving definite hypergeometric summation and transformation formulae constitutes a recent breakthrough in symbolic computation. An excellent and detailed account of this theory can be found in the book of Petkovšek, Wilf and Zeilberger [1996].

Zeilberger also was the first who observed that these algorithms can be carried over to the q-hypergeometric case. Despite the fact that there already exist implementations in Maple, written by Zeilberger (cf. Petkovšek, Wilf and Zeilberger [1996]) and Koornwinder [1993], we felt the need to come up with another implementation `qZeil`, in Mathematica, which is able to deal with various important features not covered by the ones mentioned. A similar project was carried out for the ordinary hypergeometric case; see the Mathematica implementation by Paule and Schorn [1995].

1991 *Mathematics Subject Classification.* Primary: 33D15, 33D20, 68Q40; Secondary: 05A30, 68R05.

The work of the first author was supported in part by grant P7220 of the Austrian FWF.

Another aspect concerns the theoretical foundation of a q-analogue of Gosper's algorithm. Besides Karr's [1981] approach which covers indefinite q-hypergeometric summation in the general frame of his theory of difference field extensions, up to now it had been kind of a surprise that Gosper's algorithm can be carried over to the q-case almost word by word; cf. Koornwinder [1993]. It turns out that the new algebraic concept of "greatest factorial factorization", introduced by Paule [1995], provides an algebraic explanation not only of Gosper's algorithm, but also of its analogue for q-hypergeometric telescoping. An extension to the bibasic hypergeometric case has been done by Riese [1996].

Summarizing, the general goals of the paper are: (i) to introduce our Mathematica implementation `qZeil` of a q-analogue of Zeilberger's (and Gosper's) algorithm to potential users; (ii) to explain the usage of `qZeil` by illustrating examples that range from simple to nontrivial; (iii) to present an algebraically motivated approach to q-hypergeometric telescoping the q-analogue Zeilberger algorithm is based on. Since the sections to some extent can be read independently from each other, we shortly comment on the table of contents.

In Section 2, "q-greatest factorial factorization" (qGFF) of polynomials is introduced. It is a q-analogue of the concept introduced by Paule [1995] and provides the crucial tool for analyzing q-hypergeometric telescoping.

In Section 3, qGFF is used to explain the mechanism of the algorithm for solving the q-hypergeometric telescoping problem, i.e., finding a q-hypergeometric sequence $\langle g_k \rangle$ such that $f_k = g_{k+1} - g_k$ for a given q-hypergeometric sequence $\langle f_k \rangle$. In our Mathematica implementation this algorithm is called "qTelescope".

In Section 4, a detailed description of the usage of the Mathematica package `qZeil` is given. All notions needed for the understanding of the various program features are described in short, for instance: qWZ-certification, qWZ-pairs, qWZ-dualization and qWZ-companion identities. The section concludes with a brief comparison with Koornwinder's implementation.

Section 5 deals with nontrivial applications. As in Paule and Schorn [1995], a special emphasis is put on illustrating how to apply the package to concrete problems. Besides the examples one finds in Section 4, in this section we present: an example where a computer gave the first proof of a (human) conjecture, a detailed discussion of "creative symmetrizing" for reducing the order of the q-Zeilberger output recurrence, the observation that L.J. Rogers' classical finite version of Euler's pentagonal number theorem is nothing but the dual identity of a limiting case of q-Chu-Vandermonde, and a computer proof of an infinite q-series identity recently discovered by R. J. McIntosh.

2 q-Greatest Factorial Factorization

In this section, "q-greatest factorial factorization" (qGFF) of polynomials is introduced. It is a q-analogue of a new canonical form representation (GFF) introduced by Paule [1995], defined with respect to the q-shift operator ϵ instead of the shift $(Ep)(x) = p(x+1)$ as for GFF.

2.1 Basic Definitions By $\mathbf{N}$ we understand the set of all nonnegative integers. We assume $\mathbf{K}$ to be a *field* of *characteristic zero*, for instance, the complex numbers. Its transcendental extension by the indeterminate q is denoted by $\mathbf{F}$, i.e., $\mathbf{F} = \mathbf{K}(q)$. For implementations and for algorithm specification, $\mathbf{K}$ is assumed to be effectively computable.

As usual we shall assume the result of any gcd (greatest common divisor) computation in $\mathbf{K}[x]$ or $\mathbf{F}[x]$ as being normalized to a *monic* polynomial p, i.e., the leading coefficient $\mathrm{lcf}(p) = 1$.

By ϵ we denote the q-shift operator on $\mathbf{F}[x]$, i.e., $(\epsilon p)(x) = p(qx)$ for any $p \in \mathbf{F}[x]$. The extension of this shift operator to the rational function field $\mathbf{F}(x)$, the quotient field of $\mathbf{F}[x]$, will be also denoted by ϵ.

A polynomial $p \in \mathbf{F}[x]$ is said to be *q-monic* if $p(0) = 1$. Note that this property is invariant with respect to the q-shift operator, i.e., $(\epsilon p)(0) = p(0) = 1$, whereas usual monicness is invariant with respect to the shift operator E, i.e., $\mathrm{lcf}(Ep) = \mathrm{lcf}(p) = 1$.

The polynomial degree (in x) of any $p \in \mathbf{F}[x]$, $p \neq 0$, is denoted by $\deg(p)$. We define $\deg(0) := -1$.

Evidently, any polynomial $p \in \mathbf{F}[x]$ has a unique factorization, the *q-monic decomposition*, in the form

$$p = z \cdot x^{\alpha} \cdot \widehat{p}$$

where $z \in \mathbf{F}$, $\alpha \in \mathbf{N}$, and $\widehat{p} \in \mathbf{F}[x]$ q-monic. The q-monic decomposition can be computed easily; note also that α simply is the multiplicity of the root 0.

Besides the usual normalization with respect to monicness, it will be convenient to introduce normalization also with respect to q-monicness. In these instances we write "gcd_q" instead of "gcd", indicating that the gcd_q of two q-monic polynomials is understood to be q-monic.

More generally, if $p_1 = z_1 \cdot x^{\alpha_1} \cdot \widehat{p_1}$ and $p_2 = z_2 \cdot x^{\alpha_2} \cdot \widehat{p_2}$ are the q-monic decompositions of $p_1, p_2 \in \mathbf{F}[x]$, we define

$$\mathrm{gcd}_q(p_1, p_2) := \gcd(x^{\alpha_1}, x^{\alpha_2}) \cdot \mathrm{gcd}_q(\widehat{p_1}, \widehat{p_2}),$$

where the q-monic part is treated as described above.

Example 1 *Let* $p_1 = q\,x^2\,(1-x)(1-qx)$ *and* $p_2 = \epsilon p_1 = q^3\,x^2\,(1-qx)(1-q^2x)$, *then*

$$\mathrm{gcd}_q(p_1, p_2) = x^2(1-qx),$$

whereas $\gcd(p_1, p_2) = x^2(x - 1/q)$.

□

Note that for any $p_1, p_2 \in \mathbf{F}[x]$:

$$\gcd(p_1, p_2) = 1 \iff \mathrm{gcd}_q(p_1, p_2) = 1.$$

Definition 1 *For any q-monic polynomial $p \in \mathbf{F}[x]$ and $k \in \mathbf{N}$ the* k-th falling q-factorial $[p]_{q}^{\underline{k}}$ *of p is defined as*

$$[p]_{q}^{\underline{k}} := \prod_{i=0}^{k-1} \epsilon^{-i} p.$$

Note that by the null convention $\prod_{i \in \emptyset} p_i = 1$ we have $[p]_{q}^{\underline{0}} = 1$. In view of the commonly used q-shifted factorial, i.e., $(x;q)_k = [1 - q^{k-1}x]_{q}^{\underline{k}}$ it would be more natural to introduce the product in terms of rising powers of ϵ. The reason why we still stick to the "reverse" definition is that Paule's [1995] theory of greatest factorial factorization was presented exactly this way with E instead of ϵ, and we want to keep the parallels to the q-case as close as possible.

Polynomials arising in q-hypergeometric summation, in general have several different representations in terms of q-factorials.

Example 2 *Let* $p = (1-q^3x)(1-q^2x)(1-qx)^2(1-x)$, *then* $p = [1-qx]_q^{\underline{1}}\,[(1-q^3x)(1-qx)]_q^{\underline{2}} = [(1-qx)(1-x)]_q^{\underline{1}}\,[1-q^3x]_q^{\underline{3}} = [(1-q^3x)(1-qx)]_q^{\underline{1}}\,[1-q^2x]_q^{\underline{3}} = [1-qx]_q^{\underline{2}}\,[1-q^3x]_q^{\underline{3}} = [1-qx]_q^{\underline{1}}\,[1-q^3x]_q^{\underline{4}}$, *etc.*

□

From all these possibilities the last one, which takes care of maximal chains is of particular importance. Informally, it can be obtained as follows: One selects irreducible factors of p in such a way that their product, say

$$q_1(x)\; q_1(q^{-1}x)\cdots q_1(q^{-k+1}x),$$

forms a falling q-factorial $[q_1]_q^{\underline{k}}$ of maximal length k. For the remaining irreducible factors of p this procedure is applied again in order to find all k-th falling q-factorial divisors $[q_1]_q^{\underline{k}}$, $[q_2]_q^{\underline{k}}$, etc., of that type. Then $[q_1 \cdot q_2 \cdots]_q^{\underline{k}}$ forms the q-factorial factor of p of maximal length k. Iterating this procedure one gets a factorization of p in terms of "greatest" q-factorial factors.

Definition 2 *We say that* $\langle p_1, \ldots, p_k\rangle$, $p_i \in \mathbf{F}[x]$, *is a* qGFF-form *of a* q*-monic polynomial* $p \in \mathbf{F}[x]$ *if the following conditions hold:*
(qGFF1) $p = [p_1]_q^{\underline{1}} \cdots [p_k]_q^{\underline{k}}$,
(qGFF2) each p_i *is* q*-monic, and* $k > 0$ *implies* $\deg(p_k) > 0$,
(qGFF3) $i \le j \Rightarrow \gcd_q([p_i]_q^{\underline{i}}, \epsilon p_j) = 1 = \gcd_q([p_i]_q^{\underline{i}}, \epsilon^{-j} p_j)$.

Note that, due to the null convention, $\langle\rangle$ is the qGFF-form of $1 \in \mathbf{F}[x]$. Condition (qGFF3) intuitively can be understood as prohibiting "overlaps" of q-factorials that violate length maximality.

The following theorem explicitly states the fact that the qGFF-form provides a canonical form. For instance, $\langle 1-qx, 1, 1, 1-q^3x\rangle$ is *the* qGFF-form of the polynomial p from the example above.

Theorem 1 *If* $\langle p_1, \ldots, p_k\rangle$ *and* $\langle q_1, \ldots, q_l\rangle$ *are* qGFF*-forms of a* q*-monic* $p \in \mathbf{F}[x]$ *then* $k = l$ *and* $p_i = q_i$ *for all* $i \in \{1, \ldots, k\}$.

Proof The proof is entirely analogous to that of Theorem 2.1 in Paule [1995], and is thus left to the reader.

□

From algorithmic point of view it is important to note that the qGFF form can be computed in an iterative manner essentially involving only gcd computations; see Section 3.2.

If $\langle p_1, \ldots, p_k\rangle$ is the qGFF-form of a q-monic $p \in \mathbf{F}[x]$ we also denote this fact for short by $q\mathrm{GFF}(p) = \langle p_1, \ldots, p_k\rangle$.

2.2 The Fundamental qGFF Lemma In hypergeometric summation (i.e., $q = 1$) the gcd of a polynomial p and its shift Ep plays a fundamental role. The same is true for q-hypergeometric summation with respect to the q-shift operator ϵ instead of E. The qGFF-concept, as GFF in case $q = 1$, takes special care of that observation. The mathematical and algorithmic essence lies in the following lemma.

Lemma 1 ("Fundamental qGFF Lemma") *Given a q-monic polynomial $p \in \mathbf{F}[x]$ with qGFF-form $\langle p_1, \ldots, p_k \rangle$. Then*

$$\gcd_q(p, \epsilon p) = [p_1]_q^{\underline{0}} \cdots [p_k]_q^{\underline{k-1}}.$$

Proof Proceeding by induction on k the case $k = 0$ is trivial. For $k > 0$,

$$\begin{aligned}\gcd_q&(p, \epsilon p)\\ &= [p_k]_q^{\underline{k-1}} \cdot \gcd_q([p_1]_q^{\underline{1}} \cdots [p_{k-1}]_q^{\underline{k-1}} \cdot \epsilon^{-k+1} p_k, \epsilon([p_1]_q^{\underline{1}} \cdots [p_{k-1}]_q^{\underline{k-1}} \cdot p_k))\\ &= [p_k]_q^{\underline{k-1}} \cdot \gcd_q([p_1]_q^{\underline{1}} \cdots [p_{k-1}]_q^{\underline{k-1}}, \epsilon([p_1]_q^{\underline{1}} \cdots [p_{k-1}]_q^{\underline{k-1}})).\end{aligned}$$

The first equality is obvious, the second is a consequence of (qGFF3) because for $i < k$ we have

$$\gcd_q([p_i]_q^{\underline{i}}, \epsilon p_k) = \gcd_q(\epsilon^{-k+1} p_k, \epsilon [p_i]_q^{\underline{i}}) = \epsilon(\gcd_q(\epsilon^{-k} p_k, [p_i]_q^{\underline{i}})) = 1$$

together with $\gcd_q(\epsilon^{-k+1} p_k, \epsilon p_k) | \gcd_q([p_k]_q^{\underline{k}}, \epsilon p_k) = 1$. The rest follows from applying the induction hypothesis.

□

In other words, from the qGFF-form of p, i.e., $q\text{GFF}(p) = \langle p_1, \ldots, p_k \rangle$ one directly can extract the qGFF-form of its "$\gcd_q$-shift", i.e., $q\text{GFF}(\gcd_q(p, \epsilon p)) = \langle p_2, \ldots, p_k \rangle$.

Example 3 *From $q\text{GFF}(p) = \langle 1-qx, 1, 1, 1-q^3x \rangle$ one immediately gets by Lemma 1 that $\gcd_q(p, \epsilon p) = [1 - q^3x]_q^{\underline{3}}$ and $q\text{GFF}(\gcd_q(p, \epsilon p)) = \langle 1, 1, 1 - q^3x \rangle$.*

□

Note that dividing p with $q\text{GFF}(p) = \langle p_1, \ldots, p_k \rangle$ by $\epsilon^{-1} \gcd_q(p, \epsilon p)$ or $\gcd_q(p, \epsilon p)$ results in separating the product of the first, respectively last, falling q-factorial entries:

$$\frac{p}{\epsilon^{-1} \gcd_q(p, \epsilon p)} = p_1 p_2 \cdots p_k \quad \text{and} \quad \frac{p}{\gcd_q(p, \epsilon p)} = p_1 (\epsilon^{-1} p_2) \cdots (\epsilon^{-k+1} p_k).$$

Remark 1 *For $q = 1$ the Fundamental GFF Lemma is formulated with respect to the shift operator E. In this version (see Paule [1995]) it forms a perfect analogue to the fundamental property*

$$\gcd(p, Dp) = p_1^0 \, p_2^1 \cdots p_k^{k-1}$$

of the derivation operator D used in square-free factorization, i.e., for computing $p = p_1^1 \, p_2^2 \cdots p_k^k$, the square-free factorization of $p \in \mathbf{K}[x]$.

□

3 q-Hypergeometric Telescoping

In this section we consider q-hypergeometric telescoping. In the general frame of his theory of difference field extensions, Karr [1981] was the first who provided an algorithmic answer to the problem. However, the theoretical setting is mighty and complex enough to deal also with other, non-hypergeometric types of indefinite summation problems, and, up to now, no working implementation is available.

Zeilberger (e.g., Petkovšek, Wilf and Zeilberger [1996]) has been the first who observed that the problem can be treated along the same lines as Gosper's algorithm. This approach was followed by Koornwinder [1993], who remarked "Surprisingly, Gosper's and Zeilberger's algorithms can be carried over to the q-case almost unchanged". Using GFF, defined with respect to the shift operator E, we presented a new and algebraically motivated approach to Gosper's algorithm; see Paule [1995]. Defining this type of polynomial factorization with respect to the q-shift operator ϵ, one is led to qGFF which enables to apply essentially the same argumentation, resulting in Algorithm qTelescope, now solving the problem of q-hypergeometric telescoping. In other words, the GFF concept, together with the translation into its q-version qGFF, somehow spoils Koornwinder's surprise, at least with respect to Gosper's algorithm.

3.1 The Algorithm qTelescope In this section we present the algorithm qTelescope which can be viewed as a q-analogue of Gosper's algorithm. After specifying the problem, we demonstrate that all what is needed consists only in a slight adoption of the arguments presented in Section 5 of Paule [1995].

A sequence $\langle f_k \rangle_{k\geq 0}$ is called *q-hypergeometric* over $\mathbf{F}$ if there exists a rational function $\rho \in \mathbf{F}(x)$ such that $f_{k+1}/f_k = \rho(q^k)$ for all $k \in \mathbf{N}$. Given q-hypergeometric $\langle f_k \rangle_{k\geq 0}$, the problem of *$q$-hypergeometric telescoping* is to find a q-hypergeometric solution $\langle g_k \rangle_{k\geq 0}$ of

$$g_{k+1} - g_k = f_k. \tag{1}$$

Assume that a q-hypergeometric solution $\langle g_k \rangle_{k\geq 0}$ of (1) exists. Let $\sigma \in \mathbf{F}(x)$ be such that $g_{k+1}/g_k = \sigma(q^k)$ for all $k \in \mathbf{N}$, then evidently

$$g_k = \tau(q^k) \cdot f_k, \tag{2}$$

where $\tau(x) = 1/(\sigma(x) - 1) \in \mathbf{F}(x)$.

For any integer α we define $\alpha_+ := \max(\alpha, 0)$, and $\alpha_- := \max(-\alpha, 0)$.

By relation (2), Equation (1) is equivalent to

$$z \cdot x^{\alpha_+} \cdot a \cdot \epsilon\tau - x^{\alpha_-} \cdot b \cdot \tau = x^{\alpha_-} \cdot b, \tag{3}$$

where $\rho = z \cdot x^{\alpha} \cdot a/b$ with $z \in \mathbf{F}$, α integer, and $a, b \in \mathbf{F}[x]$ relatively prime and q-monic.

Vice versa, any rational solution $\tau \in \mathbf{F}(x)$ of (3) gives rise to a q-hypergeometric solution $g_k := \tau(q^k) \cdot f_k$ of (1). This means, q-hypergeometric telescoping is equivalent to finding a *rational* solution τ of (3).

Any $\tau \in \mathbf{F}(x)$ can be represented as the quotient of relatively prime polynomials in the form $\tau = u/v$ where $u, v \in \mathbf{F}[x]$ with $v = x^{\beta} \cdot \widehat{v}$ the q-monic decomposition of v. In case such a solution τ of (3) exists, assume we know v or a multiple $V \in \mathbf{F}[x]$ of v. Then by clearing denominators in

$$z \cdot x^{\alpha_+} \cdot a \cdot \frac{\epsilon U}{\epsilon V} - x^{\alpha_-} \cdot b \cdot \frac{U}{V} = x^{\alpha_-} \cdot b,$$

the problem reduces further to finding a polynomial solution $U \in \mathbf{F}[x]$ of the resulting difference equation with polynomial coefficients,

$$z \cdot x^{\alpha_+} \cdot a \cdot V \cdot \epsilon U - x^{\alpha_-} \cdot b \cdot (\epsilon V) \cdot U = x^{\alpha_-} \cdot b \cdot V \cdot \epsilon V. \tag{4}$$

(Note that at least one polynomial solution, namely $U = u \cdot V/v$, exists.) Furthermore, equations of that type simplify by canceling $\gcd_q$'s. For instance, in order to get more information about the denominator v, let $v_i := \epsilon^i v/\gcd_q(v, \epsilon v)$, $i \in \{0, 1\}$. Then (3) is equivalent to

$$z \cdot x^{\alpha_+} \cdot a \cdot v_0 \cdot \epsilon u - x^{\alpha_-} \cdot b \cdot v_1 \cdot u = x^{\alpha_-} \cdot b \cdot v_0 \cdot v_1 \cdot \gcd_q(v, \epsilon v). \tag{5}$$

Now, if $\langle p_1, \ldots, p_m \rangle$, $m \geq 0$, is the qGFF-form of $\widehat{v}$, it follows from $\gcd(u, v) = 1 = \gcd(v_0, v_1)$ and the Fundamental qGFF Lemma that

$$v_0 = (\epsilon^0 p_1) \cdots (\epsilon^{-m+1} p_m) | b \quad \text{and} \quad v_1 = q^\beta \cdot (\epsilon p_1) \cdots (\epsilon p_m) | a.$$

This observation gives rise to a simple and straightforward algorithm for computing a multiple $\widehat{V} := [P_1]_q^{\frac{1}{}} \cdots [P_n]_q^{\frac{n}{}}$ of $\widehat{v}$. For instance, if $P_1 := \gcd_q(\epsilon^{-1} a, b)$ then obviously $p_1 | P_1$. Indeed, we shall see below that by exploiting properties (qGFF1), (qGFF2), and (qGFF3) from Section 2.1, one can iteratively extract q-monic p_i-multiples P_i such that $\epsilon P_i | a$ and $\epsilon^{-i+1} P_i | b$:

Algorithm $\widehat{\mathbf{V}}$MULT. INPUT: relatively prime and q-monic polynomials $a, b \in \mathbf{F}[x]$ that constitute the q-monic quotient of $\rho = z \cdot x^\alpha \cdot a/b \in \mathbf{F}(x)$; OUTPUT: q-monic polynomials $\langle P_1, \ldots, P_n \rangle$ such that $\widehat{V} := [P_1]_q^{\frac{1}{}} \cdots [P_n]_q^{\frac{n}{}}$ is a multiple of $\widehat{v}$, the q-monic part of the denominator $v = x^\beta \cdot \widehat{v}$ of $\tau \in \mathbf{F}(x)$.

(i) Compute $n = \min\{j \in \mathbf{N} | \gcd(\epsilon^{-1} a, \epsilon^{k-1} b) = 1 \text{ for all integers } k > j\}$.

(ii) Set $a_0 = a$, $b_0 = b$, and compute for i from 1 to n:

$$\begin{aligned} P_i &= \gcd_q(\epsilon^{-1} a_{i-1}, \epsilon^{i-1} b_{i-1}), \\ a_i &= a_{i-1}/\epsilon P_i, \\ b_i &= b_{i-1}/\epsilon^{-i+1} P_i. \end{aligned}$$

The following lemma tells that the output polynomials P_i indeed are multiples of the p_i's.

Lemma 2 *Let $\tau \in \mathbf{F}(x)$ be a rational function solution of Equation* (3) *such that $\tau = u/v$ with relatively prime $u, v \in \mathbf{F}[x]$ and $v = x^\beta \cdot \widehat{v}$ the q-monic decomposition of v. Let $q\text{GFF}(\widehat{v}) = \langle p_1, \ldots, p_m \rangle$. If n and $\langle P_1, \ldots, P_n \rangle$ are computed as in Algorithm $\widehat{\text{V}}$MULT, then:*

$$n \geq m \text{ and } p_i | P_i \text{ for all } i \in \{1, \ldots, m\}.$$

Proof In its essence the proof only uses properties (qGFF1), (qGFF2), and (qGFF3); almost word by word the argumentation can be carried over from the case $q = 1$. For that see Algorithm VMULT and the proof of Lemma 5.1 in Paule [1995]. Hence we leave the steps of the verification to the reader.

□

With the multiple $\widehat{V}$ of $\widehat{v}$ in hands, all what is left for solving (3), and thus the q-hypergeometric telescoping problem (1), is to determine an appropriate multiplicity γ such that

$$V = x^\gamma \cdot \widehat{V} \text{ is a multiple of } v = x^\beta \cdot \widehat{v}.$$

For that we consider Equation (5) again: (i) Assume that either $\alpha_- \neq 0$ or $\alpha_+ \neq 0$. In the first case we have $\alpha_+ = 0$ and $x^{\alpha_-} | u$, hence β must be 0 because of

$\gcd(u,v) = 1$. This means, we can choose $\gamma := 0$. In the second case we have $\alpha_- = 0$ and $x^{\min(\alpha_+,\beta)}|u$, because of $\gcd_q(v,\epsilon v) = x^\beta \cdot \gcd_q(\widehat{v},\epsilon\widehat{v})$. Again β must be 0, and again we can choose $\gamma := 0$. (ii) Assume that $\alpha = 0$. In this case Equation (5) evaluated at $x = 0$ turns into

$$(z - q^\beta)\, u(0) = q^\beta \cdot \delta_{0,\beta},$$

because $v_1 = q^\beta \cdot (\epsilon p_1) \cdots (\epsilon p_m)$; $\delta_{0,\beta}$ denotes the Kronecker symbol. This means, if $\beta > 0$ we obtain, observing that $u(0) \neq 0$ in this case, as a condition for β that $z = q^\beta$. Hence in case $\alpha = 0$, we choose $\gamma := \log_q(z)$ if z is a positive integer power of q, or $\gamma := 0$ otherwise.

Summarizing, q-hypergeometric telescoping can be decided constructively as follows:

Algorithm qTelescope. INPUT: $\rho \in \mathbf{F}(x)$ such that $f_{k+1}/f_k = \rho(q^k)$ for all $k \in \mathbf{N}$; OUTPUT: a q-hypergeometric solution $\langle g_k \rangle_{k\geq 0}$ of (1); in case such a solution does not exist, the algorithm stops.

(i) Decompose ρ into the form $\rho = z \cdot x^\alpha \cdot a/b$ such that $z \in \mathbf{F}$, α integer, and $a, b \in \mathbf{F}[x]$ relatively prime and q-monic.

(ii) With respect to input a, b compute q-monic polynomials $\langle P_1, \ldots, P_n \rangle$ by Algorithm $\widehat{\text{V}}$MULT and set $\widehat{V} := [P_1]_{\frac{1}{q}} \cdots [P_n]_{\frac{n}{q}}$.

(iii) Determine the zero-root multiplicity γ as follows:

$$\gamma := \begin{cases} \log_q(z) & \text{if } \alpha = 0 \text{ and } z \text{ a positive integer power of } q, \\ 0 & \text{otherwise.} \end{cases}$$

(iv) Take $V := x^\gamma \cdot \widehat{V}$. If Equation (4) can be solved for a polynomial $U \in \mathbf{F}[x]$ then $g_k := f_k \cdot U(q^k)/V(q^k)$ solves (1), if Equation (4) admits no polynomial solution U then no q-hypergeometric solution of (1) exists.

We want to note that in practical computations Equation (4) is solved in a more economical but equivalent form, which is obtained by dividing out gcd's; see Equation (7). A standard way to solve, for $Y \in \mathbf{F}[x]$, such a q-difference equation $Q \cdot \epsilon Y - R \cdot Y = P$, with polynomial coefficients from $\mathbf{F}[x]$, is the method of undetermined coefficients. For that one needs a careful analysis of possible degree bounds for Y which has been done for the Mathematica implementation of qTelescope.

3.2 Remarks on Canonical Forms Algorithm $\widehat{\text{V}}$MULT is the key ingredient of Algorithm qTelescope. The $q = 1$ version of $\widehat{\text{V}}$MULT is essentially the same algorithm Petkovšek [1992] came up with in order to compute a canonical "Gosper-form" representation; the q-version described by Abramov, Paule and Petkovšek [1995] was used to formulate a q-analogue of this form. Lemma 2 focuses on the $p_i|P_i$ property; the following lemma lists additional facts about Algorithm $\widehat{\text{V}}$MULT which can be proved in a similar fashion:

Lemma 3 *Let n, a_n, b_n, and the tuple $\langle P_1, \ldots, P_n \rangle$ be computed as in Algorithm $\widehat{\text{V}}$MULT, then:*

(i) $a = (\epsilon P_1) \cdots (\epsilon P_n) \cdot a_n,$

(ii) $b = (\epsilon^0 P_1) \cdots (\epsilon^{-n+1} P_n) \cdot b_n,$

(iii) $\forall k \in \mathbf{N}$: $\gcd(a_n, \epsilon^k b_n) = 1$,

(iv) $\forall i \in \{1, \ldots, n\}$: $\gcd([P_i]_{q}^{\underline{i}}, a_n) = 1$,

(v) $\forall i \in \{1, \ldots, n\}$: $\gcd([P_i]_{q}^{\underline{i}}, \epsilon^{-1} b_n) = 1$,

(vi) $q\mathrm{GFF}([P_1]_{q}^{\underline{1}} \cdots [P_n]_{q}^{\underline{n}}) = \langle P_1, \ldots, P_n \rangle$.

Proof Left to the reader; cf. Paule [1995] or Abramov, Paule and Petkovšek [1995]. □

The intended scope of this paper prohibits a more detailed discussion of the numerous interesting features of Algorithm $\widehat{\mathrm{V}}$MULT; we briefly touch two which are of particular interest.

By Lemma 3 we have

$$\frac{a}{b} = \frac{\epsilon \widehat{V}}{\widehat{V}} \cdot \frac{a_n}{b_n}. \tag{6}$$

Using this representation for a/b, the task to solve eq. (4) for U reduces to solve

$$z \cdot q^{-\gamma} \cdot x^{\alpha_+} \cdot a_n \cdot \epsilon W - x^{\alpha_-} \cdot (\epsilon^{-1} b_n) \cdot W = x^{\gamma + \alpha_-} \cdot \widehat{V} \tag{7}$$

for $W \in \mathbf{F}[x]$; then $U = (\epsilon^{-1} b_n) \cdot W$.

More generally, any nonzero $\rho \in \mathbf{F}(x)$ has a *unique* representation of the form

$$\rho = \frac{\epsilon P}{P} \cdot \frac{Q}{\epsilon R}, \tag{8}$$

where the polynomials $P, Q, R \in \mathbf{F}[x]$, being normalized in a certain way, are such that $\gcd(P, Q) = \gcd(P, R) = 1$, and $\gcd(Q, \epsilon^k R) = 1$ for all $k \geq 1$. A representation of this type which provides a perfect q-analogue of Petkovšek's canonical Gosper-form, was introduced by Paule and Strehl [1995]; Equation (6) simply is its restriction to the q-monic quotient of ρ. Another normalization of P, Q, and R was chosen by Abramov, Paule and Petkovšek [1995].

Paule and Strehl [1995] described, without giving detailed proofs, that the form (8), as in case $q = 1$ contains all the information needed for solving telescoping as follows: If

$$P = Q \cdot \epsilon Y - R \cdot Y \tag{9}$$

admits a solution $Y \in \mathbf{F}[x]$, then $C/D = ((\epsilon Y)/Y) \cdot (Q/R) \in \mathbf{F}(x)$ is the rational representation of the q-hypergeometric solution, i.e.,

$$g_k = \frac{D(q^k)}{C(q^k) - D(q^k)} \cdot f_k$$

solves (1).

One can easily verify that Algorithm qTelescope is equivalent to this procedure; for instance, Equation (7) is equivalent to Equation (9). It is this canonical form version of qTelescope which was taken for our Mathematica implementation, i.e., computing the form (8) with normalization as explained by Paule and Strehl [1995]; see also Section 4.2.7.

We want to conclude this section by pointing to another important computational feature of Algorithm $\widehat{\text{V}}$MULT. Namely, in view of Lemma 3 and representation (6), it can be used to compute the (canonical) qGFF-form $q\text{GFF}(P) = \langle P_1, \ldots, P_n \rangle$ for given q-monic $P \in \mathbf{F}[x]$: simply set $a = \epsilon P / \gcd_q(P, \epsilon P)$ and $b = P / \gcd_q(P, \epsilon P)$.

4 A q-Analogue of Zeilberger's Algorithm

As demonstrated in Wilf and Zeilberger [1992b], Zeilberger's [1990] fast algorithm for proving terminating hypergeometric identities generalizes to the q-case.

In this section we first give a short account of the underlying mechanism, especially of the fundamental notion of "proof certification"; for further details the reader is referred to the beautiful book by Petkovšek, Wilf and Zeilberger [1996]. This is followed by a description of the usage of the package `qZeil`, a q-analogue of Zeilberger's algorithm in the computer algebra system Mathematica. Its implementation, based on q-hypergeometric telescoping introduced in Section 3, was carried out by A. Riese in the frame of a diploma thesis supervised by the first named author of this paper. The section concludes by a brief comparison of `qZeil` with Koornwinder's [1993] Maple package.

4.1 qWZ-Certification Analogous to Zeilberger's [1990] algorithm its q-analogue takes terminating q-hypergeometric sums as input; the output is a linear recurrence that is satisfied by the input sum, together with a rational function which serves as the proof certificate. It is important to note that the proof certificate enables a verification of the output recurrence merely by checking a rational function identity. This means, the algorithm itself supplies complete information for a correctness check which works independently of the steps in which the output recurrence was manufactured.

Let $f := \langle f_{n,k} \rangle$ be a double-indexed sequence with values in $\mathbf{F}$. We shall consider only sequences where n runs through $\mathbf{N}$, whereas the second parameter k might run through all integers.

The sequence f is called q-hypergeometric in both parameters if both quotients

$$\frac{f_{n,k}}{f_{n-1,k}} \quad \text{and} \quad \frac{f_{n,k}}{f_{n,k-1}}$$

are rational functions in q^n and q^k over $\mathbf{F}$ for all n and k where the quotients are well-defined.

Recall the standard definition of the q-shifted factorial of $a \in \mathbf{F}$:

$$(a;q)_k = \begin{cases} (1-a)(1-aq)\cdots(1-aq^{k-1}), & \text{if } k > 0, \\ 1, & \text{if } k = 0, \\ 1/((1-aq^{-1})(1-aq^{-2})\cdots(1-aq^{k})), & \text{if } k < 0, \end{cases}$$

and

$$(a;q)_\infty = \prod_{k=0}^{\infty} (1 - aq^k).$$

Then the sequence of Gaussian polynomials $\left[{n \atop k}\right]_q = (q;q)_n / ((q;q)_k (q;q)_{n-k})$, if $0 \le k \le n$, and $\left[{n \atop k}\right]_q = 0$ otherwise, is q-hypergeometric in n and k. We say the sequence f has finite support with respect to k, if the following is true for each n fixed: $f_{n,k} \neq 0$ for all k from a finite integer interval I_n, and $f_{n,k} = 0$ for all k outside I_n; for example, $f_{n,k} := \left[{n \atop k}\right]_q$ with $I_n = \{0, 1, \ldots, n\}$.

Given $f = \langle f_{n,k} \rangle$ q-hypergeometric in n and k, one can prove under mild side-conditions, as demonstrated in Wilf and Zeilberger [1992b], that for a certain integer $d \geq 0$ and $n \geq d$ there exists a linear recurrence

$$c_0(n) f_{n,k} + c_1(n) f_{n-1,k} + \cdots + c_d(n) f_{n-d,k} = g_{n,k} - g_{n,k-1}, \tag{10}$$

where the coefficients are polynomials in q^n not depending on k, and where $g_{n,k}$ is a rational function multiple of $f_{n,k}$ and thus also q-hypergeometric in n and k. Given the order d, which in general is not known a priori, $g_{n,k}$ and also the coefficient polynomials $c_i(n)$ are determined by q-hypergeometric telescoping, i.e., by Algorithm qTelescope.

Assume that f has finite support with respect to k. Then summing both sides of (10) over all k results in

$$c_0(n) S_n + c_1(n) S_{n-1} + \cdots + c_d(n) S_{n-d} = 0, \tag{11}$$

a recurrence for the sum sequence $S_n := \sum_k f_{n,k}$, a finite sum due to the finite support property. We use the convention that the summation parameter k runs through all the integers, in case the summation range is not specified explicitly.

Now the qWZ-*certificate* (for short: certificate) of recurrence (10) or (11), respectively, by definition is the rational function $cert(n,k)$, rational in q^n and q^k, such that

$$g_{n,k} = cert(n,k)\ f_{n,k}.$$

Evidently, with the certificate in hands the verification of (10), and therefore (11), reduces to checking the *rational function identity*

$$r(n,k) = cert(n,k) - cert(n,k-1) \frac{f_{n,k-1}}{f_{n,k}}, \tag{12}$$

where $r(n,k)$, rational in q^n and q^k, comes from rewriting the left-hand side of (10) as $r(n,k)\ f_{n,k}$. The computation of $r(n,k)$ is straightforward, because any $f_{n-i,k}$ can be written as a rational function multiple of $f_{n,k}$, for instance, $f_{n-1,k} = (f_{n-1,k}/f_{n,k}) \cdot f_{n,k}$.

We conclude this section with a remark on the inhomogeneous case which also arises in applications. Assume that f does not have finite support, or, that one is interested in summation with bounds not naturally induced by the finite support, for instance,

$$S_n := \sum_{k=ln+m}^{on+p} f_{n,k}$$

with fixed integers l, m, o, and p. In this case the package delivers an output recurrence which is inhomogeneous of type

$$c_0(n) S_n + c_1(n) S_{n-1} + \cdots + c_d(n) S_{n-d} = g_{n,on+p} - g_{n,ln+m-1} + ct(n),$$

where the corresponding correction term $ct(n)$ is defined as

$$ct(n) := \sum_{j=1}^{d} c_j(n)[ct_1(j,n) - ct_2(j,n)],$$

with

$$ct_1(j,n) := \sum_{k=l(n-j)+m}^{ln+m-1} f_{n-j,k} \quad \text{and} \quad ct_2(j,n) := \sum_{k=o(n-j)+p+1}^{on+p} f_{n-j,k},$$

employing the extended sum-definition

$$\sum_{k=a}^{b} f_k = \begin{cases} f_a + f_{a+1} + \cdots + f_b, & a \le b, \\ 0, & a = b+1, \\ -[f_{b+1} + f_{b+2} + \cdots + f_{a-1}], & a \ge b+2. \end{cases}$$

An example for a nontrivial application of this "general bounds" feature of **qZeil** is presented in Section 5.4.

4.2 The Package qZeil

4.2.1 *Installation* The package consists of five files named **qZeil.m**, **qTelescope.m**, **qInput.m**, **qSimplify.m** and **LinSolve.m**, which have to be copied into one directory. The files are available by email request to

Peter.Paule risc.uni-linz.ac.at.

After starting a Mathematica session from this directory and typing <<qZeil.m all files are loaded automatically. In addition to these files containing the code for the algorithm, the ASCII-file **qZeilExamples.txt**, consisting of about 250 identities at the moment, can be used as a source of examples. Most of these identities are terminating summations or transformations from the book by Gasper and Rahman [1990]. The package **qZeil** already helped to improve this outstanding q-hypergeometric database; see Riese [1995].

4.2.2 *Interfaces* The package has two interfaces. Concerning indefinite summation one can run **qTelescope** for q-hypergeometric telescoping, or **qZeil**, the q-analogue of Zeilberger's algorithm, to come up with a recurrence, homogeneous or inhomogeneous, for a definite q-hypergeometric sum. The corresponding commands are given by

```
qTelescope[SUMMAND, RANGE, <INTCONST>]
```

and

```
qZeil[SUMMAND, RANGE, RECVAR, ORDER, <INTCONST>],
```

where **<PARAMETER>** denotes an optional argument. Before we will give a detailed description of the parameters, let us first present some illustrating examples.

4.2.3 *Warm-up Examples*

1. Load the package:

```
In[1]:= <<qZeil.m
Out[1]= Axel Riese's q-Zeilberger implementation version
        1.4 loaded
```

2. Compute the closed form for a special case of the q-Chu-Vandermonde summation formula

$$\sum_{k=0}^{n} \frac{(b;q)_k}{(q;q)_k} q^k = \frac{(bq;q)_n}{(q;q)_n}.$$

```
In[2]:= qTelescope[qfac[b,q,k] q^k / qfac[q,q,k], {k,0,n}]
        qfac[b q, q, n]
Out[2]= ---------------
         qfac[q, q, n]
```

3. The so-called Rogers-Szegö polynomials

$$r_n(x,a) := \sum_{k=0}^{n} \left[\begin{array}{c} u \\ k \end{array} \right]_q a^{n-k} x^k$$

are known to satisfy a recurrence of order 2:

```
In[3]:= qZeil[qBinomial[n,k,q] a^(n-k) x^k, {k,0,n}, n, 2]
                           -1 + n
Out[3]= SUM[n] == a (-1 + q      ) x SUM[-2 + n] +

         (a + x) SUM[-1 + n]
```

4. For Jackson's q-analogue of the Pfaff-Saalschütz formula

$$_3\phi_2(q^{-n}, a, b; c, abc^{-1}q^{1-n}; q, q) = \frac{(c/a, c/b; q)_n}{(c, c/ab; q)_n}$$

we obtain:

```
In[4]:= qZeil[qfac[q^(-n),q,k] qfac[a,q,k] qfac[b,q,k] q^k /
            (qfac[c,q,k] qfac[a b / c q^(1-n),q,k] *
             qfac[q,q,k]), {k,0,n}, n, 1]
                          -1 + n         -1 + n
                       c q            c q
                  (1 - ---------) (1 - ---------) SUM[-1 + n]
                           a              b
Out[4]= SUM[n] == -------------------------------------------
                                                 -1 + n
                                 -1 + n       c q
                        (1 - c q      ) (1 - ---------)
                                                 a b
```

4.2.4 *The Input Summand* As summand we allow any q-hypergeometric term of the form

$$f_{n,k} = \frac{\prod_{r=1}^{rr} (A_r\, q^{(d_r i_r)n+(e_r i_r)k+l_r}; q^{i_r})_{a_r n+b_r k+c_r}}{\prod_{s=1}^{ss} (B_s\, q^{(f_s j_s)n+(g_s j_s)k+m_s}; q^{j_s})_{u_s n+v_s k+w_s}} \cdot R(q^n, q^k) \cdot q^{\alpha\binom{k}{2}+(\beta n+\gamma)k} \cdot z^k, \qquad (13)$$

with

A_r, B_s	power products in $\mathbf{K}$,
a_r, b_r, u_s, v_s	specific integers (i.e., integers free of any parameters),
c_r, w_s	integers, which may depend on parameters free of n and k,
d_r, e_r, f_s, g_s	specific integers,
l_r, m_s	integers free of n and k,
i_r, j_s	specific nonzero integers,
R	a rational function in $\mathbf{F}(q^n, q^k)$ such that the denominator factors

completely into a product of terms of the form $(1 - Bq^{fn+gk+m})$,
α, β, γ specific integers, and
z a rational function in $\mathbf{F}$.

The q-shifted factorial $(a; q^i)_k$ has to be typed as `qfac[a,q^i,k]`. In addition we allow terms of the form `qBrackets[a,q]` for $[a]_q := (1 - q^a)/(1 - q)$, `qFactorial[a,q]` for $[a]_q! := [1]_q [2]_q \cdots [a]_q$, and `qBinomial[a,b,q]` for $\left[\begin{smallmatrix} a \\ b \end{smallmatrix} \right]_q$, provided that those expressions can be translated correctly — with respect to (13) — into terms of q-shifted factorials as with `qfac` also for these forms powers `q^i` are admitted.

4.2.5 *The Summation Range* The range of summation has to be specified in the form

$$\texttt{RANGE := \{SUMVAR, LOW, UPP\}}.$$

In `qTelescope`, `LOW` and `UPP` may be arbitrary integers or integer parameters free of `SUMVAR` satisfying `LOW` $\le$ `UPP`. In `qZeil`, `LOW` and `UPP` are linear integer functions in `RECVAR` being free of `SUMVAR` such that `LOW` $\le$ `UPP`.

In `qZeil` the user may specify one or both bounds to be $\pm$`Infinity`. In this case, the bounds are assumed to be naturally induced by the finite support. The algorithm runs considerably faster in this *Turbo*-mode, since no inhomogeneous part and no correction terms of the recurrence have to be computed.

4.2.6 *The Optional Argument* `INTCONST` Since Mathematica is not able to handle typed variables, it is necessary to simulate them by telling the system explicitly which indeterminates should be treated as nonnegative integer constants. If one assigns to the optional argument `INTCONST` a list of Mathematica symbols representing those indeterminates, the program will assume them to be nonnegative integers. This also improves the simplification abilities of the program.

Consider the following example. Suppose we want to find a closed form for the indefinite sum

$$\sum_{k=0}^{n} \left[\begin{matrix} m+k \\ k \end{matrix} \right]_q q^k.$$

Without any knowledge about m the program is not able to recognize m and $m + k$ in $(q; q)_m$ and $(q; q)_{m+k}$, respectively, as integers. The problem disappears if we make the assignment `INTCONST := {m}`.

```
In[6]:= qTelescope[qBinomial[m+k,k,q] q^k, {k, 0, n}, {m}]
Out[6]= qBinomial[1 + m + n, 1 + m, q]
```

Note that in `qZeil` and `qTelescope` all indeterminates appearing in the bounds as well as the recursion variable `RECVAR` (in `qZeil`) are assumed to be elements of `INTCONST` automatically.

4.2.7 *The Certificate and other Global Variables* The (simplified) certificate $cert(n, k)$, i.e., the rational function from Section 4.1 such that $g_{n,k} = cert(n, k) \cdot f_{n,k}$, is delivered by calling the function `Cert` without any parameters. For example, after obtaining `Out[4]` type

```
In[5]:= Cert
          1 - k + n             k              k     k     n
       c q          (-1 + a q ) (-1 + b q ) (q  - q )
```

```
Out[5]= ----------------------------------------------
                 n           n            1 + k        n
        (-1 + q ) (-q + c q ) (-(a b q       ) + c q )
```

and the proof of Jackson's q-analogue of the Pfaff-Saalschütz formula reduces to checking the initial values at $n = 0$ and verifying the rational function equation (12) for this value of $cert(n, k)$.

The output behavior of the program can be influenced by the global Boolean variables `Talk` and `Output`.

If `Talk` is set to `True`, the user can see explicitly which step of the algorithm is executed at the moment. This is mainly thought for time-consuming examples. Default value for `Talk` is `False`.

If `Output` is set to `True`, then running `qTelescope` or `qZeil` generates the file `GoOut`, where some intermediate results of the actual computation are written to. Default value for `Output` is `True`.

The entries `P-factor P_fac(T)`, `Q-numerator Q_num(T)` and `R-denominator R_den(T)` in `GoOut`, where T is used as an abbreviation for q^k, correspond to polynomials $\bar{P}, \bar{Q}$ and $\bar{R}$, respectively, constituting a slightly modified version

$$\frac{f_k}{f_{k-1}} = \frac{\bar{P}}{\epsilon^{-1}\bar{P}} \frac{\bar{Q}}{\bar{R}},$$

of the q-analogue (8) of Petkovšek's canonical Gosper-form, normalized as explained in Paule and Strehl [1995]. This is because in practice it turns out to be more convenient to start out with the rational function f_k/f_{k-1}, and to solve $f_k = g_k - g_{k-1}$ instead of the version (1). Hence in the actual implementation of qTelescope, Equation (9) is solved in the correspondingly modified version.

Finally, by setting the global variable `Simp` to `False` one can suppress the automatic simplification of the telescoping solution sequence $g_{n,k}$ and the correction terms. By default, the program applies the rules listed in the file `qSimplify.m` to those expressions. Since the size of the result may grow enormously, this should be done only in case of emergency.

4.2.8 *Computing Companion Identities* As WZ-pairs in the hypergeometric case (see Petkovšek, Wilf and Zeilberger [1996]), qWZ-pairs play an important role in q-certification. Analogous to case $q = 1$, one can use qWZ-pairs to get new identities "for free", i.e., without too much additional effort. Following Wilf and Zeilberger [1992a], one of these identities "for free" is called the *companion identity.* It is based on the symmetry of f and g in the qWZ-equation

$$f_{n,k} - f_{n-1,k} = g_{n,k} - g_{n,k-1}.$$

Double-indexed sequences of this type, which in addition are q-hypergeometric in both indices, are called qWZ-pairs. As in case $q = 1$, one has: If f and g form a qWZ-pair satisfying the following conditions:

(F) for each integer k, the limit $f_k := \lim_{n\to\infty} f_{n,k}$ exists and is finite,

(G) $\lim_{k\to-\infty} \sum_{n\geq 0} g_{n+1,k} = 0$,

then the companion identity is given by

$$\sum_{n\geq 0} g_{n+1,k} = \sum_{j\leq k} (f_j - f_{0,j}) .$$

The program computes the companion identity, if the global variable `Companion` is set to `True`, and f and g in fact form a qWZ-pair. To compute f_k we might have to make the assumption $|q| < 1$, or, to take the limit w.r.t. sequences of formal power (or Laurent) series. The condition (G) has to be checked by the user. Note that (G) is satisfied automatically if f has finite support with respect to k. Default value for `Companion` is `False`, the result is assigned to the variable `CompId`.

For example, for the q-Chu-Vandermonde identity

$$\sum_{k=0}^{n} \begin{bmatrix} n \\ k \end{bmatrix}_q \begin{bmatrix} b \\ k \end{bmatrix}_q q^{k^2} = \begin{bmatrix} b+n \\ n \end{bmatrix}_q ,$$

we obtain the following result:

```
In[7] := Companion = True;
         qZeil[qBinomial[n,k,q] qBinomial[b,k,q] q^(k^2) /
               qBinomial[b+n,n,q], {k,0,n}, n, 1, {b}]
Out[8]= SUM[n] == 1

In[9]:= CompId
                                  2
                     1 + k + k   + n                             2
Out[9]= Sum[-((q                      qBinomial[n, k, q] qfac[q, q, b]

             qfac[q, q, n]) / (qfac[q, q, -1 + b - k] qfac[q, q, k]

              qfac[q, q, 1 + b + n])), {n, 0, Infinity}] ==

                                 2
                               jj
                              q    qBinomial[b, b - jj, q] qfac[q, q, b]
     -(k >= 0) + Sum[----------------------------------------,
                                           qfac[q, q, jj]

        {jj, -Infinity, k}]
```

Here we follow the convention that for any true-or-false predicate *pred* we define $(pred) := 1$ if *pred* is true, and $(pred) := 0$ if *pred* is false.

Hence, for $b \geq 0$ and $k \geq 0$ the companion identity reads as (cf. Riese [1995])

$$\frac{q^{1+k+k^2}(1-q^{k+1})}{(1-q^{b+1})} \begin{bmatrix} b \\ k+1 \end{bmatrix}_q \cdot \sum_{n\geq k} \frac{\left[\begin{smallmatrix} n \\ k \end{smallmatrix}\right]_q}{\left[\begin{smallmatrix} n+b+1 \\ n \end{smallmatrix}\right]_q} q^n = 1 - \sum_{j=0}^{k} \frac{\left[\begin{smallmatrix} b \\ j \end{smallmatrix}\right]_q (q;q)_b}{(q;q)_j} q^{j^2} .$$

For the special case $b = n$ of the q-Chu-Vandermonde identity we get the result spelled out in Wilf and Zeilberger [1992b].

4.2.9 *Computing Dual Identities* Another method for discovering new identities is based on the fact that to any qWZ-pair one can associate a dual pair that may produce new identities.

As in the $q = 1$ case (cf. Wilf and Zeilberger [1992a], or Gessel [1995] who made a systematic investigation of dual identity production), one introduces the operation of *shadowing.* Let, for instance, $a_n = (q;q)_n$ for $n \geq 0$. Then the defining property of a_n is that it satisfies the recurrence equation $a_n = (1-q^n)\,a_{n-1}$ together with the initial condition $a_0 = 1$. Trying to extend this sequence to the "opposite side", one could ask for a sequence $\bar{a}_n$ such that $\bar{a}_n = (1-q^n)\,\bar{a}_{n-1}$ holds for the negative integers. A sequence that satisfies this condition is

$$\bar{a}_n = \frac{(-1)^n q^{\binom{n+1}{2}}}{(q;q)_{-n-1}} \qquad \text{for } n \leq -1.$$

We call $\bar{a}_n$ the *shadow* of a_n. More generally, for $a_{n,k} = (\alpha; q^i)_{an+bk+c}$, where α is free of n and k, the shadow is defined by

$$\bar{a}_{n,k} = \frac{(-1)^{an+bk+c}\alpha^{an+bk+c+1} q^{i\left[\binom{an+bk+c}{2}-1\right]}}{(q^{2i}/\alpha; q^i)_{-an-bk-c-1}},$$

with the property that

$$\frac{a_{n,k}}{a_{n-1,k}} = \frac{\bar{a}_{n,k}}{\bar{a}_{n-1,k}} \quad \text{and} \quad \frac{a_{n,k}}{a_{n,k-1}} = \frac{\bar{a}_{n,k}}{\bar{a}_{n,k-1}}.$$

The shadow $\bar{f}_{n,k}$ of a summand term $f_{n,k}$ (see Section 4.2.4) is defined to be the result of formally replacing each factor of the form $(A;q^d)_{an+bk+c}$ in f according to the shadowing rule described above. Since $f_{n,k}/f_{n-1,k} = \bar{f}_{n,k}/\bar{f}_{n-1,k}$ and $f_{n,k}/f_{n,k-1} = \bar{f}_{n,k}/\bar{f}_{n,k-1}$, the sequences f and $\bar{f}$ are essentially equivalent. This means, they only differ in initial values, in their domain of definition, and when they vanish. Thus, it follows that, if f and g form a qWZ-pair, then so do $\bar{f}$ and $\bar{g}$.

Evidently, one is free to shadow only some of the factors of $f_{n,k}$ and $g_{n,k}$ and fixing the others, this way getting different shadow pairs. Following Wilf [1993], a choice that seems to give fruitful results in general, is to shadow only those factors $(A;q^d)_{an+bk+c}$ for which $a + b \neq 0$. Hence, we will apply this kind of "default shadowing" in the algorithm. In this case, to avoid trivial qWZ-pairs like $(0,0)$, etc., we have to cancel all q-shifted factorial expressions in $f_{n,k}$ and $g_{n,k}$ being free of n and k, which again gives us a qWZ-pair.

The final step in dualization is to pass from the shadow pair $(\bar{f},\bar{g})$ to the dual pair (f',g') by a flip of variables and sequences, transforming the domain of n back to the nonnegative integers. Finally, the dual pair is defined as

$$\left(f'_{n,k}, g'_{n,k}\right) := \left(\bar{g}_{-k,-n-1}, \bar{f}_{-k-1,-n}\right),$$

which does not influence the fact that the sequences form a qWZ-pair, but which does alter the certificate via the same change of variables.

Note that in general dualization does not commute with specialization, i.e., the dual identity of some special case of an identity is not the same as the specialization of the dual identity. However, dualization is an involution up to constant factors.

The program computes the dual qWZ-pair, if the global variable `Dual` is set to `True`, and f and g actually form a qWZ-pair. The result is assigned to the variable `DualPair`. Default value for `Dual` is `False`.

Up to now we have not made a systematic investigation of producing new q-identities by dualization, as done for $q = 1$ by Gessel [1995]. We only present a few examples explaining the use of the package; see also Section 5.3.

For the q-Chu-Vandermonde identity above one gets the following:

```
In[12]:= Dual = True;
         qZeil[qBinomial[n,k,q] qBinomial[b,k,q] q^(k^2) /
               qBinomial[b+n,n,q], {k,0,n}, n, 1, {b}]
Out[13]= SUM[n] == 1

In[14]:= DualPair
                                            2                   2
                 k + n  k/2 + b k + k /2 - n/2 - b n - n /2
Out[14]= {((-1)        q

     qBinomial[n, k, q] qfac[q, q, -1 - b + k]

     qfac[q, q, -1 - b - n] qfac[q, q, n]) / qfac[q, q, k],

                                 2                   2
        k + n  b + k/2 + b k + k /2 + n/2 - b n - n /2
   ((-1)      q

     qBinomial[-1 + n, k, q] qfac[q, q, -b + k]

     qfac[q, q, -1 - b - n] qfac[q, q, -1 + n]) / qfac[q, q, k]}
```

Hence, after replacing b by $-b-1$ the dual identity becomes

$$\sum_{k=0}^{n}(-1)^{n+k}q^{(n-k)(2b-k-n+1)/2}\begin{bmatrix} n \\ k \end{bmatrix}_q \begin{bmatrix} b+k \\ k \end{bmatrix}_q = \begin{bmatrix} b \\ n \end{bmatrix}_q,$$

which is the same as the original identity modulo a renaming of the parameters. An identity satisfying this property is called *self-dual.*

As mentioned above, for the special case $b = n$ we do not obtain just the dual identity with b replaced by n, but:

$$\sum_{k=0}^{n}\begin{bmatrix} 2k \\ k \end{bmatrix}_q \begin{bmatrix} n \\ k \end{bmatrix}_q^2 \frac{q^{n-k}+q^n-2q^k}{1+q^k} = 0$$

presented by Wilf and Zeilberger [1992b].

Next, let us consider the q-Saalschütz identity in the form

$$\sum_{k=0}^{n}\begin{bmatrix} r-s+m \\ k \end{bmatrix}_q \begin{bmatrix} s-r+n \\ n-k \end{bmatrix}_q \begin{bmatrix} s+k \\ m+n \end{bmatrix}_q q^{(n-k)(r-s+m-k)} = \begin{bmatrix} r \\ n \end{bmatrix}_q \begin{bmatrix} s \\ m \end{bmatrix}_q.$$

The program computes the following dual identity (cf. Riese [1995]):

$$\sum_{k=0}^{n} p \begin{bmatrix} m+k \\ k \end{bmatrix}_q \begin{bmatrix} s \\ r-k \end{bmatrix}_q \begin{bmatrix} m-s \\ n-k \end{bmatrix}_q q^{(n-k)(r-k)} = \begin{bmatrix} m+r-s \\ n \end{bmatrix}_q \begin{bmatrix} n+s \\ r \end{bmatrix}_q .$$

Renaming the parameters we find the q-Saalschütz identity also to be self-dual. For the special case $m = n$ and $r = s$, the process of dualization leads to the following result (cf. Riese [1995]):

$$\sum_{k=0}^{n} \frac{1+q^{-k}-2q^{n-2k}-2q^{k-s}+q^{n-s}+q^{n-s-k}}{1+q^k}$$

$$\cdot \begin{bmatrix} 2k \\ k \end{bmatrix}_q \begin{bmatrix} n \\ k \end{bmatrix}_q^2 \frac{(q;q)_{n+s-2k-1}}{(q;q)_{s-k}^2} = 0,$$

where $s \geq n+1$.

Finally, for the q-Dixon identity

$$\sum_{k} (-1)^k \begin{bmatrix} n+b \\ n+k \end{bmatrix}_q \begin{bmatrix} n+c \\ c+k \end{bmatrix}_q \begin{bmatrix} b+c \\ b+k \end{bmatrix}_q q^{k(3k-1)/2} = \begin{bmatrix} n+b+c \\ n,b,c \end{bmatrix}_q$$

we get, by "creative symmetrizing" described in Section 5.2, i.e., taking as input,

```
In[15]:= qZeil[(1+q^k)/2 (-1)^k qBinomial[n+b,n+k,q] *
            qBinomial[n+c,c+k,q] qBinomial[b+c,b+k,q] *
            q^(k(3k-1)/2) qfac[q,q,n] qfac[q,q,b] qfac[q,q,c] /
            qfac[q,q,n+b+c],
            {k, -Infinity, Infinity}, n, 1, {b, c}]
Out[15]= SUM[n] == SUM[-1 + n]
```

and then calling `DualPair`, the dual identity

$$\sum_{k=0}^{n} (-1)^k q^{\binom{k+1}{2}-n(1+k+n)} \begin{bmatrix} b+c+k \\ b-1 \end{bmatrix}_q \begin{bmatrix} b+n \\ b+k \end{bmatrix}_q \begin{bmatrix} n+k \\ k \end{bmatrix}_q = \begin{bmatrix} b+c \\ c+n+1 \end{bmatrix}_q \begin{bmatrix} c \\ n \end{bmatrix}_q ,$$

which is nothing but a specialization of the q-Saalschütz identity.

4.2.10 *Some Remarks on Run-Time and History* Concerning the run-time, the main computational part of Gosper's algorithm and its q-analogue consists in solving a system of homogeneous linear equations with polynomial coefficients. It turned out that the Mathematica functions `NullSpace` and `LinearSolve` are absolutely impracticable even for rather simple applications. To overcome this problem, E. Aichinger wrote a Mathematica function `ENullSpace` based on Gaussian elimination, which does the job excellently for most of the examples. The interface `LinSolve` was written by M. Schorn; see Paule and Schorn [1995]. In the first prototype versions of `qZeil` up to 95 percent of the run-time were spent for solving the system of equations. Meanwhile this amount has decreased to about 30–40 percent in average, mainly due to a preprocessing of the system in which all constant factors with respect to the summation variable are extracted. Furthermore, a lot of considerations had to be put into finding a powerful *and* efficient simplification procedure. As a compromise, the strategy now is based on collecting several rewrite rules into blocks which are applied one after the other.

4.3 A Comparison with Koornwinder's Package As already mentioned, Koornwinder [1993] implemented Zeilberger's algorithm and its q-analogue in Maple. Furthermore, he gives a rigorous description of the ordinary algorithm and some remarks how to carry it over to the q-case. His program implements q-hypergeometric telescoping for

$$\sum_{k=0}^{n} \frac{(\alpha_1;q)_k\,(\alpha_2;q)_k\cdots(\alpha_r;q)_k}{(q;q)_k\,(\beta_1;q)_k\cdots(\beta_s;q)_k}\left((-1)^k q^{\binom{k}{2}}\right)^{1+s-r}\zeta^k$$

and the q-Zeilberger algorithm for

$$\sum_{k=0}^{n} \frac{(q^{-n};q)_k\,(q^{i_2 n}\alpha_2;q)_k\cdots(q^{i_r n}\alpha_r;q)_k}{(q;q)_k\,(q^{j_1 n}\beta_1;q)_k\cdots(q^{j_s n}\beta_s;q)_k}\left((-1)^k q^{\binom{k}{2}}\right)^{1+s-r}(q^{n\nu}\zeta)^k,$$

where $\alpha_1,\ldots,\alpha_r,\beta_1,\ldots,\beta_s$ and ζ are rational functions over the rational number field in a fixed number of indeterminates including q (but not q^k), and $i_2,\ldots,i_r$, $j_1,\ldots,j_s,\nu$ are integers such that

$\log_q(\beta_t)$ noninteger if $j_t=-1,-2,\ldots$; $\quad\log_q(\beta_t)\neq 0,-1,-2,\ldots$ if $j_t=0$;
$\log_q(\alpha_t)$ noninteger if $i_t=0$; $\quad\zeta\neq 0$.

Since the input specification described above is confined to basic hypergeometric series in base q, which is too restrictive in practice, with our implementation we tried to overcome some of these shortcomings which are:

- The summation range cannot be changed to an interval different from $[0,n]$ like $[-n,n]$ or $[0,2n]$, etc. So, for instance, it is not possible to find a recurrence for Jackson's terminating q-analogue of Dixon's sum

$$_3\phi_2\left[\begin{matrix}q^{-2n},b,c\\ q^{1-2n}/b,q^{1-2n}/c\end{matrix};q,\frac{q^{2-n}}{bc}\right]=\frac{(b,c;q)_n\,(q,bc;q)_{2n}}{(q,bc;q)_n\,(b,c;q)_{2n}}.$$

- Since $(q^{-n};q)_k$ and $(q;q)_k^{-1}$ have to be factors of the summand, the program always assumes finite support. Therefore no inhomogeneous recurrences can be dealt with.
- Concerning the bases of the q-shifted factorials, no powers of q are accepted. In addition, it is impossible to split q-shifted factorials of the form $(\alpha;q^m)_k$ for $m>1$ into $(\alpha^{(1)},\ldots,\alpha^{(m)};q)_k$, because in general the $\alpha^{(i)}$ are the *complex* roots of α. Furthermore, no rational powers of indeterminates are allowed.
- The index in q-shifted factorial expressions is restricted to be k, and the α's and β's must be free of k. One often has to apply expensive transformations to achieve this form, or even worse, $(\alpha;q)_{2k}$, for instance, cannot be split into $(\alpha;q)_k\,(\alpha q^k;q)_k$, etc.
- No polynomial part can be specified. Hence, "creative symmetrizing" from Section 5.2 in general cannot be applied.

Finally, let us compare the run-time for some typical examples all taken from the Gasper and Rahman [1990] book, which fit also to Koornwinder's input specification:

1. the q-Chu-Vandermonde sum, Equation (II.6) in Appendix II,
2. the q-Dixon sum, Equation (II.13) in Appendix II,

3. the sum of a very-well-poised ${}_6\phi_5$ series, Equation (II.21) in Appendix II,
4. Jackson's q-analogue of Dougall's ${}_7F_6$ sum, Equation (II.22) in Appendix II,
5. Watson's transformation formula, Equation (III.17) in Appendix III,
6. Bailey's ${}_{10}\phi_9$ transformation formula, Equation (III.28) in Appendix III,
7. a transformation formula of a nearly-poised ${}_5\phi_4$ series, Equation (III.25) in Appendix III.

The following table of run-times (in seconds) refers to tests on a Pentium 100 with 16 MB memory using Mathematica 2.2 for Windows and Maple V.2 for Windows, respectively. To have a fair competition, we eliminated a bug in Koornwinder's degree bound computation. Also minor adjustments for the input terms were made due to syntax restrictions of the Maple package. For the entry "Turbo-qZeil" recall the specification of the Turbo-mode from Section 4.2.5.

#	Series	Order	qZeil	Turbo-qZeil	Koornwinder
1	${}_2\phi_1$	1	1,59	0,88	0,55
2	${}_4\phi_3$	1	4,23	1,76	1,23
3	${}_6\phi_5$	1	5,33	2,25	1,80
4	${}_8\phi_7$	1	14,17	7,80	12,03
5	${}_4\phi_3$	2	13,07	9,39	4,59
5	${}_8\phi_7$	2	17,74	9,72	8,95
6	${}_{10}\phi_9$	2	235	224	out of memory[1]
6	${}_{10}\phi_9$	2	214	194	out of memory[2]
7	${}_5\phi_4$	3	47	38	113
7	${}_{12}\phi_{11}$	3	out of memory	out of memory	out of memory

5 Applications

In this section we present nontrivial examples illustrating various features of our package as well as the wide range of its applicability. For each subsection we assume that the `qZeil.m` package has been loaded as explained in Section 4.2.1.

5.1 An Identity Conjectured by D. Stanton Dennis Stanton [1995] conjectured the following identity,

$$\sum_k (-1)^k q^{4k^2} \begin{bmatrix} 2n \\ n-4k \end{bmatrix}_q = \sum_k q^{2k^2} \begin{bmatrix} n \\ 2k \end{bmatrix}_{q^2} (-q;q^2)_{n-2k}(-1;q^4)_k. \quad (14)$$

As demonstrated below, for the package `qZeil` the proof of (14) is no problem at all. This computer proof was the first proof that was given for Stanton's conjecture.

At this occasion we want to point out that we tried to make the package as user-friendly as possible. This means, the translation from q-series notation to Mathematica input for `qZeil` should cause only minimal effort from side of the user.

Note also that for summations over finite support, as in (14), it is convenient to take as summation range `{k,-Infinity,Infinity}` which saves computing time.

For the left-hand side of (14) one has

[1] after 5600 seconds

[2] after 6200 seconds

```
In[16]:=  qZeil[(-1)^k q^(4 k^2) qBinomial[2n,n-4k,q],
                  {k,-Infinity,Infinity}, n, 3]
Out[16]=
               -3 + 2 n        -5 + 2 n         -4 + 2 n
  SUM[n] == q          (1 - q         ) (1 - q         ) SUM[-3 + n] +

      -7 + 2 n   5    2 n    1 + 2 n    2 + 2 n
     q          (q  - q    - q        - q        ) SUM[-2 + n] +

       3    2 n    1 + 2 n    2 + 2 n
     (q  + q    + q        + q        ) SUM[-1 + n]
     ---------------------------------------------
                          3
                         q
```

and the proof certificate $cert(n,k)$ comes in the nicely factored form:

```
In[17]:= Cert
             -12 k + n   1 + 4 k    n     4 k    n     2 + 4 k    n
Out[17]= (q            (q        - q ) (-q    + q ) (-q        + q )

           3 + 4 k    n
        (-q        + q )) /

           n         n         n         n        2 n   3    2 n
     ((q - q ) (-1 + q ) (1 + q ) (q + q ) (q - q   ) (q  - q   ))
```

For the right-hand side with input

```
In[18]:= qZeil[q^(2 k^2) qBinomial[n,2k,q^2] qfac[-q,q^2,n-2k] *
                qfac[-1,q^4,k], {k,-Infinity,Infinity}, n, 3];
```

we get the same recurrence with the proof certificate, also nicely factored:

```
In[19]:= Cert
Out[19]=
   6 + 4 k + 2 n                 4 k     2 k    n     2 k    n
  (q              (1 + q) (1 + q   ) (-q    + q ) (q    + q )

        1 + 2 k    n     1 + 2 k    n
     (-q        + q ) (q        + q )) /

           n         n         n         n     1 + 4 k    2 n
     ((q - q ) (-1 + q ) (1 + q ) (q + q ) (q        + q   )

        3 + 4 k    2 n     5 + 4 k    2 n
      (q        + q   ) (q        + q   ))
```

Finally, the proof of (14) is completed by checking the identity at the initial values $n = 0$, $n = 1$, and $n = 2$.

5.2 Creative Symmetrizing In the hypergeometric case it is well-known that Zeilberger's algorithm not always delivers the minimal recurrence; see Andrews [1996], Paule and Schorn [1995], or Petkovšek, Wilf and Zeilberger [1996]. In the q-case the situation is similar, but one can observe that more standard identities like q-Dixon (cf. Gasper and Rahman [1990, (II.15)]) are infected by the nonminimality-virus. However, creative symmetrizing removes this infection in most of these cases. Creative symmetrizing also explains the remarkable observation that in almost all of these instances the $q = 1$ specialization behaves nice with respect to minimality.

Consider the following identity due to L.J. Rogers:

$$\sum_k \frac{(-1)^k q^{k(3k-1)/2}}{(q;q)_{n+k}(q;q)_{n-k}} = \frac{1}{(q;q)_n}. \tag{15}$$

This identity not only is a finite version of Euler's pentagonal number theorem (cf. Andrews [1976]), but also plays a fundamental role in connection with the Rogers-Ramanujan identities and, more general, in the context of Bailey chains; an excellent account is given in Andrews [1986].

Trying to prove (15) with `qZeil` results in a surprise: the program succeeds only with a recurrence of order 3 instead of expected order 1:

```
In[20]:= f[n_,k_] := (-1)^k q^(k(3k-1)/2) /
                     (qfac[q,q,n+k] qfac[q,q,n-k])

In[21]:= qZeil[f[n,k], {k,-n,n}, n, 1]
Out[21]= No solution: Increase order by 1

In[22]:= qZeil[f[n,k], {k,-n,n}, n, 2]
Out[22]= No solution: Increase order by 1

In[23]:= qZeil[f[n,k], {k,-n,n}, n, 3]
                        3
                       q  SUM[-3 + n]
Out[23]= SUM[n] == -------------------------- -
                        2 n          -1 + 2 n
                  (1 - q   ) (1 - q          )

   2    3    4    2 n
 (q  + q  + q  + q    ) SUM[-2 + n]
 --------------------------------- -
             2 n          -1 + 2 n
    q (1 - q   ) (1 - q          )

    2    3    4    3 n     2 + 2 n     1 + 3 n
 (-q  - q  - q  + q    + q        + q        ) SUM[-1 + n]
 ----------------------------------------------------------
                  2       2 n          -1 + 2 n
                 q  (1 - q   ) (1 - q          )
```

Hence in situations like that we are faced with three problems:

(i) How to find the minimal recurrence?

(ii) The computing time and size of the proof certificate might increase drastically.

(iii) One cannot apply the qWZ-pair machinery in order to get dual and companion identities "for free".

A solution to problem (i) is provided by the q-analogue of Petkovšek's algorithm **Hyper**; see Abramov, Paule and Petkovšek [1995]. Nevertheless, *creative symmetrizing* in many instances solves all of these problems.

Creative symmetrizing has been introduced by Paule [1994] for certifying finite versions of Rogers-Ramanujan type identities, for which the same problem as for (15) was observed. As found by Petkovšek, the method also can be applied in analogous $q = 1$ situations; see the book by Petkovšek, Wilf, and Zeilberger [1996], in which the term "creative antisymmetrizing" was introduced for a special case of the method (i.e., if $R(n,k) = -1$ in the lemma below).

Creative symmetrizing is based on the following essentially trivial lemma.

Lemma 4 ("Creative Symmetrizing") *For fixed $n \in \mathbf{N}$ let I_n be an integer interval with bounds α and β, both being free of k. If for all $k \in I_n$,*

$$f_{n,\alpha+\beta-k} = R(n,k) \cdot f_{n,k},$$

then

$$\sum_{k=\alpha}^{\beta} f_{n,k} = \frac{1}{2} \sum_{k=\alpha}^{\beta} (1 + R(n,k)) \cdot f_{n,k}.$$

Proof The proof is an immediate consequence of symmetrizing around $(\alpha+\beta)/2$, i.e.,

$$2 \sum_{k=\alpha}^{\beta} f_{n,k} = \sum_{k=\alpha}^{\beta} (f_{n,k} + f_{n,\alpha+\beta-k}).$$

□

In the applications below, $R(n,k)$ will be either a polynomial or a rational function in q^n and q^k. The point of creative symmetrizing lies in the observation that applying `qZeil` to the symmetrized summand $(1 + R(n,k)) \cdot f_{n,k}$ instead of $f_{n,k}$, for certain types of input sequences increases the chances to get a minimal output recurrence.

As mentioned above there are cases, like q-Dixon, with minimal order of the Zeilberger recurrence in the ordinary case, but with nonminimal order for their q-analogue. If in these instances creative symmetrizing succeeds in minimal order reduction, the explanation lies in the fact that for $q = 1$ the rational function $R(n,k)$ reduces to a constant. See, for instance, the q-Dixon input `In[15]` in Section 4.2.9.

With respect to (15) creative symmetrizing works as follows: Denote the summand by $f_{n,k}$, then evidently $f_{n,-k} = q^k \cdot f_{n,k}$, i.e., $R(n,k) = q^k$ and $\alpha = -n$, $\beta = n$. Applying `qZeil` to the symmetrized summand $(1 + q^k) \cdot f_{n,k}$ now indeed delivers the expected minimal recurrence:

```
In[24]:= qZeil[(1+q^k) f[n,k], {k,-n,n}, n, 1]
                         SUM[-1 + n]
Out[24]= SUM[n] == -----------
                              n
                          1 - q

In[25]:= Cert
          k + n    k     n
         q      (-q  + q )
Out[25]= -----------------
```

```
              k         n
     (1 + q ) (-1 + q )
```

(Note that the proof certificate is particularly nice.)

The next example — one Bailey chain step from (15) and its companion (see Andrews [1986]) — concerns finite versions of the Rogers-Ramanujan identities. For instance, a finite version of the Rogers-Ramanujan identity related to 2 and 3 (mod 5) is

$$\sum_k \frac{q^{k^2+k}}{(q;q)_k(q;q)_{n-k}} = \sum_k \frac{(-1)^k q^{k(5k+3)/2}}{(q;q)_{n+k+1}(q;q)_{n-k}}, \tag{16}$$

which was stated first in this form by Andrews [1974], see also Bressoud [1981].

With qZeil, one proves that the left-hand side satisfies a recurrence of order 2, namely:

```
In[26]:= qZeil[q^(k^2+k) / (qfac[q,q,k] qfac[q,q,n-k]),
                   {k,0,n}, n, 2]

Out[26]=
                                                  n     2 n
                  q SUM[-2 + n]     (1 + q - q   + q    ) SUM[-1 + n]
     SUM[n] == -(-------------) + -------------------------------
                          n                             n
                     1 - q                          1 - q

In[27]:= Cert
          -k + 2 n    k     n
         q           (q  - q )
Out[27]= ------------------
                      n
               -1 + q
```

Now the surprise comes with respect to the right-hand side. Namely, the program only finds a recurrence of order 5 (!) with a sufficiently lengthy certificate. Creative symmetrizing also resolves this problem: Denote by $f_{n,k}$ the corresponding summand, and observe that with $\alpha = -n-1$ and $\beta = n$ we have $f_{n,-1-k} = (-q^{2k+1}) \cdot f_{n,k}$, i.e., $R(n,k) = -q^{2k+1}$. Applying qZeil to the symmetrized summand $(1-q^{2k+1}) \cdot f_{n,k}$ produces the expected minimal recurrence of the same form as the one for the left-hand side:

```
In[28]:= f[n_,k_] := (-1)^k q^(k(5k+3)/2) /
                        (qfac[q,q,n+k+1] qfac[q,q,n-k])

In[29]:= qZeil[(1-q^(2k+1)) f[n,k], {k,0,n}, n, 2]
Out[29]=
                                                  n     2 n
                  q SUM[-2 + n]     (1 + q - q   + q    ) SUM[-1 + n]
     SUM[n] == -(-------------) + -------------------------------
                          n                             n
                     1 - q                          1 - q

In[30]:= Cert
          2 + 2 k + 2 n          1 + k     k     n
```

```
           q              (-1 + q        ) (q   - q )
Out[30]= ------------------------------------------
                              1 + 2 k          n
                    (-1 + q           ) (1 - q )
```

Checking equality at the initial values at $n = 0$ and $n = 1$ completes the proof of (16). (Note that the proof certificates for both sides again are sufficiently nice, so that the verification of (12) can be easily done by a human.) The `qZeil` proof of the companion identity to (16) can be found in Paule [1994].

We want to remark that identity (16) and its companion are the specializations $a = 1$ and $a = q$ of

$$\sum_k \frac{a^k q^{k^2}}{(q;q)_k (q;q)_{n-k}} = \sum_k \frac{(-1)^k a^{2k} q^{k(5k-1)/2} (aq;q)_{k-1}}{(q;q)_{n-k} (aq;q)_{n+k} (q;q)_k} \cdot (1 - aq^{2k}), \tag{17}$$

a terminating version of Watson's q-analogue of Whipple's theorem; see Gasper and Rahman [1990, (III.18)].

Indeed, for $a = q$ we find that the left-hand side of (17) equals the *unilateral* sum

$$\sum_{k \geq 0} (1 - q^{2k+1}) f_{n,k}.$$

Hence, in the classical unilateral version the symmetrizing factor, which we have introduced into the bilateral summation, is already there! This can be taken as a classical explanation, why creative symmetrizing can be successfully applied in so many cases.

Since Watson's identity is a transformation formula for a terminating very-well poised ${}_8\phi_7$ series, this gives rise to the conjecture that creative symmetrizing could be applied successfully in very-well poised context. This is strongly confirmed, also for well- or nearly-poised series, by inspection of numerous examples from various sources; see the file `qZeilExamples.txt`.

We conclude this section by having a look at a summation that arose in work of Andrews and Jackson [1990], namely

$${}_3\phi_2 \left[\begin{matrix} q^{-2n}, b, c \\ q^{-2n}/b, q^{-2n}/c, \end{matrix} ; q, \frac{q^{-n}}{bc} \right] = \frac{(q^{-2n};q)_n (q^{-2n}/bc;q)_n}{(q^{-2n}/b;q)_n (q^{-2n}/c;q)_n}. \tag{18}$$

Independently this *almost poised* series identity was proved by Bressoud [1987], who introduced this new notion together with numerous interesting examples.

Applying `qZeil` directly to the summand $f_{n,k}$ of the series in (18), the output recurrence order is 3.

With respect to Lemma 4 we have $\alpha = 0$, $\beta = 2n$, and $f_{n,2n-k} = R(n,k) \cdot f_{n,k}$ with

$$R(n,k) = q^{n-k} \frac{(1 - bq^k)(1 - cq^k)}{(1 - bq^{2n-k})(1 - cq^{2n-k})}.$$

Applying `qZeil` to the symmetrized summand $(1 + R(n,k)) \cdot f_{n,k}$ produces the expected minimal recurrence:

```
In[31]:= f[n_,k_] := qfac[q^(-2n),q,k] qfac[b,q,k] qfac[c,q,k] /
         (qfac[q^(-2n)/b,q,k] qfac[q^(-2n)/c,q,k] qfac[q,q,k]) *
         (q^(-n)/(b c))^k
```

```
In[32]:= R[n_,k_] := q^(n-k) (1-b q^k)(1-c q^k) /
                     ((1-b q^(2n-k))(1-c q^(2n-k)))
In[33]:= Factor[1+R[n,k]]
            k    n    k          3 n      k + n       k + n
Out[33]= ((q  + q ) (q  + b c q    - b q      - c q       +

          2 k + n        k + 2 n        k       2 n     k       2 n
    b c q        - b c q        )) / ((-q  + b q   ) (-q  + c q   ))

In[34]:= qZeil[Factor[1+R[n,k]] f[n,k], {k,0,2n}, n, 1]
                             n         n         n              2 n
Out[34]= SUM[n] == ((1 + q ) (1 - b q ) (1 - c q ) (1 - b c q    )

              -1 + 2 n          -1 + 2 n
       (1 - q        ) (1 - b c q        ) SUM[-1 + n]) /

                   n         2 n          2 n           -1 + 2 n
     ((1 - b c q ) (1 - b q   ) (1 - c q   ) (1 - b q         )

                -1 + 2 n
       (1 - c q         ))
```

(In this case the certificate is about the size of one page. In such instances the verification of (12) is left to the computer.)

5.3 An Identity of L.J. Rogers as a Dual Identity As already mentioned, creative symmetrizing enables to apply the qWZ-pair machinery. As an example let us consider the finite form of Euler' pentagonal number theorem (15) due to L.J. Rogers.

We start out with the symmetrized version as input:

```
In[35]:= Dual = True;
         qZeil[(1+q^k) (-1)^k q^(k(3k-1)/2) qfac[q,q,n] /
              (qfac[q,q,n+k] qfac[q,q,n-k]), {k,-n,n}, n, 1]
```

and obtain as expected:

```
Out[36]= SUM[n] == 2
```

The dual pair reads as:

```
In[37]:= DualPair
                                 2
             -k + n - k n + n
            q                    qfac[q, q, k + n]
Out[37]= {-----------------------------------------,
             qfac[q, q, k] qfac[q, q, -k + n]

                       2
           -n - k n + n          n
          q                 (1 + q ) qfac[q, q, k + n]
        -(---------------------------------------------)}
           qfac[q, q, k] qfac[q, q, -1 - k + n]
```

Summing the first entry we obtain:

```
In[38]:= qZeil[DualPair[[1]], {k,0,n}, n, 1]
Out[38]= SUM[n] == 1
```

which in usual notation is

$$\sum_{k=0}^{n} q^{(n-k)(n+1)} \frac{(q;q)_{n+k}}{(q;q)_k (q;q)_{n-k}} = 1. \tag{19}$$

Dualization of (19) brings us back to the symmetrized version of (15):

```
In[39]:= DualPair
                                      2
                k   -k/2 + (3 k )/2          k
            (-1)   q                     (1 + q ) qfac[q, q, n]
Out[39]= {-----------------------------------------------------,
                qfac[q, q, -k + n] qfac[q, q, k + n]

                                    2
            k   (3 k)/2 + (3 k )/2 + n
        (-1)   q                          qfac[q, q, -1 + n]
        ------------------------------------------------------}
            qfac[q, q, -1 - k + n] qfac[q, q, k + n]
```

Summarizing, because identity (19) is the limiting case $c \to \infty$ of Equation (II.7) in Gasper and Rahman [1990] with $a = q^{n+1}$, we have obtained Rogers' finite version (15) of Euler's pentagonal number theorem as the dual of a limiting case of Chu-Vandermonde (reversed).

We want to conclude this section by computing the WZ-companion identity with respect to the symmetrized version of (15):

```
In[40]:= Companion = True;

In[41]:= qZeil[(1+q^k) (-1)^k q^(k(3k-1)/2) qfac[q,q,n] /
              (qfac[q,q,n+k] qfac[q,q,n-k]), {k,-n,n}, n, 1]
Out[41]= SUM[n] == 2

In[42]:= CompId
                                                2
                   k   1 + (3 k)/2 + (3 k )/2 + n
               (-1)   q                              qfac[q, q, n]
Out[42]= Sum[-------------------------------------------------------,
                  qfac[q, q, -k + n] qfac[q, q, 1 + k + n]

          {n, 0, Infinity}] == -2 (k >= 0) +

                                     2
                  jj   -jj/2 + (3 jj )/2       jj
              (-1)    q                    (1 + q  )
          Sum[----------------------------------------, {jj, -Infinity, k}]
```

```
qfac[q, q, Infinity]
```

5.4 An Identity Discovered by R. J. McIntosh Comparing asymptotic expansions associated with different q-series, R. J. McIntosh [1995] conjectured that

$$\sum_{k=0}^{\infty} \frac{q^{(n+2k)(n+2k+1)/2}}{(q^2;q^2)_k} = (-q;q)_\infty \sum_{k=0}^{\infty} \frac{(-1)^k q^{k(k+1)/2-nk}}{(q^2;q^2)_k}, \tag{20}$$

where n is any integer. This conjecture was communicated to G.E. Andrews who later provided a proof. Using the package `qZeil` the verification of (20) is mere routine.

First let us assume n to be a *nonnegative* integer. Define for nonnegative integer m:

$$L_n(m) := \sum_{k=0}^{m} \frac{q^{(n+2k)(n+2k+1)/2}}{(q^2;q^2)_k} \quad \text{and} \quad R_n(m) := \sum_{k=0}^{m} \frac{(-1)^k q^{k(k+1)/2-nk}}{(q^2;q^2)_k}.$$

Now make use of the "general bounds" feature of `qZeil`, explained in Section 4.1,

```
In[43]:= qZeil[q^((n+2k)(n+2k+1)/2) / qfac[q^2,q^2,k],
                {k,0,m}, n, 2]
```

and obtain

```
                                  2                     2
                         m + 2 m   + n/2 + 2 m n + n /2
                        q
Out[43]= SUM[n] == ------------------------------ + SUM[-2 + n] -
                                    2   2
                              qfac[q , q , m]

        1 - n
       q        SUM[-1 + n]
```

with

```
In[44]:= Cert
Out[44]= 1
```

This means, $cert(n,k) = 1$ is the proof certificate of the inhomogeneous recurrence $(n \geq 2)$

$$L_n(m) = \frac{q^{\binom{n+1}{2}+m+2mn+2m^2}}{(q^2;q^2)_m} - q^{-n+1} L_{n-1}(m) + L_{n-2}(m). \tag{21}$$

Analogously for R_n,

```
In[45]:= qZeil[(-1)^k q^(k(k+1)/2 - n k) / qfac[q^2,q^2,k],
                {k,0,m}, n, 2]
                                                    2
                           m   1 + (3 m)/2 + m /2 - n - m n
                       (-1)   q
Out[45]= SUM[n] == ---------------------------------- +
                                        2   2
                                  qfac[q , q , m]
```

```
                           1 - n
      SUM[-2 + n] - q            SUM[-1 + n]
In[46]:= Cert
             1 + k - n
Out[46]= q
```

which means, $cert(n,k) = q^{1+k-n}$ is the proof certificate of the inhomogeneous recurrence ($n \geq 2$)

$$R_n(m) = \frac{q^{1-n(m+1)+(m+3)m/2}}{(q^2;q^2)_m} - q^{-n+1}R_{n-1}(m) + R_{n-2}(m). \tag{22}$$

Since for fixed n and $m \to \infty$ the inhomogeneous parts of (21) and (22) vanish, the sequences $L_n(\infty)$, $R_n(\infty)$, and also $(-q;q)_\infty \cdot R_n(\infty)$ satisfy the same *homogeneous* recurrence of order 2. Therefore, all what is left to prove (20) for $n \geq 0$ is to check (20) at the initial values $n = 0$ and $n = 1$.

For *negative* integer n the proof is entirely analogous; one simply replaces n by $-n$ in the defining sums for $L_n(m)$ and $R_n(m)$, and the initial value check at $n = 0$ and $n = -1$ completes the proof.

The initial value check is an easy consequence of a classical identity which is due to L.J. Rogers:

$$\sum_{k=0}^{\infty} \frac{q^{k(2k+1)}t^{2k}}{(q^2;q^2)_k} = (tq;q)_\infty \sum_{k=0}^{\infty} \frac{q^{\binom{k+1}{2}}t^k}{(q;q)_k(tq;q)_k}. \tag{23}$$

This is Equation (3.22) in Andrews and Askey [1977], where one finds an elegant proof together with some background information.

Eq. (23) with $t = -1$ gives $L_0(\infty) = R_0(\infty)$. Eq. (23) with $t = -1/q$ results in

$$L_{-1}(\infty) = R_1(\infty) + R_0(\infty). \tag{24}$$

Hence, by equality at $n = 0$ we have

$$R_1(\infty) = L_{-1}(\infty) - L_0(\infty) = \sum_{k\geq 0} \frac{q^{k(2k-1)}}{(q^2;q^2)_k}(1-q^{2k}) = L_1(\infty).$$

Finally,

$$R_1(\infty) - R_{-1}(\infty) = (-q;q)_\infty \sum_{k\geq 0}(-1)^k \frac{q^{k(k-1)/2}}{(q^2;q^2)_k}(1-q^{2k}) = -R_0(\infty),$$

which implies $R_{-1}(\infty) = R_1(\infty) + R_0(\infty) = L_{-1}(\infty)$, where the last equality follows from (24).

Acknowledgement We wish to thank Doron Zeilberger for inspiring comments.

References

Abramov, S. A., Paule, P. and Petkovšek, M. [1995], *q-Hypergeometric Solutions of q-Difference Equations*, preprint.

Andrews, G. E. [1974], *Problem 74-12*, SIAM Rev., **16**, 390.

Andrews, G. E. [1976], *The Theory of Partitions*, Encyclopedia of Mathematics and its Applications, **2** (G.-C. Rota, ed.), Addison-Wesley, Reading (Reissued: Cambridge University Press, London and New York, 1985).

Andrews, G. E. [1986], *q-Series: Their Development and Application in Analysis, Number Theory, Combinatorics, Physics, and Computer Algebra*, CBMS Regional Conference Series, No. 66, Amer. Math. Soc., Providence.

Andrews, G. E. [1996], *Pfaff's Method (III): Comparison with the WZ Method*, preprint, Elec. J. Combinatorics, **3**(2), R21.

Andrews, G. E. and Askey, R. [1977], *Enumeration of partitions: The role of Eulerian series and q-orthogonal polynomials*, Higher Combinatorics (M. Aigner, ed.), Reidel Publ. Co., Dordrecht, Holland, 3–26.

Andrews, G. E. and Jackson, D. M. [1990], *An algebraically derived q-analog of a character sum associated with a class of semiregular permutations*, Pac. J. Math., **14**, 207–218.

Bressoud, D. M. [1981], *Solution of Problem 74-12*, SIAM Rev., **23**, 101–104.

Bressoud, D. M. [1987], *Almost Poised Basic Hypergeometric Series*, Proc. Indian Acad. Sci. (Math. Sci.), **96**, 61–66.

Gasper, G. and Rahman, M. [1990], *Basic Hypergeometric Series*, Encyclopedia of Mathematics and its Applications (G.-C. Rota, ed.), **35**, Cambridge University Press, London and New York.

Gessel, I. [1995], *Finding identities with the WZ method*, J. Symbolic Computation, **20**, 537-566.

Gosper, R. W. [1978], *Decision procedures for indefinite hypergeometric summation*, Proc. Natl. Acad. Sci. U.S.A., **75**, 40–42.

Karr, M. [1981], *Summation in finite terms*, Journal of the ACM, **28**, 305–350.

Koornwinder, T. H. [1993], *On Zeilberger's Algorithm and its q-analogue*, J. of Comput. and Appl. Math., **48**, 91–111.

McIntosh, R. J. [1995], *Some Asymptotic Formulae for q-Hypergeometric Series*, preprint, to appear in: J. London Math. Soc.

Paule, P. [1994], *Short and Easy Computer Proofs of the Rogers-Ramanujan Identities and of Identities of Similar Type*, Elec. J. Combinatorics, **1**, R10.

Paule, P. [1995], *Greatest Factorial Factorization and Symbolic Summation*, J. Symbolic Computation, **20**, 235-268.

Paule, P. and Schorn, M. [1993], *A Mathematica version of Zeilberger's algorithm for proving binomial coefficient identities*, J. Symbolic Computation, **20**, 673-698.

Paule, P. and Strehl, V. [1995], *Symbolic summation — some recent developments*, Computeralgebra in science and engineering — algorithms, systems, applications (J. Fleischer, J. Grabmeier, F. Hehl, and W. Kuechlin, eds.), World Scientific, Singapore, 138–162.

Petkovšek, M. [1992], *Hypergeometric solutions of linear recurrences with polynomial coefficients*, J. Symbolic Computation, **14**, 243–264.

Petkovšek, M., Wilf, H. S. and Zeilberger, D. [1996], $A = B$, A. K. Peters.

Riese, A. [1995], *A Mathematica q-analogue of Zeilberger's algorithm for proving q-hypergeometric identities*, diploma thesis, J. Kepler University, Linz.

Riese, A. [1996], *A Generalization of Gosper's Algorithm to Bibasic Hypergeometric Summation*, Elec. J. Combinatorics, **3**, R19.

Stanton, D. [1995], *Talk* at the Workshop "Special Functions, q-Series and Related Topics", organized by the Fields Institute for Research in Mathematical Sciences at University College, 12–23 June 1995, Toronto, Ontario.

Wilf, H. S. [1993], *Identities and their computer proofs*, "SPICE" lecture notes, available via anonymous `ftp` of the file `pub/wilf/lecnotes.ps` at the site `ftp.cis.upenn.edu`.

Wilf, H. S. and Zeilberger, D. [1992a], *Rational function certification of multisum/integral/ 'q'identities*, Bull. Amer. Math. Soc. (N.S.), **27**, 148–153.

Wilf, H. S. and Zeilberger, D. [1992b], *An algorithmic proof theory for hypergeometric (ordinary and "q") multisum/integral identities*, Invent. Math., **108**, 575–633.

Zeilberger, D. [1990], *A fast algorithm for proving terminating hypergeometric identities*, Discr. Math., **80**, 207–211.

Fields Institute Communications
Volume **14**, 1997

Orthogonal Polynomials in the Complex Plane and on the Real Line

Walter Van Assche
Department of Mathematics
Katholieke Universiteit Leuven
B-3001 Heverlee (Leuven), Belgium

Abstract. We present a short introduction into general orthogonal polynomials in the complex plane, with special attention for the real line and the unit circle. Most of the material is classical and available in different textbooks (see the references for relevant literature). This introduction brings together the analysis in the complex plane, the real line and the unit circle, which should be useful for those initiating their research in this field. The emphasis is on extremal properties, location of zeros, recurrence relation, and quadrature rather than on asymptotic results.

1 Orthogonal polynomials in the complex plane

1.1 Preliminaries Let μ be a positive Borel measure in the complex plane and consider the Hilbert space $L_2(\mu)$ of measurable functions ϕ for which

$$\int |\phi(z)|^2 \, d\mu(z) < \infty.$$

In $L_2(\mu)$ we have the inner product

$$\langle \phi, \psi \rangle = \int \phi(z)\overline{\psi(z)} \, d\mu(z), \qquad \phi, \psi \in L_2(\mu).$$

Suppose $\phi_0, \phi_1, \phi_2, \ldots$ is a system of linearly independent functions in $L_2(\mu)$. Quite often it is much more convenient to transform this system of linearly independent functions into another system of linearly independent functions $\varphi_0, \varphi_1, \varphi_2, \ldots$ such that φ_n is a linear combination of the $n+1$ functions $\phi_0, \phi_1, \ldots, \phi_n$ and

$$\langle \varphi_n, \varphi_m \rangle = \int \varphi_n(z)\overline{\varphi_m(z)} \, d\mu(z) = 0, \qquad m \neq n.$$

1991 *Mathematics Subject Classification.* Primary: 42C05; Secondary: 33C45.
Senior Research Associate of the Belgian National Fund for Scientific Research.

This new system of functions is then said to be a system of *orthogonal functions* with respect to the measure μ. If moreover

$$\|\varphi_n\|^2 = \int |\varphi_n(z)|^2 \, d\mu(z) = 1, \qquad n \geq 0,$$

then φ_n are orthonormal functions.

A closed expression can be obtained for these orthogonal functions in terms of the original functions $\phi_n(x)$ and the *Gram matrix*

$$G_n = \begin{pmatrix} \langle \phi_0, \phi_0 \rangle & \langle \phi_1, \phi_0 \rangle & \cdots & \langle \phi_n, \phi_0 \rangle \\ \langle \phi_0, \phi_1 \rangle & \langle \phi_1, \phi_1 \rangle & \cdots & \langle \phi_n, \phi_1 \rangle \\ \vdots & \vdots & \cdots & \vdots \\ \langle \phi_0, \phi_n \rangle & \langle \phi_1, \phi_n \rangle & \cdots & \langle \phi_n, \phi_n \rangle \end{pmatrix}.$$

Such a Gram matrix is always Hermitian and positive definite, since $\langle \phi_k, \phi_l \rangle = \overline{\langle \phi_l, \phi_k \rangle}$ and

$$(x_0, x_1, \dots, x_n) \; G_n \begin{pmatrix} \overline{x_0} \\ \overline{x_1} \\ \vdots \\ \overline{x_n} \end{pmatrix} = \sum_{k=0}^{n} \sum_{l=0}^{n} x_k \overline{x_l} \langle \phi_l, \phi_k \rangle,$$

and if we write $\psi_n = \sum_{k=0}^{n} \overline{x_k} \phi_k$ then this becomes

$$(x_0, x_1, \dots, x_n) \; G_n \begin{pmatrix} \overline{x_0} \\ \overline{x_1} \\ \vdots \\ \overline{x_n} \end{pmatrix} = \langle \psi_n, \psi_n \rangle = \|\psi_n\|^2$$

which is always non-negative and can only be zero when ψ_n is zero μ-always everywhere. As a consequence the determinant

$$\Delta_n = \begin{vmatrix} \langle \phi_0, \phi_0 \rangle & \langle \phi_1, \phi_0 \rangle & \cdots & \langle \phi_n, \phi_0 \rangle \\ \langle \phi_0, \phi_1 \rangle & \langle \phi_1, \phi_1 \rangle & \cdots & \langle \phi_n, \phi_1 \rangle \\ \vdots & \vdots & \cdots & \vdots \\ \langle \phi_0, \phi_n \rangle & \langle \phi_1, \phi_n \rangle & \cdots & \langle \phi_n, \phi_n \rangle \end{vmatrix}$$

is real and positive.

Theorem 1.1 *The orthonormal functions φ_n are given by*

$$\varphi_n(z) = \frac{1}{\sqrt{\Delta_n \Delta_{n-1}}} \begin{vmatrix} \langle \phi_0, \phi_0 \rangle & \langle \phi_1, \phi_0 \rangle & \cdots & \langle \phi_n, \phi_0 \rangle \\ \langle \phi_0, \phi_1 \rangle & \langle \phi_1, \phi_1 \rangle & \cdots & \langle \phi_n, \phi_1 \rangle \\ \vdots & \vdots & \cdots & \vdots \\ \langle \phi_0, \phi_{n-1} \rangle & \langle \phi_1, \phi_{n-1} \rangle & \cdots & \langle \phi_n, \phi_{n-1} \rangle \\ \phi_0(z) & \phi_1(z) & \cdots & \phi_n(z) \end{vmatrix}. \tag{1.1}$$

Proof For the orthogonality, the linearity implies that it is sufficient to show

$$\langle \varphi_n, \phi_k \rangle = 0, \qquad k = 0, 1, \dots, n-1.$$

By expanding the determinant along the last row this easily gives

$$\langle \varphi_n, \phi_k \rangle = \frac{1}{\sqrt{\Delta_n \Delta_{n-1}}} \begin{vmatrix} \langle \phi_0, \phi_0 \rangle & \langle \phi_1, \phi_0 \rangle & \cdots & \langle \phi_n, \phi_0 \rangle \\ \langle \phi_0, \phi_1 \rangle & \langle \phi_1, \phi_1 \rangle & \cdots & \langle \phi_n, \phi_1 \rangle \\ \vdots & \vdots & \cdots & \vdots \\ \langle \phi_0, \phi_{n-1} \rangle & \langle \phi_1, \phi_{n-1} \rangle & \cdots & \langle \phi_n, \phi_{n-1} \rangle \\ \langle \phi_0, \phi_k \rangle & \langle \phi_1, \phi_k \rangle & \cdots & \langle \phi_n, \phi_k \rangle \end{vmatrix}.$$

When $0 \leq k \leq n-1$ it follows that the last row is equal to the $(k+1)$th row so that the determinant is zero. Furthermore we see that

$$\langle \varphi_n, \phi_n \rangle = \sqrt{\frac{\Delta_n}{\Delta_{n-1}}}.$$

Observe that

$$\varphi_n(z) = \sqrt{\frac{\Delta_{n-1}}{\Delta_n}} \phi_n(z) + \psi_{n-1}(z)$$

where ψ_{n-1} is a linear combination of the first n functions $\phi_0, \phi_1, \ldots, \phi_{n-1}$, hence

$$\langle \varphi_n, \varphi_n \rangle = \sqrt{\frac{\Delta_{n-1}}{\Delta_n}} \langle \varphi_n, \phi_n \rangle = 1,$$

which shows the orthonormality.

□

Our interest lies in the linearly independent system of monomials $1, z, z^2, z^3, \ldots$ and from this basis in the linear space of polynomials we want to obtain a system of orthogonal polynomials with respect to a given positive measure μ. The system of orthonormal polynomials is then unique if we impose that the leading coefficient is positive. We thus have a system of polynomials φ_n such that

$$\int \varphi_n(z) \overline{\varphi_m(z)}\, d\mu(z) = \delta_{m,n}, \qquad m, n \geq 0,$$

with $\varphi_n(z) = \kappa_n z^n + \cdots$ and $\kappa_n > 0$.

1.2 Extremal properties One of the main advantages of working with orthonormal functions is that one can easily obtain Fourier expansions of functions in $L_2(\mu)$. Let $f \in L_2(\mu)$, then we can write the formal expansion

$$f(z) \sim \sum_{k=0}^{\infty} f_k \varphi_k(z).$$

The coefficients f_k are then called *Fourier coefficients.* By multiplying both sides of the expansion by $\overline{\varphi_n(z)}$ and then integrating with respect to μ, we find a simple expression for these Fourier coefficients:

$$f_n = \int f(z) \overline{\varphi_n(z)}\, d\mu(z) = \langle f, \varphi_n \rangle. \tag{1.2}$$

If we use the Cauchy-Schwarz inequality, then we find that

$$|f_n| \leq \left(\int |f(z)|^2\, d\mu(z) \right)^{1/2} \left(\int |\varphi_n(z)|^2\, d\mu(z) \right)^{1/2} = \|f\|,$$

so that all the Fourier coefficients are finite. The partial sums of this Fourier series then have a very useful extremal property: the nth partial sum

$$f_n(z) = \sum_{k=0}^{n} f_k \varphi_k(z)$$

is the best $L_2(\mu)$ approximation to f with polynomials of degree at most n.

Theorem 1.2 *Let $f \in L_2(\mu)$, then the integral*

$$\int |f(z) - q_n(z)|^2 \, d\mu(z)$$

attains its minimum taken over all polynomials q_n of degree at most n when q_n is the nth partial sum f_n of the Fourier series. The minimum is

$$\|f\|^2 - \sum_{k=0}^{n} |f_k|^2. \tag{1.3}$$

Proof One easily obtains

$$\int |f(z)-q_n(z)|^2 \, d\mu(z) = \int |f(z)|^2 \, d\mu(z) + \int |q_n(z)|^2 \, d\mu(z) - 2\Re \int f(z)\overline{q_n(z)} \, d\mu(z).$$

If we write $q_n(z) = \sum_{k=0}^{n} a_k \varphi_k(z)$, then

$$\int |q_n(z)|^2 \, d\mu(z) = \sum_{k=0}^{n} \sum_{j=0}^{n} a_k \overline{a_j} \langle \varphi_k, \varphi_j \rangle = \sum_{k=0}^{n} |a_k|^2,$$

and

$$\int f(z)\overline{q_n(z)} \, d\mu(z) = \sum_{k=0}^{n} \overline{a_k} \langle f, \varphi_k \rangle = \sum_{k=0}^{n} f_k \overline{a_k}.$$

Therefore

$$\begin{aligned} \int |f(z) - q_n(z)|^2 \, d\mu(z) &= \|f\|^2 + \sum_{k=0}^{n} |a_k|^2 - 2\Re \sum_{k=0}^{n} f_k \overline{a_k} \\ &= \|f\|^2 - \sum_{k=0}^{n} |f_k|^2 + \sum_{k=0}^{n} |a_k - f_k|^2. \end{aligned}$$

Clearly this expression is minimal whenever $a_k = f_k$ in which case the polynomial q_n becomes the nth partial sum of the Fourier series for f, and the minimum is given by (1.3).

□

As a consequence we always have the inequality

$$\sum_{k=0}^{n} |f_k|^2 \leq \|f\|^2,$$

and as $n \to \infty$ this gives *Bessel's inequality*

$$\sum_{k=0}^{\infty} |f_k|^2 \leq \|f\|^2.$$

The case when the Bessel inequality becomes an equality for every $f \in L_2(\mu)$ is very important and occurs when polynomials are dense in $L_2(\mu)$.

If we use the formula (1.2) for the Fourier coefficients, then the partial sum for the Fourier expansion becomes

$$f_n(z) = \int K_n(z,\xi) f(\xi)\, d\mu(\xi),$$

where the kernel $K_n(z,\xi)$ is given by

$$K_n(z,\xi) = \sum_{k=0}^{n} \varphi_k(z)\overline{\varphi_k(\xi)}.$$

If f is a polynomial of degree at most n, then the partial sum f_n coincides with f and we have the property that for every polynomial q_n of degree at most n

$$q_n(z) = \int K_n(z,\xi) q_n(\xi)\, d\mu(\xi).$$

For this reason we often call $K_n(z,\xi)$ the *reproducing kernel.*

A particular case of the extremal property described in Theorem 1.2 gives an interesting interpretation of the leading coefficient of φ_n.

Theorem 1.3 *Suppose that the support of μ contains at least n points. The minimum of the $L_2(\mu)$-norm*

$$\|q_n\| = \left(\int |q_n(z)|^2\, d\mu(z) \right)^{1/2}$$

taken over all monic polynomials q_n of degree n is attained when $q_n = \varphi_n/\kappa_n$. The minimum is equal to $1/\kappa_n$.

Proof Each monic polynomial q_n can be written as

$$q_n(z) = \hat{\varphi}_n(z) + r_{n-1}(z),$$

where $\hat{\varphi}_n = \varphi_n/\kappa_n$ is the monic orthogonal polynomial of degree n and r_{n-1} is some polynomial of degree at most $n-1$. We then have

$$\begin{aligned} \int |q_n(z)|^2\, d\mu(z) &= \int |\hat{\varphi}_n(z)|^2\, d\mu(z) + \int |r_{n-1}(z)|^2\, d\mu(z) \\ &\quad + 2\Re \int \hat{\varphi}_n(z)\overline{r_{n-1}(z)}\, d\mu(z). \end{aligned}$$

The last term vanishes by orthogonality, hence

$$\int |q_n(z)|^2\, d\mu(z) = \int |\hat{\varphi}_n(z)|^2\, d\mu(z) + \int |r_{n-1}(z)|^2\, d\mu(z),$$

and this expression is minimal when

$$\int |r_{n-1}(z)|^2\, d\mu(z) = 0.$$

This integral is zero if and only if $r_{n-1} = 0$, since r_{n-1} has at most $n-1$ zeros and μ has at least n points in its support. Hence the minimum is attained when $q_n(z) = \hat{\varphi}_n(z)$ and since $\varphi_n(z)$ has norm one, we find that the norm of $q_n(z)$ is $1/\kappa_n$. □

Another way of expressing this theorem is as follows: the best $L_2(\mu)$-approximation to z^n by polynomials of degree $< n$ is given by the polynomial $z^n - \hat{\varphi}_n(z)$ and the error is given by $1/\kappa_n$. Hence the leading coefficient of the orthonormal polynomial of degree n indicates how well a polynomial of degree n can be approximated in $L_2(\mu)$ by polynomials of lower degree.

The reproducing kernel $K_n(z,\xi)$ also has an extremal property.

Theorem 1.4 *The minimum of the integral*

$$\int |q_n(z)|^2 \, d\mu(z)$$

taken over all polynomials q_n of degree at most n for which $q_n(\xi) = 1$, is attained for the polynomial

$$\frac{\sum_{k=0}^n \varphi_k(z)\overline{\varphi_k(\xi)}}{\sum_{k=0}^n |\varphi_k(\xi)|^2} = \frac{K_n(z,\xi)}{K_n(\xi,\xi)}.$$

The minimum is equal to $[K_n(\xi,\xi)]^{-1}$.

Proof We can express the polynomial q_n as a Fourier series in the first $n+1$ orthonormal polynomials φ_k $(k = 0, 1, \ldots, n)$:

$$q_n(z) = \sum_{k=0}^n a_k \varphi_k(z).$$

The integral then becomes

$$\int |q_n(z)|^2 \, d\mu(z) = \sum_{k=0}^n \sum_{j=0}^n a_k \overline{a_j} \int \varphi_k(z) \overline{\varphi_j(z)} \, d\mu(z) = \sum_{k=0}^n |a_k|^2,$$

and the constraint $q_n(\xi) = 1$ becomes

$$\sum_{k=0}^n a_k \varphi_k(\xi) = 1.$$

By the Cauchy-Schwarz inequality we have

$$1 = \left| \sum_{k=0}^n a_k \varphi_k(\xi) \right|^2 \leq \left(\sum_{k=0}^n |a_k|^2 \right) \left(\sum_{k=0}^n |\varphi_k(\xi)|^2 \right), \tag{1.4}$$

so that

$$\sum_{k=0}^n |a_k|^2 \geq \frac{1}{\sum_{k=0}^n |\varphi_k(\xi)|^2}.$$

This lower bound for the required integral is attained when there is equality in (1.4), i.e., when $a_k = C\overline{\varphi_k(\xi)}$, with C some complex constant. The constraint thus becomes

$$C \sum_{k=0}^n |\varphi_k(\xi)|^2 = 1,$$

so that $C = 1/\sum_{k=0}^n |\varphi_k(\xi)|^2$. The minimizing polynomial is therefore given by $K_n(z,\xi)/K_n(\xi,\xi)$ and the minimum is $1/K_n(\xi,\xi)$, as asserted.

□

The reproducing kernel with equal arguments $K_n(\xi,\xi)$ thus indicates how small the $L_2(\mu)$-norm of polynomials of degree n can be made, if these polynomials are forced to take the specific value 1 at ξ. The reciprocal of the reproducing kernel with equal arguments is usually called the *Christoffel function*, and one often writes

$$\lambda_n(z) = \frac{1}{K_n(z,z)} = \left(\sum_{k=0}^{n} |\varphi_k(z)|^2\right)^{-1}.$$

A dual version of this extremal property is

Theorem 1.5 *Let $\xi \in \mathbb{C}$ be fixed, then the maximum of $|q_n(\xi)|^2$ taken over all polynomials q_n of degree at most n with unit $L_2(\mu)$-norm*

$$\int |q_n(z)|^2 \, d\mu(z) = 1,$$

is attained for

$$q_n(z) = e^{i\theta} \frac{K_n(z,\xi)}{\sqrt{K_n(\xi,\xi)}}$$

for any $\theta \in [0, 2\pi)$, and the maximum is $K_n(\xi,\xi)$.

Proof Consider the polynomial $\tilde{q}_n(z) = q_n(z)/q_n(\xi)$, then $\tilde{q}_n(\xi) = 1$ and

$$\int |\tilde{q}_n(z)|^2 \, d\mu(z) = \frac{1}{|q_n(\xi)|^2}.$$

Hence the maximum of $|q_n(\xi)|^2$ taken over all polynomials q_n of degree at most n with $\|q_n\| = 1$ corresponds to the reciprocal of the minimum of $\|\tilde{q}_n\|$ taken over all polynomials $\tilde{q}_n$ of degree at most n for which $\tilde{q}_n(\xi) = 1$. We already know that this minimum is attained for $\tilde{q}_n(z) = K_n(z,\xi)/K_n(\xi,\xi)$ and is equal to $1/K_n(\xi,\xi)$, hence the required maximum is $K_n(\xi,\xi)$ and is attained for $q_n(z) = q_n(\xi)\tilde{q}_n(z) = e^{i\theta}\sqrt{K_n(\xi,\xi)}K_n(z,\xi)/K_n(\xi,\xi)$, which proves the theorem.

□

1.3 Properties of the zeros When studying polynomials, one of the most relevant quantities to consider are the zeros. The fundamental theorem of algebra asserts that any polynomial of degree n has exactly n zeros (counting multiplicities). When dealing with orthogonal polynomials, one can say much more about their location.

The support of a measure μ is given by

$$\text{supp}(\mu) = \{z \in \mathbb{C} : \mu(B_{z,\epsilon}) > 0 \text{ for every } \epsilon > 0\},$$

where $B_{z,\epsilon}$ is the open disk with center z and radius ϵ. One easily verifies that $\text{supp}(\mu)$ is a closed set. By the convex hull $\text{Co}(A)$ of a set $A \subset \mathbb{C}$ we mean the smallest convex set containing A, and a set $G \subset \mathbb{C}$ is convex if for each pair of points $x, y \in G$ the line connecting x and y is a subset of G. Clearly

$$\text{Co}(A) = \bigcap_{\substack{A \subset G \subset \mathbb{C} \\ G \text{ convex}}} G,$$

and if A is closed then $\text{Co}(A)$ is the intersection of all closed half-planes containing A.

Theorem 1.6 (Fejér) *If φ_n is the orthogonal polynomial of degree n with respect to the measure μ, then all its zeros lie in* $\mathrm{Co}(\mathrm{supp}(\mu))$.

Proof Suppose that z_0 is a zero outside $\mathrm{Co}(\mathrm{supp}(\mu))$, then we write

$$\hat{\varphi}_n(z) = (z - z_0)p_{n-1}(z),$$

where p_{n-1} is a monic polynomial of degree $n-1$. Since $z_0 \notin \mathrm{Co}(\mathrm{supp}(\mu))$, there exists a closed half-plane G such that $\mathrm{supp}(\mu) \subset G$ and $z_0 \notin G$. The boundary of this half-plane is a line L separating $\mathrm{supp}(\mu)$ and z_0. Let z_1 be the point on L closest to z_0, then for every $z \in \mathrm{supp}(\mu)$

$$|z - z_0| = |z - z_1| + |\hat{z} - z_0|,$$

where $\hat{z}$ is the intersection of the the line connecting z and z_0 and the circle with center z and radius $|z - z_1|$.

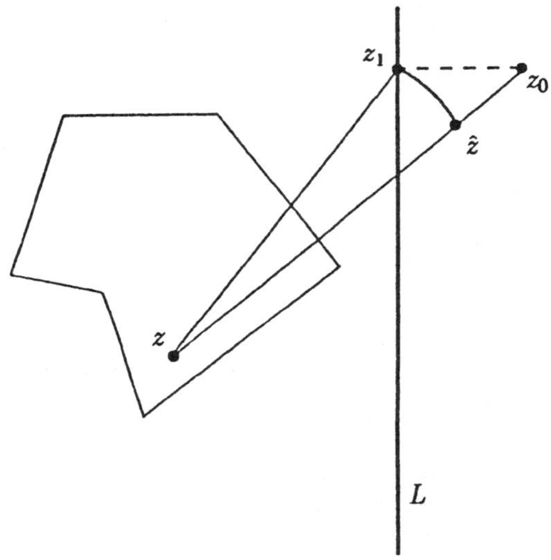

We thus have

$$|z - z_0| > |z - z_1|, \qquad \text{for all } z \in \mathrm{supp}(\mu).$$

Therefore

$$\int |(z - z_1)p_{n-1}(z)|^2 \, d\mu(z) < \int |\hat{\varphi}_n(z)|^2 \, d\mu(z),$$

which is in contradiction with the extremal property given in Theorem 1.3. Therefore we must conclude that there are no zeros outside $\mathrm{Co}(\mathrm{supp}(\mu))$.

□

More turns out to be true (Saff [1990]):

Theorem 1.7 *If* $\mathrm{Co}(\mathrm{supp}(\mu))$ *is not a line segment and* $\mathrm{supp}(\mu)$ *is an infinite set, then the zeros of the orthogonal polynomials for the measure μ can not be on the boundary of the convex hull of* $\mathrm{supp}(\mu)$.

Proof Suppose z_0 is a zero of ϕ_n on the boundary of the convex hull of $E = \mathrm{supp}(\mu)$. Then there is a line L containing z_0 such that E is contained in the closed half-plane bounded by this line L. Let L be the set $\{z = x + iy \in \mathbb{C} : ax + by = c\}$, where $a, b, c \in \mathbb{R}$. Without loss of generality we can assume that E is contained in the half-plane $\{z = x + iy \in \mathbb{C} : ax + by \leq c\}$.

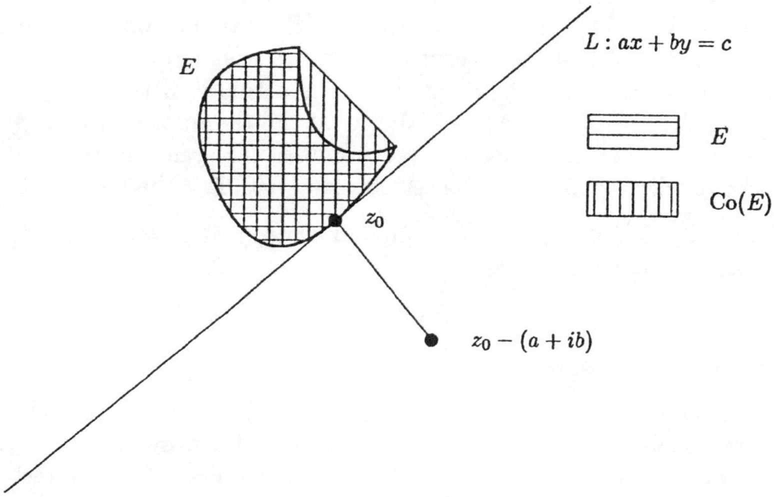

Since $z_0 = x_0 + iy_0 \in L$ we have $ax_0 + by_0 = c$, hence the point $z_0 - (a + ib)$ belongs to the complementary half-plane $\{z = x + iy \in \mathbb{C} : ax + by < c\}$. Since z_0 is a zero of φ_n, we have for the monic polynomial $\hat{\varphi}_n(z) = (z - z_0)p_{n-1}(z)$, where p_{n-1} is a monic polynomial of degree $n - 1$. Consider the monic polynomial

$$q_n(z) = [z - z_0 + t(a + ib)]p_{n-1}(z),$$

with $t \in \mathbb{R}$. Then by the extremal property in Theorem 1.3, the integral

$$I(t) = \int_E |q_n(z)|^2 \, d\mu(z) = \int_E [(x - x_0 + ta)^2 + (y - y_0 + tb)^2]|p_{n-1}(z)|^2 \, d\mu(z)$$

is minimal when $t = 0$. Since this is a real function of the real variable t, this implies that $I'(0) = 0$. Clearly

$$I'(0) = 2 \int_E [a(x - x_0) + b(y - y_0)]|p_{n-1}(z)|^2 \, d\mu(z),$$

but $ax_0 + by_0 = c$ and for every $z = x + iy \in E$ we have $ax + by \geq c$, hence

$$I'(0) = 2 \int_E [ax + by - c]|p_{n-1}(z)|^2 \, d\mu(z)$$

and $I'(0) = 0$ thus implies that $[ax + by - c]|p_{n-1}(z)|^2 = 0$ μ-almost everywhere. This means that the support of μ must be a subset of the set $L \cup Z_n$, where Z_n are the zeros of p_{n-1} and thus also zeros of φ_n. There are two possibilities: either all the points of E are on the line L or at least one point of E is in Z_n. The first possibility must be excluded since then the convex hull of E would be a line segment. For the second possibility the convex hull of E will be a polygon with one side on L and vertices at the points of E which are in Z_n. Repeat the previous argument with one of the vertices z_1 in Z_n, then we conclude that there will only be a finite number of points of E in the open half-plane bounded by a line L' through z_1. But this would imply that E contains a finite number of points. This contradiction proves that there can be no zeros on the boundary of Co(E).

□

The convex hull of the support of μ therefore gives a good global idea of where the zeros of the orthogonal polynomials are located. Even though one expects that the zeros are on the support of the measure, this is not true, since there may be zeros on the set $\mathrm{Co}(E) \setminus E$, where $E = \mathrm{supp}(\mu)$. Furthermore, if the set E has holes, then there may be zeros in these holes, and in fact all the zeros may be in the holes, as is shown by considering orthogonal polynomials on the unit circle. It turns out that most of the zeros will be on the set obtained by filling the holes of E.

Definition 1.8 *Let D_∞ be the connected component of the complement of E that contains ∞, then D_∞ is open and*

$$\mathrm{Pc}(E) = \mathbb{C} \setminus D_\infty$$

is the polynomial convex hull of E.

Clearly $\mathrm{Pc}(E) \subset \mathrm{Co}(E)$. The terminology comes from several complex variables (Hörmander [1966]). For a set $A \subset \mathbb{C}^n$ the polynomial convex hull is the set of all $x \in \mathbb{C}^n$ for which $|f(x)| \leq \sup_{z \in A} |f(z)|$ for all polynomials f. If we restrict f to polynomials of degree 1, then this gives the convex hull. For one complex variable the polynomial convex hull of $A \subset \mathbb{C}$ coincides with the complement of the outer component of A.

We will now show that most of the zeros of orthogonal polynomials are in the polynomial convex hull of the support of μ (Stahl and Totik [1992], Theorem 2.1.1):

Theorem 1.9 (Widom) *Suppose K is a compact set in D_∞. Then the number of zeros of φ_n in K is bounded by a constant independent of n.*

Proof Suppose that the number of zeros of φ_n in K tends to infinity as $n \to \infty$, then there are infinitely many zeros in K and thus there exists a point $z_0 \in K$ such that for any neighborhood $D(z_0)$ of z_0 the number of zeros of φ_n is unbounded as $n \to \infty$. Let $f(z) = 1/(z - z_0)$, then f is analytic in $\mathbb{C} \setminus D(z_0)$. If $E = \mathrm{supp}(\mu)$, then $f(\mathrm{Pc}(E))$ is a compact set such that $0 \notin f((\mathrm{Pc}(E))$. Let $\epsilon > 0$ be given, then Runge's theorem[1] shows that there exists a polynomial q_{m-1} of degree $m-1$ such that

$$|\frac{1}{x} - q_{m-1}(x)| \leq \epsilon, \qquad \forall x \in f(\mathrm{Pc}(E)),$$

so that

$$|z - z_0 - q_{m-1}(f(z))| \leq \epsilon, \qquad \forall z \in E \subset \mathrm{Pc}(E).$$

Let $p_{m-1}(z) = z^{m-1} q_{m-1}(1/z)$, then p_{m-1} is a polynomial of degree $m-1$ and we have

$$|z - z_0 - f(z)^{m-1} p_{m-1}(z - z_0)| \leq \epsilon, \qquad \forall z \in E.$$

If we choose $\epsilon \leq d(z_0, E)/2$, then we find

$$\left| \frac{(z - z_0)^m - p_{m-1}(z - z_0)}{(z - z_0)^m} \right| \leq 1/2, \qquad \forall z \in E.$$

[1]Let K be a compact set in the complex plane so that $\overline{\mathbb{C}} \setminus K$ is connected, and f be an analytic function on Ω, where Ω is some open set containing K. Then there is a sequence of polynomials P_n such that $P_n \to f$ uniformly on K (see, e.g., Rudin [1978], p. 290).

The polynomial $Q_m(z) = (z - z_0)^m - p_{m-1}(z - z_0)$ is therefore a polynomial of degree m satisfying for all $z \in E$

$$|Q_m(z)/(z - z_0)^m| \leq 1/2.$$

Let $D_\delta(z_0)$ be an open disk with center z_0 and radius δ, then the number of zeros of φ_n in $D_\delta(z_0)$ is unbounded as $n \to \infty$, and hence there exists an integer n such that φ_n has at least m zeros in $D_\delta(z_0)$. Denote m of these zeros by $z_1, z_2, \dots, z_m$, then

$$\left| \frac{Q_m(z)}{\prod_{k=1}^m (z - z_k)} \right| \leq \frac{1}{2} \prod_{k=1}^m \left| \frac{z - z_0}{z - z_k} \right|,$$

and since for every $z \in E$

$$\left| \frac{z - z_0}{z - z_k} \right| \leq \left| \frac{z - z_0}{z - z_0 - z_k + z_0} \right| \leq \frac{1}{1 - \delta/d(z_0, E)},$$

the choice $\delta = d(z_0, E)(1 - a)$ gives

$$\left| \frac{Q_m(z)}{\prod_{k=1}^m (z - z_k)} \right| \leq \frac{1}{2a^m}, \qquad \forall z \in E.$$

We thus have

$$\int_E \left| \frac{\hat{\varphi}_n(z) Q_m(z)}{\prod_{k=1}^m (z - z_k)} \right|^2 d\mu(z) \leq \frac{1}{4a^{2m}} \int_E |\hat{\varphi}_n(z)|^2 \, d\mu(z),$$

which for $2a^m > 1$ is in contradiction with the extremal property of Theorem 1.3, and thus we conclude that the number of zeros of $\varphi_n(z)$ in K remains bounded.

□

1.4 Examples

1.4.1 *Orthogonal polynomials on a disk* Let μ be Lebesgue measure (uniform measure) on the disk $D_R = \{z \in \mathbb{C} : |z| \leq R\}$, where $R > 0$, so that

$$\int_{D_R} f(z) \, d\mu(z) = \iint_{D_R} f(x + iy) \, dx \, dy = \int_0^{2\pi} \int_0^R f(re^{i\theta}) \, r \, dr \, d\theta.$$

The entries of the Gram matrix are then given by

$$\begin{aligned} \langle z^k, z^l \rangle &= \int_{D_R} z^k \overline{z^l} \, d\mu(z) \\ &= \int_0^{2\pi} \int_0^R r^{k+l} e^{i(k-l)\theta} \, r \, dr \, d\theta \\ &= \frac{R^{k+l+2}}{k + l + 2} \int_0^{2\pi} e^{i(k-l)\theta} \, d\theta. \end{aligned}$$

The (complex) trigonometrical system $e^{ik\theta}$ $(k \in \mathbb{Z})$ is well-known to be orthogonal and a simple computation gives

$$\langle z^k, z^l \rangle = \delta_{k,l} \frac{2\pi R^{2k+2}}{2k + 2}.$$

The Gram matrix G_n therefore becomes a diagonal matrix with determinant

$$\Delta_n = \frac{\pi^{n+1} R^{(n+1)(n+2)}}{(n+1)!}.$$

The monic orthogonal polynomials are thus

$$\hat{\varphi}_n(z) = z^n,$$

and the leading coefficient κ_n is

$$\kappa_n = \sqrt{\frac{\Delta_{n-1}}{\Delta_n}} = \sqrt{\frac{n+1}{\pi}} R^{-n-1},$$

so that the orthonormal polynomials are

$$\varphi_n(z) = \sqrt{\frac{n+1}{\pi R^2}} \left(\frac{z}{R}\right)^n.$$

This shows that monomials are orthogonal polynomials on a disk, and the radius R of the disk only shows up in the normalization.

1.4.2 *Orthogonal polynomials on a curve* Let C be a finite number of rectifiable Jordan arcs. For the measure μ we take some measure which is absolutely continuous with respect to arc length, so that the inner product between two functions f and g becomes

$$\langle f, g \rangle = \int_C f(z)\overline{g(z)}\, |dz|.$$

As a special case we can consider a circle C_R with center at the origin and radius R. If we take Lebesgue measure, so that

$$\int_{C_R} f(z)\, d\mu(z) = R \int_0^{2\pi} f(Re^{i\theta})\, d\theta,$$

then the entries of the Gram matrix are

$$\langle z^k, z^l \rangle = R^{k+l+1} \int_0^{2\pi} e^{i(k-l)\theta}\, d\theta = 2\pi R^{2k+1} \delta_{k,l},$$

hence the Gram matrix G_n is diagonal and its determinant is $\Delta_n = (2\pi)^{n+1} R^{(n+1)^2}$. The monic orthogonal polynomials on the circle C_R are thus given by

$$\hat{\varphi}_n(z) = z^n,$$

and the leading coefficient $\kappa_n = \sqrt{\Delta_{n-1}/\Delta_n}$ is equal to $R^{-n}/\sqrt{2\pi R}$ so that the orthonormal polynomials are

$$\varphi_n(z) = \frac{1}{\sqrt{2\pi R}} \left(\frac{z}{R}\right)^n.$$

Note the similarity with the orthogonal polynomials on the disk D_R.

2 Orthogonal polynomials on the real line

2.1 Hankel matrices Suppose a positive Borel measure μ on the real line is given. We assume that all the moments

$$m_n = \int x^n \, d\mu(x) \tag{2.1}$$

are finite and that the support of μ

$$\operatorname{supp}(\mu) = \{x \in \mathbb{R} : \mu(x-\epsilon, x+\epsilon) > 0 \text{ for every } \epsilon > 0\}$$

contains infinitely many points. Also, without loss of generality, we assume that we are dealing with a probability measure so that $\mu(\mathbb{R}) = 1$. Orthogonalizing the linear space of polynomials, we obtain a system of orthonormal polynomials on the real line which will be denoted by p_n $(n = 0, 1, 2, \ldots)$. We are now dealing with real-valued functions so that we do not need complex conjugation in the inner product, and so

$$\int_{-\infty}^{\infty} p_n(x) p_m(x) \, d\mu(x) = \delta_{m,n}, \qquad m, n \geq 0,$$

and this system is unique if we impose the leading coefficients, denoted in the real case by γ_n, to be positive:

$$p_n(x) = \gamma_n x^n + \delta_n x^{n-1} + \cdots, \qquad \gamma_n > 0.$$

The Gram matrix now becomes

$$H_n = \begin{pmatrix} m_0 & m_1 & \cdots & m_n \\ m_1 & m_2 & \cdots & m_{n+1} \\ \vdots & \vdots & \cdots & \vdots \\ m_n & m_{n+1} & \cdots & m_{2n} \end{pmatrix},$$

and matrices of this type, with constant values along anti-diagonals, are known as *Hankel matrices.* The orthonormal polynomial of degree n is thus given by

$$p_n(x) = \frac{1}{\sqrt{\Delta_n \Delta_{n-1}}} \begin{vmatrix} m_0 & m_1 & m_2 & \cdots & m_n \\ m_1 & m_2 & m_3 & \cdots & m_{n+1} \\ \vdots & \vdots & \vdots & \cdots & \vdots \\ m_{n-1} & m_n & m_{n+1} & \cdots & m_{2n-1} \\ 1 & x & x^2 & \cdots & x^n \end{vmatrix},$$

where $\Delta_n = \det H_n$ is the determinant of the Hankel matrix H_n and the leading coefficient is given by a ratio of two Hankel determinants

$$\gamma_n = \sqrt{\frac{\Delta_{n-1}}{\Delta_n}}.$$

2.2 Recurrence relation The representation in terms of Hankel matrices is not very useful for the actual computation of the orthogonal polynomials since this requires the evaluation of determinants. For orthogonal polynomials on the real line there is a much more efficient and linear way for computation by means of a three-term recurrence relation.

Theorem 2.1 *Orthonormal polynomials $p_n(x) = \gamma_n x^n + \delta_n x^{n-1} + \cdots$ on the real line satisfy a three-term recurrence relation*

$$xp_n(x) = a_{n+1}p_{n+1}(x) + b_n p_n(x) + a_n p_{n-1}(x), \tag{2.2}$$

with initial conditions $p_0 = 1$ and $p_{-1} = 0$. The recurrence coefficients are given by

$$a_n = \int xp_{n-1}(x)p_n(x)\,d\mu(x) = \frac{\gamma_{n-1}}{\gamma_n} > 0, \tag{2.3}$$

$$b_n = \int xp_n^2(x)\,d\mu(x) = \frac{\delta_n}{\gamma_n} - \frac{\delta_{n+1}}{\gamma_{n+1}}. \tag{2.4}$$

The monic orthogonal polynomials $\hat{p}_n = p_n/\gamma_n$ satisfy the three-term recurrence relation

$$\hat{p}_{n+1}(x) = (x - b_n)\hat{p}_n(x) - a_n^2 \hat{p}_{n-1}(x), \tag{2.5}$$

with initial conditions $\hat{p}_0 = 1$ and $\hat{p}_{-1} = 0$.

Proof The polynomial $xp_n(x)$ has degree $n+1$ and can therefore be expressed as a Fourier series using the first $n+2$ orthonormal polynomials. This gives

$$xp_n(x) = \sum_{k=0}^{n+1} c_{k,n} p_k(x),$$

where the Fourier coefficients $c_{k,n}$ are given by

$$c_{k,n} = \int xp_n(x)p_k(x)\,d\mu(x).$$

Observe that $xp_k(x)$ is a polynomial of degree $k+1$, hence if $k+1 < n$ then the integral for $c_{k,n}$ vanishes because of the orthogonality. Therefore the Fourier series for $xp_n(x)$ only contains three terms. If we put $c_{n-1,n} = a_n$ and $c_{n,n} = b_n$ and observe that $c_{n+1,n} = c_{n,n+1} = a_{n+1}$, then the three-term recurrence relation for the orthonormal polynomials follows. By comparing the coefficient of x^{n+1} on both sides of (2.2) we find $\gamma_n = a_{n+1}\gamma_{n+1}$ from which $a_n = \gamma_{n-1}/\gamma_n$ follows. Comparing the coefficients of x^n gives $\delta_n = a_{n+1}\delta_{n+1} + b_n\gamma_n$, which gives

$$b_n = \frac{\delta_n}{\gamma_n} - \frac{\delta_{n+1}}{\gamma_{n+1}}.$$

If we substitute p_n by $\gamma_n \hat{p}_n$ then the recurrence relation for the monic orthogonal polynomials follows.

□

This theorem also has a converse, known as Favard's theorem (Favard [1935]), which says that a system of polynomials satisfying a three-term recurrence relation as in (2.2), with $a_{n+1} > 0$ and $b_n \in \mathbb{R}$ for all $n \in \mathbb{N}$, will always be an orthonormal system with respect to some positive measure μ on the real line. This measure will in general not be unique, but a sufficient condition for unicity is that the recurrence coefficients are bounded. The result is much older than Favard's paper and was already known to Chebyshev when the measure μ has finite support.

From the recurrence relation we may obtain some formulas which will be useful. If we multiply both sides of the recurrence relation (2.2) by $p_n(y)$ then

$$xp_n(x)p_n(y) = a_{n+1}p_{n+1}(x)p_n(y) + b_np_n(x)p_n(y) + a_np_{n-1}(x)p_n(y).$$

Interchanging x and y gives

$$yp_n(x)p_n(y) = a_{n+1}p_{n+1}(y)p_n(x) + b_np_n(x)p_n(y) + a_np_{n-1}(y)p_n(x).$$

Now subtract both equations, then

$$\begin{aligned}(x-y)p_n(x)p_n(y) &= a_{n+1}[p_{n+1}(x)p_n(y) - p_n(x)p_{n+1}(y)] \\ &\quad - a_n[p_n(x)p_{n-1}(y) - p_{n-1}(x)p_n(y)].\end{aligned}$$

By summing over n, the right-hand side becomes a telescoping series, so that

$$\sum_{k=0}^{n-1} p_k(x)p_k(y) = a_n \frac{p_n(x)p_{n-1}(y) - p_{n-1}(x)p_n(y)}{x-y}. \tag{2.6}$$

This formula is known as the *Christoffel-Darboux* identity. When y tends to x we obtain its confluent form

$$\sum_{k=0}^{n-1} p_k^2(x) = a_n[p_n'(x)p_{n-1}(x) - p_{n-1}'(x)p_n(x)]. \tag{2.7}$$

2.3 Properties of the zeros Orthogonal polynomials on the real line have zeros with nice properties.

Theorem 2.2 *The zeros of p_n are all real and simple. If the support of μ is a subset of the interval $[a,b]$ then all the zeros of p_n are in $[a,b]$.*

Proof Since $\mathrm{Co}(\mathrm{supp}(\mu)) \subset [a,b]$, Theorem 1.6 implies that all the zeros are in $[a,b]$. Alternatively we can also reason as follows: suppose there are k zeros $x_1,\dots,x_k$ outside $[a,b]$, then $p_n(x)/(x-x_1)\cdots(x-x_k)$ is a polynomial of degree $n-k$ with all its zeros inside $[a,b]$. If $k>0$ then by orthogonality

$$\int_a^b \frac{p_n^2(x)}{(x-x_1)\cdots(x-x_k)}\, d\mu(x) = 0,$$

but $p_n^2(x)/(x-x_1)\cdots(x-x_k)$ does not change sign on $[a,b]$ and $\mathrm{supp}(\mu)$ contains infinitely many points, so that this integral can never vanish. Therefore the assumption $k>0$ cannot be true and we conclude that $k=0$. Hence all the zeros of p_n are real and in $[a,b]$. If x_0 is a zero of multiplicity >1, then $p_n(x)/(x-x_0)^2$ is a polynomial of degree $n-2$ and by orthogonality

$$\int_a^b \frac{p_n^2(x)}{(x-x_0)^2}\, d\mu(x) = 0,$$

but $p_n^2(x)/(x-x_0)^2$ is positive on $[a,b]$ and $\mathrm{supp}(\mu)$ contains infinitely many points, so that the integral cannot vanish. Therefore there are no zeros of multiplicity >1.

□

The reality of the zeros can also be obtained from the recurrence relation. In matrix form we can write (2.2) as

$$\begin{pmatrix} b_0 & a_1 & 0 & 0 & \cdots & 0 \\ a_1 & b_1 & a_2 & 0 & \cdots & 0 \\ 0 & a_2 & b_2 & a_3 & & 0 \\ \vdots & & \ddots & \ddots & \ddots & \vdots \\ 0 & 0 & & a_{n-2} & b_{n-2} & a_{n-1} \\ 0 & 0 & \cdots & 0 & a_{n-1} & b_{n-1} \end{pmatrix} \begin{pmatrix} p_0(x) \\ p_1(x) \\ p_2(x) \\ \vdots \\ p_{n-2}(x) \\ p_{n-1}(x) \end{pmatrix}$$

$$= x \begin{pmatrix} p_0(x) \\ p_1(x) \\ p_2(x) \\ \vdots \\ p_{n-2}(x) \\ p_{n-1}(x) \end{pmatrix} - a_n p_n(x) \begin{pmatrix} 0 \\ 0 \\ 0 \\ \vdots \\ 0 \\ 1 \end{pmatrix}.$$

Therefore if we take $x_{j,n}$ so that $p_n(x_{j,n}) = 0$ $(j = 1, \ldots, n)$ then $x_{j,n}$ turns out to be an eigenvalue of the matrix

$$J_n = \begin{pmatrix} b_0 & a_1 & 0 & 0 & \cdots & 0 \\ a_1 & b_1 & a_2 & 0 & \cdots & 0 \\ 0 & a_2 & b_2 & a_3 & \cdots & 0 \\ \vdots & \vdots & \ddots & \ddots & \ddots & \vdots \\ 0 & 0 & & a_{n-2} & b_{n-2} & a_{n-1} \\ 0 & 0 & \cdots & 0 & a_{n-1} & b_{n-1} \end{pmatrix}$$

with eigenvector

$$\begin{pmatrix} p_0(x_{j,n}) \\ p_1(x_{j,n}) \\ p_2(x_{j,n}) \\ \vdots \\ p_{n-2}(x_{j,n}) \\ p_{n-1}(x_{j,n}) \end{pmatrix}.$$

The matrix J_n is tridiagonal and symmetric and is known as a *(finite) Jacobi matrix*. The symmetry of J_n immediately implies that the eigenvalues (zeros) are real. The simplicity of the zeros also follows from the recurrence relation by considering the confluent form of the Christoffel-Darboux formula (2.7). If x_0 is a zero of multiplicity > 1 then (2.7) yields

$$\sum_{k=0}^{n-1} p_k^2(x_0) = 0,$$

which is impossible since $p_0(x_0)$ is a non-zero constant.

The zeros of orthogonal polynomials also have the property that the zeros of p_n interlace with the zeros of p_{n-1}:

Theorem 2.3 *Suppose $x_{1,n} < x_{2,n} < \cdots < x_{n,n}$ are the zeros of p_n, then*

$$x_{j,n} < x_{j,n-1} < x_{j+1,n}.$$

Proof Since all the zeros of p_n are real and the leading coefficient of p_n is positive, we find that the sign of $p_n'(x_{j,n})$ is $(-1)^{n-j}$. By using (2.7) at the zeros $x_{j,n}$ we find $p_n'(x_{j,n})p_{n-1}(x_{j,n}) > 0$, so that the sign of $p_{n-1}(x_{j,n})$ is also $(-1)^{n-j}$. The polynomial p_{n-1} thus changes sign when going from $x_{j,n}$ to $x_{j+1,n}$, so there has to be at least one zero in between these two zeros of p_n. This can be said for $j = 1$ until $j = n-1$, and since p_{n-1} has $n-1$ zeros, we have therefore exactly one zero of p_{n-1} between two consecutive zeros of $p_n(x)$. Also it is not possible that p_n and p_{n-1} have a common zero x_0, because then (2.7) gives $\sum_{k=0}^n p_k^2(x_0) = 0$, which is impossible since p_0 is a non-zero constant.

□

This interlacing property can also be proved by using the Jacobi matrix J_n. Here we need the *inclusion principle* for the eigenvalues of an Hermitian matrix (see e.g., Horn and Johnson [1985], Theorem 4.3.15 on p. 189):

Theorem 2.4 *Let A be an $n \times n$ Hermitian matrix and let A_r be any principal $r \times r$ submatrix obtained from A by deleting $n-r$ rows and the corresponding $n-r$ columns. If $\lambda_k(A)$ are the eigenvalues of A in increasing order, then*

$$\lambda_k(A) \leq \lambda_k(A_r) \leq \lambda_{k+n-r}(A).$$

The interlacing property for orthogonal polynomials on the real line is the case where $A = J_n$ and A_r is the principal submatrix J_{n-1} obtained from J_n by deleting the last row and column. Note that for the zeros of orthogonal polynomials we have the stronger conclusion with strict inequalities. If we delete from J_n the first row and column, then the resulting matrix $J_{n-1}^{(1)}$ is a Jacobi matrix containing the recurrence coefficients with index shifted by one. The corresponding orthonormal polynomials are known as *associated polynomials* and we denote them by $p_n^{(1)}$ $(n = 0, 1, 2, \dots)$. They satisfy the shifted recurrence relation

$$xp_n^{(1)}(x) = a_{n+2}p_{n+1}^{(1)}(x) + b_{n+1}p_n^{(1)}(x) + a_{n+1}p_{n-1}^{(1)}(x),$$

with initial values $p_0^{(1)} = 1$ and $p_{-1}^{(1)} = 0$. It is therefore clear that $p_{n-1}^{(1)}$ $(n \in \mathbb{N})$ is another solution of the three-term recurrence relation (2.2). Since we are dealing with linear recurrence relations, it is worthwhile to compute the *Wronskian* or *Casorati determinant* of two solutions u_n and v_n $(n \in \mathbb{N})$. This is given by

$$W(u_n, v_n) = a_{n+1} \begin{vmatrix} u_n & v_n \\ u_{n+1} & v_{n+1} \end{vmatrix} = a_{n+1}(u_n v_{n+1} - u_{n+1} v_n).$$

If u_n and v_n are solutions of (2.2) then

$$xu_n = a_{n+1}u_{n+1} + b_n u_n + a_n u_{n-1}, \tag{2.8}$$

$$xv_n = a_{n+1}v_{n+1} + b_n v_n + a_n v_{n-1}. \tag{2.9}$$

Multiply (2.8) by v_n and (2.9) by u_n and subtract both equations, then it follows that $W(u_n, v_n) = W(u_{n-1}, v_{n-1})$ so that this Wronskian is independent of n and thus also equal to $W(u_0, v_0)$. For the two solutions $p_n(x)$ and $p_{n-1}^{(1)}(x)$ we then find

$$a_{n+1}\left(p_n(x)p_n^{(1)}(x) - p_{n+1}(x)p_{n-1}^{(1)}(x)\right) = a_1. \tag{2.10}$$

The Wronskian of these two solutions is different from zero, hence these are two linearly independent solutions of the recurrence relation (2.2) and every solution of the recurrence relation can be written as a linear combination of these two solutions. By taking $x = x_{j,n}$ in (2.10) we see that the sign of $p_{n-1}^{(1)}(x_{j,n})$ is equal to $-\operatorname{sign} p_{n+1}(x_{j,n})$ which by the interlacing property is equal to $(-1)^{n-j}$, from which we find

$$x_{j,n} < x_{j,n-1}^{(1)} < x_{j+1,n},$$

with $x_{j,n-1}^{(1)}$ $(1 \leq j \leq n-1)$ the zeros of $p_{n-1}^{(1)}$ in increasing order.

2.4 Gaussian quadrature Numerical quadrature consists of approximating the integral of a function f

$$I(f) = \int f(x)\, d\mu(x)$$

by a finite sum which uses only n function evaluations

$$I_n(f) = \sum_{j=1}^{n} f(x_{j,n})\beta_{j,n}.$$

Here the n nodes $x_{j,n}(1 \leq j \leq n)$ and the n quadrature coefficients $\beta_{j,n}(1 \leq j \leq n)$ have to be chosen properly so that the quadrature formula is correct for as many functions f as possible. We say that the degree of accuracy of the quadrature rule is k when all polynomials of degree k are integrated correctly but there is a polynomial of degree $k+1$ which is not integrated correctly. If we fix the n points $x_{j,n}$, then the equations

$$\sum_{j=0}^{n} \beta_{j,n} x_{j,n}^{k} = m_k, \qquad k = 0, 1, \dots, n-1,$$

where m_k are the moments given by (2.1), form a system of n linear equations for the n unknowns $\beta_{j,n}$. The matrix of this system is the Vandermonde matrix

$$\begin{pmatrix} 1 & 1 & \cdots & 1 \\ x_{1,n} & x_{2,n} & \cdots & x_{n,n} \\ x_{1,n}^2 & x_{2,n}^2 & \cdots & x_{n,n}^2 \\ \vdots & \vdots & \cdots & \vdots \\ x_{1,n}^{n-1} & x_{2,n}^{n-1} & \cdots & x_{n,n}^{n-1} \end{pmatrix},$$

which is non-singular whenever all the nodes $x_{j,n}$ $(1 \leq j \leq n)$ are distinct. Hence every choice of n distinct nodes has a unique set of quadrature coefficients $\beta_{j,n}$ $(1 \leq j \leq n)$ such that the corresponding quadrature rule has degree of accuracy at least $n-1$.

These quadrature rules are closely related to interpolation. If we approximate the function f by a polynomial interpolating f at the points $x_{j,n}$, then the Lagrange interpolating polynomial is

$$L_n(x) = \sum_{j=1}^{n} f(x_{j,n})\ell_{j,n}(x),$$

where the fundamental polynomials of Lagrange interpolation are given by

$$\ell_{j,n}(x) = \frac{q_n(x)}{q_n'(x_{j,n})(x - x_{j,n})}, \quad q_n(x) = \prod_{j=1}^{n}(x - x_{j,n}).$$

Since each $\ell_{j,n}$ has degree $n-1$ we see that L_n has degree $n-1$, and the property $\ell_{j,n}(x_{k,n}) = \delta_{k,j}$ shows that this polynomial indeed interpolates: $L_n(x_{j,n}) = f(x_{j,n})$ $(1 \le j \le n)$. Integrating L_n gives

$$\int L_n(x)\, d\mu(x) = \sum_{j=1}^{n} f(x_{j,n}) \int \ell_{j,n}(x)\, d\mu(x).$$

When f is a polynomial of degree $n-1$, then $L_n = f$, so we conclude that the quadrature coefficients are given by

$$\beta_{j,n} = \int \ell_{j,n}(x)\, d\mu(x).$$

Such interpolatory quadrature rules always have degree of accuracy at least $n-1$. We will show that choosing the nodes as the zeros of the orthogonal polynomial of degree n leads to much higher accuracy.

Theorem 2.5 (Gauss-Jacobi-Christoffel quadrature) *If we choose the n nodes of the quadrature rule as the n zeros $x_{j,n}$ of the orthonormal polynomial p_n for the measure μ and if we denote the corresponding quadrature coefficients by $\lambda_{j,n}$ $(1 \le j \le n)$, then for every polynomial π_{2n-1} of degree less than or equal to $2n-1$*

$$\sum_{j=1}^{n} \lambda_{j,n}\pi_{2n-1}(x_{j,n}) = \int \pi_{2n-1}(x)\, d\mu(x).$$

The quadrature coefficients $\lambda_{j,n}$ are positive and are given by

$$\lambda_{j,n} = \frac{1}{a_n p_{n-1}(x_{j,n})p_n'(x_{j,n})} = \frac{-1}{a_{n+1}p_{n+1}(x_{j,n})p_n'(x_{j,n})},$$

or

$$\lambda_{j,n} = \left(\sum_{k=0}^{n-1} p_k^2(x_{j,n}) \right)^{-1}. \tag{2.11}$$

Proof If L_n is the interpolating polynomial for π_{2n-1} at the nodes $x_{j,n}$, then

$$\sum_{j=1}^{n} \lambda_{j,n}\pi_{2n-1}(x_{j,n}) - \int \pi_{2n-1}(x)\, d\mu(x) = \int [L_n(x) - \pi_{2n-1}(x)]\, d\mu(x).$$

The integrand $L_n - \pi_{2n-1}$ is a polynomial of degree at most $2n-1$ which vanishes at the nodes $x_{j,n}$ $(1 \le j \le n)$, hence

$$L_n(x) - \pi_{2n-1}(x) = p_n(x) r_{n-1}(x),$$

where r_{n-1} is a polynomial of degree at most $n-1$. If p_n is the orthonormal polynomial of degree n for the measure μ, then the orthogonality gives

$$\int [L_n(x) - \pi_{2n-1}(x)]\, d\mu(x) = 0,$$

which shows that the degree of accuracy is at least $2n-1$. If π_{2n} is a polynomial of degree $2n$ and L_n its interpolating polynomial, then

$$L_n(x) - \pi_{2n}(x) = p_n(x) r_n(x),$$

with r_n a polynomial of degree n, and the integral is then different from zero. This shows that the degree of accuracy is exactly $2n-1$.

The corresponding quadrature coefficients are

$$\lambda_{j,n} = \int \ell_{j,n}(x)\,d\mu(x).$$

Observe that

$$\int \ell_{j,n}^2(x)\,d\mu(x) - \int \ell_{j,n}(x)\,d\mu(x) = \frac{1}{p_n'(x_{j,n})} \int p_n(x)\frac{\ell_{j,n}(x)-1}{x-x_{j,n}}\,d\mu(x) = 0,$$

because $(\ell_{j,n}(x)-1)/(x-x_{j,n})$ is a polynomial of degree $n-2$, hence

$$\lambda_{j,n} = \int \ell_{j,n}^2(x)\,d\mu(x),$$

which shows that these coefficients are positive. Consider the Christoffel-Darboux formula (2.6) with $y = x_{j,n}$ then

$$\sum_{k=0}^{n-1} p_k(x)p_k(x_{j,n}) = a_n \frac{p_n(x)p_{n-1}(x_{j,n})}{x-x_{j,n}},$$

hence

$$\begin{aligned}\int \ell_{j,n}(x)\,d\mu(x) &= \frac{1}{p_n'(x_{j,n})p_{n-1}(x_{j,n})} \int \frac{p_{n-1}(x_{j,n})p_n(x)}{x-x_{j,n}}\,d\mu(x) \\ &= \frac{1}{a_n p_n'(x_{j,n})p_{n-1}(x_{j,n})} \int \sum_{k=0}^{n-1} p_k(x)p_k(x_{j,n})\,d\mu(x),\end{aligned}$$

and by the orthogonality

$$\int p_k(x)p_k(x_{j,n})\,d\mu(x) = \delta_{k,0},$$

giving the desired expression for $\lambda_{j,n}$. The Christoffel-Darboux formula (2.7) shows that $\lambda_{j,n}$ can also be expressed in terms of $p_{n+1}(x_{j,n})$. Finally using the confluent form (2.7) of the Christoffel-Darboux formula, with $x = x_{j,n}$ gives (2.11).

□

2.5 Extremal properties Apart from the extremal properties which we have already seen for general orthogonal polynomials, we can also work out some new extremal properties which give a nice interpretation of the extreme zeros.

Theorem 2.6 (Chebyshev) *The maximum of the ratio*

$$\frac{\int q_n^2(x)x\,d\mu(x)}{\int q_n^2(x)\,d\mu(x)}$$

taken over all polynomials q_n *of degree at most* n *is equal to the largest zero* $x_{n+1,n+1}$ *of the orthogonal polynomial* p_{n+1} *for the measure* μ*. The minimum is equal to the smallest zero* $x_{1,n+1}$ *of* p_{n+1}*.*

Proof If we use the Gauss-Jacobi-Christoffel formula with the zeros of p_{n+1} then the ratio becomes

$$\frac{\sum_{j=1}^{n+1} \lambda_{j,n+1} q_n^2(x_{j,n+1}) x_{j,n+1}}{\sum_{j=1}^{n+1} \lambda_{j,n+1} q_n^2(x_{j,n+1})}.$$

Clearly we have $x_{1,n+1} \le x_{j,n+1} \le x_{n+1,n+1}$, which shows that $x_{1,n+1}$ is a lower bound and $x_{n+1,n+1}$ an upper bound for this ratio. The lower bound is obtained if we take $q_n(x) = p_{n+1}(x)/(x - x_{1,n+1})$ and the upper bound for $q_n(x) = p_{n+1}(x)/(x - x_{n+1,n+1})$.

□

An alternative proof, using Jacobi matrices, is in terms of the Fourier expansion

$$q_n(x) = \sum_{k=0}^{n} c_k p_k(x),$$

so that

$$\int q_n^2(x) x \, d\mu(x) = \sum_{k=0}^{n} \sum_{j=0}^{n} c_k c_j \int x p_k(x) p_j(x) \, d\mu(x) = \sum_{k=0}^{n} \sum_{j=0}^{n} c_k c_j (J_{n+1})_{k,j},$$

where J_{n+1} is the Jacobi matrix of order $n+1$. The ratio is thus

$$\frac{c^* J_{n+1} c}{c^* c},$$

where $c^* = (c_0, c_1, \dots, c_n)$. The maximum of this ratio is equal to the largest eigenvalue of J_{n+1} and the minimum is the smallest eigenvalue of J_{n+1}.

2.6 Examples The most important orthogonal polynomials on the real line are the 'very' classical orthogonal polynomials. They are characterized by

- a second order linear differential equation of the form

  $$\sigma(x) y''(x) + \tau(x) y'(x) + \lambda_n y(x) = 0,$$

  where σ is a polynomial of degree at most two and τ a polynomial of degree one, both independent of n. If we want a solution y which is a polynomial of degree n, then comparing the leading coefficients shows that $\lambda_n = -n(n-1)\sigma''/2 - n\tau'$. For these specific values of λ_n, the polynomial solution y_n will consist of orthogonal polynomials;

- a formula involving the nth derivative of a function, known as the *Rodrigues formula*:

  $$y_n(x) = \frac{1}{C_n w(x)} \left(\frac{d}{dx}\right)^n [w(x) \sigma^n(x)],$$

  where C_n is independent of x and w is non-negative on a certain real interval, σ is a polynomial of degree at most 2, independent of n and $w'/w = [\tau - \sigma']/\sigma$ is a rational function, with τ a polynomial of degree at most one. With these conditions one obtains polynomials y_n of degree n which are orthogonal;

- a property involving the derivatives: for the very classical orthogonal polynomials, the derivatives are also orthogonal polynomials.

2.6.1 *Jacobi polynomials* If we choose $w(x) = (1-x)^\alpha(1+x)^\beta$, then w is non-negative on the interval $[-1,1]$. Clearly

$$\frac{w'(x)}{w(x)} = \frac{-\alpha}{1-x} + \frac{\beta}{1+x},$$

hence the choice $\sigma(x) = x^2-1$ and $\tau(x) = (2+\alpha+\beta)x+\alpha-\beta$ gives the differential equation

$$(x^2-1)y_n''(x) + [(2+\alpha+\beta)x+\alpha-\beta]y_n'(x) = n[n+1+\alpha+\beta]y_n(x)$$

and for $C_n = 2^n n!$ the Rodrigues formula

$$(1-x)^\alpha(1+x)^\beta y_n(x) = \frac{(-1)^n}{2^n n!}\left(\frac{d}{dx}\right)^n [(1-x)^{\alpha+n}(1+x)^{\beta+n}].$$

The polynomial y_n is usually denoted by $P_n^{(\alpha,\beta)}$ and is known as the *Jacobi polynomial* of degree n. Integration by parts easily shows that

$$\begin{aligned}\int_{-1}^1 & P_n^{(\alpha,\beta)}(x)P_m^{(\alpha,\beta)}(x)(1-x)^\alpha(1+x)^\beta\,dx \\ &= \frac{2^{\alpha+\beta+1}}{n!(2n+\alpha+\beta+1)}\frac{\Gamma(n+\alpha+1)\Gamma(n+\beta+1)}{\Gamma(n+\alpha+\beta+1)}\delta_{n,m},\end{aligned}$$

hence the Jacobi polynomials are orthogonal on the interval $[-1,1]$ with respect to the weight function $(1-x)^\alpha(1+x)^\beta$. For integrability reasons we need to take $\alpha,\beta > -1$.

Solving the differential equation by the Frobenius method or changing the variable x to $1-2t$ so that the differential equation is transformed to the hypergeometric equation of Gauss, shows that the Jacobi polynomial of degree n is a (terminating) hypergeometric series, in particular

$$\begin{aligned}P_n^{(\alpha,\beta)}(x) &= \binom{n+\alpha}{n}{}_2F_1(-n,n+\alpha+\beta+1;\alpha+1;\frac{1-x}{2}) \\ &= \binom{n+\alpha}{n}\sum_{k=0}^n \frac{(-n)_k(n+\alpha+\beta+1)_k}{(\alpha+1)_k k!}\left(\frac{1-x}{2}\right)^k.\end{aligned}$$

Hypergeometric functions satisfy several contiguous relations, and one of these relations gives the three-term recurrence relation

$$\begin{aligned}&2n(n+\alpha+\beta)(2n+\alpha+\beta-2)P_n^{(\alpha,\beta)}(x) \\ &\quad= (2n+\alpha+\beta-1)[(2n+\alpha+\beta)(2n+\alpha+\beta-2)x+\alpha^2-\beta^2]P_{n-1}^{(\alpha,\beta)}(x) \\ &\quad\quad - 2(n+\alpha-1)(n+\beta-1)(2n+\alpha+\beta)P_{n-2}^{(\alpha,\beta)}(x).\end{aligned}$$

The normalized weight function

$$2^{-\alpha-\beta-1}\frac{\Gamma(\alpha+\beta+2)}{\Gamma(\alpha+1)\Gamma(\beta+1)}(1-x)^\alpha(1+x)^\beta$$

then has orthonormal polynomials

$$p_n(x) = \sqrt{\frac{2n+\alpha+\beta+1}{n+\alpha+\beta+1}\frac{n!(\alpha+\beta+2)_n}{(\alpha+1)_n(\beta+1)_n}}\,P_n^{(\alpha,\beta)}(x),$$

and the recurrence coefficients become

$$a_n^2 = \frac{4n(n+\alpha)(n+\beta)(n+\alpha+\beta)}{(2n+\alpha+\beta-1)(2n+\alpha+\beta)^2(2n+\alpha+\beta+1)},$$

$$b_n = \frac{\beta^2-\alpha^2}{(2n+\alpha+\beta)(2n+\alpha+\beta+2)},$$

where $b_0 = -\alpha$ in case $\alpha+\beta = 0$.

Particular cases are $\alpha = \beta = -1/2$, giving the *Chebyshev polynomials of the first kind* $T_n(x) = 2^{2n}(n!)^2/(2n)!\ P_n^{(-1/2,-1/2)}(x)$, satisfying

$$\frac{1}{\pi}\int_{-1}^{1} T_n(x)T_m(x)\frac{dx}{\sqrt{1-x^2}} = \begin{cases} 1 & n=m=0, \\ \frac{1}{2}\,\delta_{n,m} & \textit{elsewhere}. \end{cases}$$

The change of variable $x = \cos\theta$ gives

$$T_n(x) = \cos n\theta.$$

When $\alpha = \beta = 1/2$ we have the *Chebyshev polynomials of the second kind* $U_n(x) = 2^{2n+1}[(n+1)!]^2/(2n+2)!\ P_n^{(1/2,1/2)}(x)$ satisfying

$$\frac{2}{\pi}\int_{-1}^{1} U_n(x)U_m(x)\sqrt{1-x^2}\,dx = \delta_{m,n}.$$

When $x = \cos\theta$ we find

$$U_n(x) = \frac{\sin(n+1)\theta}{\sin\theta}.$$

The Chebyshev polynomials therefore express the orthogonality of the trigonometric system $\{\cos k\theta, \sin k\theta, k \in \mathbb{N}\}$.

Another important special case is when $\alpha = \beta = 0$, which corresponds to *Legendre polynomials.* The symmetric cases when $\alpha = \beta$ give rise to the *ultraspherical polynomials,* also known as *Gegenbauer polynomials.* These are usually denoted by

$$C_n^\lambda(x) = \frac{\Gamma(\lambda+1/2)\Gamma(n+2\lambda)}{\Gamma(2\lambda)\Gamma(n+\lambda+1/2)} P_n^{(\lambda-1/2,\lambda-1/2)}(x).$$

2.6.2 *Laguerre polynomials* The choice $w(x) = x^\alpha e^{-x}$ gives a weight function on $[0,\infty)$ for which $w'(x)/w(x) = (\alpha - x)/x$. Hence $\sigma(x) = x$ and $\tau(x) = \alpha+1-x$ give the differential equation

$$xy_n''(x) + (\alpha+1-x)y_n'(x) = -ny_n(x),$$

and with $C_n = n!$ we have the Rodrigues formula

$$e^{-x}x^\alpha y_n(x) = \frac{1}{n!}\left(\frac{d}{dx}\right)^n (e^{-x}x^{n+\alpha}).$$

The polynomials y_n are denoted by $L_n^{(\alpha)}$ and are known as the *Laguerre polynomials.* Integration by parts shows the orthogonality relations

$$\int_0^\infty e^{-x}x^\alpha L_n^{(\alpha)}(x)L_m^{(\alpha)}(x)\,dx = \frac{\Gamma(n+\alpha+1)}{n!}\,\delta_{m,n},$$

so that the Laguerre polynomials are orthogonal on $[0,\infty)$ with respect to the weight function $x^\alpha e^{-x}$. The integrability follows if we assume $\alpha > -1$. The differential equation now is a confluent hypergeometric equation and the Laguerre polynomial is a terminating confluent hypergeometric series

$$\begin{aligned} L_n^{(\alpha)}(x) &= \binom{n+\alpha}{n} {}_1F_1(-n;\alpha+1;x) \\ &= \sum_{j=0}^{n} \binom{n+\alpha}{n-j} \frac{(-x)^j}{j!}. \end{aligned}$$

The following recurrence relation holds:

$$nL_n^{(\alpha)}(x) = (-x+2n+\alpha-1)L_{n-1}^{(\alpha)}(x) - (n+\alpha-1)L_{n-2}^{(\alpha)}(x).$$

For the normalized weight function

$$\frac{1}{\Gamma(\alpha+1)} x^\alpha e^{-x}$$

the orthonormal polynomials are given by

$$p_n(x) = (-1)^n \binom{n+\alpha}{n}^{-1/2} L_n^{(\alpha)}(x),$$

and the recurrence coefficients are

$$a_n^2 = n(n+\alpha), \quad b_n = 2n+\alpha+1.$$

2.6.3 *Hermite polynomials* When $w(x) = e^{-x^2}$, then w is non-negative on the whole real line and $w'(x)/w(x) = -2x$. The choice $\sigma(x) = 1$ and $\tau(x) = -2x$ then gives the differential equation

$$y_n''(x) - 2xy_n'(x) = -2ny_n(x),$$

and for $C_n = (-1)^n$ we have the Rodrigues formula

$$e^{-x^2} y_n(x) = (-1)^n \left(\frac{d}{dx}\right)^n e^{-x^2}.$$

The polynomials are known as the *Hermite polynomials* and we write $y_n = H_n$. The orthogonality relations are

$$\int_{-\infty}^{\infty} e^{-x^2} H_n(x) H_m(x)\, dx = \sqrt{\pi} 2^n n! \delta_{n,m}$$

so that these polynomials are orthogonal on $(-\infty, \infty)$ with respect to the weight function e^{-x^2}. They satisfy the recurrence relation

$$H_n(x) = 2xH_{n-1}(x) - 2(n-1)H_{n-2}(x).$$

The orthonormal polynomials for the normalized weight function

$$\frac{1}{\sqrt{\pi}} e^{-x^2}$$

are given by

$$p_n(x) = (2^n n!)^{-1/2} H_n(x),$$

and the recurrence coefficients are

$$a_n^2 = n/2, \quad b_n = 0.$$

2.6.4 *The Askey table* The three examples given higher correspond to the very classical orthogonal polynomials. The characterization in terms of a second order differential equation, a Rodrigues formula, and properties of the derivatives can be generalized by also considering difference operators and q-difference operators. This will lead to a much larger class of orthogonal polynomials, nowadays known as the *classical orthogonal polynomials* (Andrews and Askey [1985], Askey and Wilson [1985]). These polynomials can be arranged in a table, which is known as the Askey table and its q-extension. It would lead us too far here to describe this table in detail, and instead we refer the interested reader to Koekoek and Swarttouw [1994].

3 Orthogonal polynomials on the unit circle

3.1 Toeplitz matrices Suppose ν is a positive Borel measure on $[0, 2\pi)$, then this measure induces a measure—also denoted by ν—on the unit circle $C = \{e^{i\theta} : \theta \in [0, 2\pi)\}$ by

$$\int_C f(z)\, d\nu(z) = \int_0^{2\pi} f(e^{i\theta})\, d\nu(\theta).$$

Without loss of generality we will assume that ν is a probability measure so that $\nu(C) = 1$. The entries of the Gram matrix are

$$\langle z^k, z^l \rangle = \int_0^{2\pi} e^{i(k-l)\theta}\, d\nu(\theta) = c_{k-l},$$

where c_k $(k \in \mathbb{Z})$ are the *trigonometric moments* of the measure ν. The Gram matrix is usually denoted by

$$T_n = \begin{pmatrix} c_0 & c_1 & c_2 & \cdots & c_n \\ c_{-1} & c_0 & c_1 & \cdots & c_{n-1} \\ c_{-2} & c_{-1} & c_0 & \cdots & c_{n-2} \\ \vdots & \vdots & \vdots & \cdots & \vdots \\ c_{-n} & c_{-n+1} & c_{-n+2} & \cdots & c_0 \end{pmatrix},$$

and this Hermitian matrix $(c_{-k} = \overline{c_k})$ has constant values along the diagonals. Such matrices are known as *Toeplitz matrices.* The orthonormal polynomials are

$$\varphi_n(z) = \frac{1}{\sqrt{\Delta_{n-1}\Delta_n}} \begin{vmatrix} c_0 & c_1 & c_2 & \cdots & c_n \\ c_{-1} & c_0 & c_1 & \cdots & c_{n-1} \\ c_{-2} & c_{-1} & c_0 & \cdots & c_{n-2} \\ \vdots & \vdots & \vdots & \cdots & \vdots \\ c_{-n+1} & c_{-n+2} & c_{-n+3} & \cdots & c_{-1} \\ 1 & z & z^2 & \cdots & z^n \end{vmatrix} = \kappa_n z^n + \cdots,$$

and the leading coefficients are again given by a ratio of two Toeplitz determinants:

$$\kappa_n = \sqrt{\frac{\Delta_{n-1}}{\Delta_n}}.$$

It is customary to denote the monic orthogonal polynomials on the unit circle by $\Phi_n = \varphi_n/\kappa_n$. It is clear that when the measure ν is symmetric in the sense that

$$\int_0^{\pi} z^k\, d\nu(\theta) = \int_{\pi}^{2\pi} \overline{z^k}\, d\nu(\theta), \qquad k \in \mathbb{N},$$

then the trigonometric moments are all real and are given by

$$c_k = 2 \int_0^{\pi} \cos k\theta \, d\nu(\theta),$$

so that the orthogonal polynomials will have real coefficients.

For the study of orthogonal polynomials on the unit circle, it is often convenient to work with reversed polynomials. If $q_n(z) = \sum_{k=0}^n a_k z^k$ is a polynomial of degree at most n, then its reversed polynomial is given by

$$q_n^*(z) = \sum_{k=0}^{n} \overline{a_k} z^{n-k} = z^n \bar{q}_n(1/z),$$

where $\bar{q}_n$ is the polynomial q_n with complex conjugated coefficients. The $*$-operation depends on the degree n so that we always have to specify for which n we are using this operation. Notice that the $*$-operation is an involution

$$[q_n^*(z)]^* = q_n(z),$$

and that on the unit circle

$$q_n^*(z) = z^n \overline{q_n(z)}, \qquad z = e^{i\theta}.$$

Theorem 3.1 *If $K_n(z,\xi) = \sum_{k=0}^n \varphi_k(z)\overline{\varphi_k(\xi)}$ is the reproducing kernel for the orthogonal polynomials φ_n on the unit circle, then*

$$K_n(z,0) = \sum_{k=0}^{n} \varphi_k(z)\overline{\varphi_k(0)} = \kappa_n \varphi_n^*(z), \tag{3.1}$$

and

$$K_n(0,0) = \sum_{k=0}^{n} |\varphi_k(0)|^2 = \kappa_n^2. \tag{3.2}$$

Proof If q_n is a monic polynomial of degree n, then q_n^* is a polynomial of degree at most n with $q_n^*(0) = 1$. The minimum of

$$\int_C |q_n(z)|^2 \, d\nu(z)$$

taken over all monic polynomials q_n is therefore equal to the minimum of

$$\int_C |q_n^*(z)|^2 \, d\nu(z)$$

taken over all polynomials q_n^* of degree at most n for which $q_n^*(0) = 1$. Combining Theorems 1.3 and 1.4 thus gives $\kappa_n^2 = K_n(0,0)$. The minimizing polynomials are related by the $*$-operation, hence $\hat{\varphi}_n^*(z) = K_n(z,0)/K_n(0,0)$, which leads to (3.1).

□

Recall that Φ_n is the monic orthogonal polynomial on the unit circle and that by Theorem 1.6 all the zeros $z_{j,n}$ of Φ_n are inside the closed unit disk, hence $|\Phi_n(0)| = \prod_{k=1}^n |z_{j,n}|$ which shows that $|\Phi_n(0)| \le 1$. The strict inequality holds, since (3.2) gives

$$\kappa_n^2 - \kappa_{n-1}^2 = |\varphi_n(0)|^2 = \kappa_n^2 |\Phi_n(0)|^2,$$

which gives

$$\left(\frac{\kappa_{n-1}}{\kappa_n}\right)^2 = 1 - |\Phi_n(0)|^2, \tag{3.3}$$

so that $|\Phi_n(0)| < 1$ for every $n \in \mathbb{N}$.

3.2 Recurrence relations We have seen earlier that orthogonal polynomials on the real line satisfy a three-term recurrence relation. Such a recurrence relation in general does not hold for orthogonal polynomials in the complex plane, but there is also a recurrence formula for orthogonal polynomials on the unit circle. In fact, on the unit circle the orthogonal polynomials satisfy a system of two recurrence relations of first order:

Theorem 3.2 *If φ_n are orthonormal polynomials on the unit circle, then*

$$\kappa_{n-1} z\varphi_{n-1}(z) = \kappa_n \varphi_n(z) - \varphi_n(0)\varphi_n^*(z) \tag{3.4}$$

$$\kappa_{n-1}\varphi_n(z) = \kappa_n z\varphi_{n-1}(z) + \varphi_n(0)\varphi_{n-1}^*(z). \tag{3.5}$$

For the monic orthogonal polynomials Φ_n one has

$$\frac{\kappa_{n-1}^2}{\kappa_n^2} z\Phi_{n-1}(z) = \Phi_n(z) - \Phi_n(0)\Phi_n^*(z), \tag{3.6}$$

$$\Phi_n(z) = z\Phi_{n-1}(z) + \Phi_n(0)\Phi_{n-1}^*(z). \tag{3.7}$$

Proof From (3.1) we find

$$\kappa_n \varphi_n^*(z) - \kappa_{n-1}\varphi_{n-1}^*(z) = \varphi_n(z)\overline{\varphi_n(0)}.$$

If we take the $*$-operation for polynomials of degree n, then this gives

$$\kappa_n\varphi_n(z) - \kappa_{n-1}z\varphi_{n-1}(z) = \varphi_n(0)\varphi_n^*(z),$$

which corresponds to (3.4). Eliminating φ_n^* from the last two equations gives

$$\kappa_n^2\varphi_n(z) - \kappa_n\kappa_{n-1}z\varphi_{n-1}(z) - \kappa_{n-1}\varphi_n(0)\varphi_{n-1}^*(z) = |\varphi_n(0)|^2\varphi_n(z).$$

Now use $\kappa_n^2 = \kappa_{n-1}^2 + |\varphi_n(0)|^2$, then (3.5) follows.

The recurrence relations for the monic polynomials follow easily by putting $\varphi_n = \kappa_n\Phi_n$.

□

If we eliminate Φ_n from (3.6) and (3.7) then we find

$$\Phi_n^*(z) = \overline{\Phi_n(0)}z\Phi_{n-1}(z) + \Phi_{n-1}^*(z). \tag{3.8}$$

The recurrence relations (3.7) and (3.8) for the monic orthogonal polynomials are easily written in matrix form as

$$\begin{pmatrix} \Phi_n(z) \\ \Phi_n^*(z) \end{pmatrix} = \begin{pmatrix} z & \Phi_n(0) \\ \overline{\Phi_n(0)}z & 1 \end{pmatrix} \begin{pmatrix} \Phi_{n-1}(z) \\ \Phi_{n-1}^*(z) \end{pmatrix}.$$

The monic orthogonal polynomials on the unit circle are therefore completely determined by the sequence $\Phi_n(0)$ ($n \in \mathbb{N}$). These numbers are thus very informative and are often called the *reflection coefficients*. These reflection coefficients also determine the orthonormal polynomials since

$$\kappa_n^2 = \frac{\kappa_{n-1}^2}{1 - |\Phi_n(0)|^2}.$$

For orthogonal polynomials on the unit circle we have a simple expression for the reproducing kernel, similar to the Christoffel-Darboux formula on the real line:

Theorem 3.3 *If φ_n are orthonormal polynomials on the unit circle, then*

$$K_{n-1}(z,\xi) = \sum_{k=0}^{n-1} \varphi_k(z)\overline{\varphi_k(\xi)} = \frac{\varphi_n^*(z)\overline{\varphi_n^*(\xi)} - \varphi_n(z)\overline{\varphi_n(\xi)}}{1 - z\bar{\xi}}. \tag{3.9}$$

Proof Eliminate φ_n from (3.4) and (3.5) then

$$\kappa_{n-1}\varphi_n^*(z) = \overline{\varphi_n(0)} z\varphi_{n-1}(z) + \kappa_n \varphi_{n-1}^*(z). \tag{3.10}$$

If we use (3.5) for z and ξ then

$$\begin{aligned} &\varphi_n(z)\overline{\varphi_n(\xi)} \\ &= \kappa_{n-1}^{-2}[\kappa_n z\varphi_{n-1}(z) + \varphi_n(0)\varphi_{n-1}^*(z)]\ [\kappa_n\bar{\xi}\ \overline{\varphi_{n-1}(\xi)} + \overline{\varphi_n(0)}\ \overline{\varphi_{n-1}^*(\xi)}] \\ &= \frac{\kappa_n^2}{\kappa_{n-1}^2} z\bar{\xi}\varphi_{n-1}(z)\overline{\varphi_{n-1}(\xi)} + \frac{|\varphi_n(0)|^2}{\kappa_{n-1}^2}\varphi_{n-1}^*(z)\overline{\varphi_{n-1}^*(\xi)} \\ &\quad + \frac{\kappa_n\overline{\varphi_n(0)}}{\kappa_{n-1}^2} z\varphi_{n-1}(z)\overline{\varphi_{n-1}^*(\xi)} + \frac{\kappa_n\varphi_n(0)}{\kappa_{n-1}^2}\bar{\xi}\varphi_{n-1}^*(z)\overline{\varphi_{n-1}(\xi)}. \end{aligned}$$

In a similar way we can use (3.10) to find

$$\begin{aligned} &\varphi_n^*(z)\overline{\varphi_n^*(\xi)} \\ &= \kappa_{n-1}^{-2}[\overline{\varphi_n(0)} z\varphi_{n-1}(z) + \kappa_n\varphi_{n-1}^*(z)]\ [\varphi_n(0)\bar{\xi}\ \overline{\varphi_{n-1}(\xi)} + \kappa_n\overline{\varphi_{n-1}^*(\xi)}] \\ &= \frac{|\varphi_n(0)|^2}{\kappa_{n-1}^2} z\bar{\xi}\varphi_{n-1}(z)\overline{\varphi_{n-1}(\xi)} + \frac{\kappa_n^2}{\kappa_{n-1}^2}\varphi_{n-1}^*(z)\overline{\varphi_{n-1}^*(\xi)} \\ &\quad + \frac{\kappa_n\overline{\varphi_n(0)}}{\kappa_{n-1}^2} z\varphi_{n-1}(z)\overline{\varphi_{n-1}^*(\xi)} + \frac{\kappa_n\varphi_n(0)}{\kappa_{n-1}^2}\bar{\xi}\varphi_{n-1}^*(z)\overline{\varphi_{n-1}(\xi)}. \end{aligned}$$

If we subtract these equations and use $\kappa_n^2 = \kappa_{n-1}^2 + |\varphi_n(0)|^2$, then we find

$$\varphi_n^*(z)\overline{\varphi_n^*(\xi)} - \varphi_n(z)\overline{\varphi_n(\xi)} = \varphi_{n-1}^*(z)\overline{\varphi_{n-1}^*(\xi)} - z\bar{\xi}\varphi_{n-1}(z)\overline{\varphi_{n-1}(\xi)},$$

and thus

$$\begin{aligned} (1 - z\bar{\xi})\varphi_{n-1}(z)\overline{\varphi_{n-1}(\xi)} &= \left(\varphi_n^*(z)\overline{\varphi_n^*(\xi)} - \varphi_n(z)\overline{\varphi_n(\xi)}\right) \\ &\quad - \left(\varphi_{n-1}^*(z)\overline{\varphi_{n-1}^*(\xi)} - \varphi_{n-1}(z)\overline{\varphi_{n-1}(\xi)}\right). \end{aligned}$$

When we sum over n, then the right-hand side becomes a telescoping sum, which can be summed easily and yields (3.9).

□

From this Christoffel-Darboux formula we have for $z = \xi$

$$K_{n-1}(z,z) = \sum_{k=0}^{n-1} |\varphi_k(z)|^2 = \frac{|\varphi_n^*(z)|^2 - |\varphi_n(z)|^2}{1-|z|^2}.$$

If z_0 is a zero of φ_n with $|z_0| = 1$, then this Christoffel-Darboux formula implies that $|\varphi_n(z_0)|^2 = |\varphi_n^*(z_0)|^2 = 0$. Use the recurrence formula (3.4) to conclude that then also $\varphi_{n-1}(z_0) = 0$. Repeating the argument then shows that $\varphi_k(z_0) = 0$ for $k \leq n$, but for $k = 0$ we have $\varphi_0 = 1$, which gives a contradiction, hence we conclude that φ_n has no zeros on the unit circle, and thus all the zeros of φ_n are in the open unit disk $\{|z| < 1\}$. Of course, this property also follows from Theorem 1.7.

3.3 Finite orthogonality Given the sequence of orthonormal polynomials φ_k $(k = 0, 1, 2, \ldots)$ with orthogonality measure ν on the unit circle, we have by definition

$$\int_0^{2\pi} \varphi_n(z)\overline{\varphi_m(z)}\, d\nu(\theta) = \delta_{m,n}, \qquad m, n \geq 0.$$

For the first $n+1$ polynomials we can find a very useful orthogonality property with respect to a measure which is absolutely continuous with respect to Lebesgue measure, for which the Radon-Nikodym derivative is the reciprocal of a trigonometric polynomial. This property will turn out to be quite useful for investigating the asymptotic behaviour.

Theorem 3.4 *Suppose φ_n $(n \in \mathbb{N})$ are polynomials satisfying the recurrence relation*

$$\kappa_{n-1}\varphi_n(z) = \kappa_n z \varphi_{n-1}(z) + \varphi_n(0)\varphi_n^*(z). \tag{3.11}$$

Then these polynomials up to degree n satisfy the orthogonality relations

$$\frac{1}{2\pi}\int_0^{2\pi} \varphi_k(z)\overline{\varphi_j(z)}\frac{d\theta}{|\varphi_n(z)|^2} = \delta_{k,j}, \qquad 0 \leq j \leq k \leq n,\ z = e^{i\theta}. \tag{3.12}$$

Proof We use (backward) induction on k. For $k = n$ we have

$$\frac{1}{2\pi}\int_0^{2\pi} |\varphi_n(z)|^2 \frac{d\theta}{|\varphi_n(z)|^2} = 1,$$

and for $0 \leq j \leq n-1$ we have

$$\begin{aligned} \frac{1}{2\pi}\int_0^{2\pi} \varphi_n(z)\overline{\varphi_j(z)}\frac{d\theta}{|\varphi_n(z)|^2} &= \frac{1}{2\pi}\int_0^{2\pi} \overline{\frac{\varphi_j(z)}{\varphi_n(z)}}\, d\theta \\ &= \frac{1}{2\pi}\int_0^{2\pi} \frac{e^{i(n-j)\theta}\varphi_j^*(e^{i\theta})}{\varphi_n^*(e^{i\theta})}\, d\theta. \end{aligned}$$

The last integral can be computed using Cauchy's theorem and vanishes since φ_n^* has no zeros inside the open unit disk. This shows that the theorem holds for $k = n$. Next we suppose that the theorem holds for some $k \leq n$ and then we will show that it also holds for $k-1$. Taking the modulus in (3.11) gives for $|z| = 1$

$$\kappa_k^2|\varphi_k(z)|^2 = \kappa_{k-1}^2|\varphi_{k-1}(z)|^2 + |\varphi_k(0)|^2|\varphi_k^*(z)|^2 + 2\kappa_{k-1}\Re\overline{\varphi_k(0)\varphi_{k-1}^*(z)}\varphi_k(z).$$

The induction hypothesis implies that φ_k is orthogonal to all polynomials of degree at most $k-1$ with respect to the measure $d\theta/|\varphi_n(z)|^2$, hence integrating with respect to this measure gives

$$\begin{aligned}\frac{\kappa_k^2}{2\pi}\int_0^{2\pi}|\varphi_k(z)|^2\frac{d\theta}{|\varphi_n(z)|^2} &= \frac{\kappa_{k-1}^2}{2\pi}\int_0^{2\pi}|\varphi_{k-1}(z)|^2\frac{d\theta}{|\varphi_n(z)|^2}\\ &\quad+\frac{|\varphi_k(0)|^2}{2\pi}\int_0^{2\pi}|\varphi_k^*(z)|^2\frac{d\theta}{|\varphi_n(z)|^2}.\end{aligned}$$

Since $|\varphi_k^*(z)|=|\varphi_k(z)|$ for $|z|=1$ and $\kappa_k^2=\kappa_{k-1}^2+|\varphi_k(0)|^2$ we thus find

$$\frac{1}{2\pi}\int_0^{2\pi}|\varphi_{k-1}(z)|^2\frac{d\theta}{|\varphi_n(z)|^2}=1.$$

For $0\le j\le k-2$ we find from (3.11)

$$\begin{aligned}\frac{\kappa_{k-1}}{2\pi}\int_0^{2\pi}\varphi_{k-1}(z)\overline{\varphi_j(z)}\frac{d\theta}{|\varphi_n(z)|^2} &= \frac{\kappa_k}{2\pi}\int_0^{2\pi}\varphi_k(z)\overline{z\varphi_j(z)}\frac{d\theta}{|\varphi_n(z)|^2}\\ &\quad-\frac{\varphi_k(0)}{2\pi}\int_0^{2\pi}\varphi_k^*(z)\overline{z\varphi_j(z)}\frac{d\theta}{|\varphi_n(z)|^2}\\ &= \frac{\varphi_k(0)}{2\pi}\int_0^{2\pi}\overline{\varphi_k(z)}\varphi_j^*(z)z^{k-j-1}\frac{d\theta}{|\varphi_n(z)|^2},\end{aligned}$$

and the latter is 0 because $\varphi_j^*(z)z^{k-j-1}$ is a polynomial of degree at most $k-1$ and thus, by the induction hypothesis, orthogonal to φ_k with respect to the measure $d\theta/|\varphi_n(z)|^2$, which completes the proof.

□

3.4 Connection with orthogonal polynomials on $[-1,1]$ Suppose μ is a positive measure on the interval $[-1,1]$, then there exists a sequence of orthonormal polynomials

$$p_n(x)=\gamma_n x^n+\delta_n x^{n-1}+\cdots,\qquad \gamma_n>0,$$

such that

$$\int_{-1}^{1}p_n(x)p_m(x)\,d\mu(x)=\delta_{m,n},\qquad m,n\ge 0.$$

We can define a measure ν on $[0,2\pi)$ by

$$d\nu(\theta)=\frac{1}{2}|d\mu(\cos\theta)|.$$

If $d\mu(x)=w(x)\,dx$ then $d\nu(\theta)=\frac{1}{2}w(\cos\theta)|\sin\theta|\,d\theta$. If μ is a probability measure, then ν is also a probability measure (this is why we introduced the factor $\frac{1}{2}$). This measure on $[0,2\pi)$ then induces a measure ν on the unit circle and there exists a sequence of orthonormal polynomials

$$\varphi_n(z)=\kappa_n z^n+\cdots,\qquad \kappa_n>0,$$

such that

$$\int_0^{2\pi}\varphi_n(z)\overline{\varphi_m(z)}\,d\nu(\theta)=\delta_{m,n},\qquad m,n\ge 0.$$

The corresponding monic polynomials are

$$\hat{p}_n = p_n/\gamma_n, \quad \Phi_n = \varphi_n/\kappa_n.$$

The polynomials on $[-1,1]$ and on the unit circle are closely related:

Theorem 3.5 (Szegő) *The orthogonal polynomials on the unit circle φ_n for the measure ν have real coefficients. If $x = (z+z^{-1})/2$, so that $x = \cos\theta$ corresponds to $z = e^{i\theta}$, then*

$$p_n(x) = \frac{\kappa_{2n}}{\sqrt{2[1+\Phi_{2n}(0)]}} \left(z^{-n}\Phi_{2n}(z) + z^n \Phi_{2n}(1/z) \right). \tag{3.13}$$

Proof The coefficients of the orthogonal polynomials are real because the measure ν is clearly symmetric. Recall the Chebyshev polynomials of the first kind, which for $x = \cos\theta$ are given by

$$T_n(x) = \cos n\theta.$$

If $x = (z+z^{-1})/2$ then one easily finds

$$T_n(x) = \frac{z^n + z^{-n}}{2},$$

since this formula is correct whenever $x = \cos\theta$. First we will show that $z^{-n}\Phi_{2n}(z) + z^n\Phi_{2n}(1/z)$ is indeed of polynomial in x of degree n. If $\Phi_{2n}(z) = \sum_{k=0}^{2n} a_k z^k$, then

$$z^{-n}\Phi_{2n}(z) + z^n\Phi_{2n}(1/z) = \sum_{k=0}^{2n} a_k (z^{k-n} + z^{n-k}) = 2\sum_{k=0}^{2n} a_k T_{|n-k|}(x),$$

which is indeed a polynomial in x of degree n. The leading coefficient of T_n is 2^{n-1}, hence the leading coefficient of $z^{-n}\Phi_{2n}(z) + z^n\Phi_{2n}(1/z)$ is $2^n(a_0 + a_{2n}) = 2^n[\Phi_{2n}(0)+1]$. Next we will show that this polynomial is orthogonal in $L_2(\mu)$ to every polynomial T_k with $k < n$. To this end we compute

$$\begin{aligned} &\int_{-1}^{1} [z^{-n}\Phi_{2n}(z) + z^n\Phi_{2n}(1/z)] T_k(x)\, d\mu(x) \\ &= \int_0^{2\pi} [z^{-n}\Phi_{2n}(z) + z^n\Phi_{2n}(1/z)] z^k \, d\nu(\theta) \\ &= \int_0^{2\pi} \Phi_{2n}(z) z^{-n+k}\, d\nu(\theta) + \int_0^{2\pi} z^{n+k}\overline{\Phi_{2n}(z)}\, d\nu(\theta), \end{aligned}$$

and these integrals are zero whenever $0 \le k < n$ by orthogonality. When $k = n$ we find

$$\int_{-1}^{1} [z^{-n}\Phi_{2n}(z) + z^n\Phi_{2n}(1/z)] T_n(x)\, d\mu(x) = \int_0^{2\pi} z^{2n}\overline{\Phi_{2n}(z)}\, d\nu(\theta) = 1/\kappa_{2n}^2.$$

Since $z^{-n}\Phi_{2n}(z) + z^n\Phi_{2n}(1/z) = 2(1+\Phi_{2n}(0))T_n(x) + q_{n-1}(x)$ we thus find that the norm of $z^{-n}\Phi_{2n}(z) + z^n\Phi_{2n}(1/z)$ is $\kappa_{2n}^{-1}\sqrt{2[1+\Phi_{2n}(0)]}$ so that the desired formula follows for $n > 0$. For $n = 0$ one easily verifies that $p_0 = 1$, as required.

□

From (3.13) one can obtain a relation between the recurrence coefficients on $[-1,1]$ and the reflection coefficients:

Theorem 3.6 *The recurrence coefficients for the orthogonal polynomials on* $[-1,1]$ *and the reflection coefficients* $\Phi_n(0)$ *are related by*

$$2a_n = \sqrt{[1-\Phi_{2n}(0)][1-\Phi_{2n-1}^2(0)][1+\Phi_{2n-2}(0)]}, \qquad (3.14)$$

$$2b_n = \Phi_{2n-1}(0)[1-\Phi_{2n}(0)] - \Phi_{2n+1}(0)[1+\Phi_{2n}(0)]. \qquad (3.15)$$

Conversely, if $R_n = \hat{p}_{n+1}/\hat{p}_n$, *then*

$$\Phi_{2n}(0) = R_n(1) - R_n(-1) - 1, \quad \Phi_{2n+1}(0) = \frac{R_n(1)+R_n(-1)}{R_n(1)-R_n(-1)}. \qquad (3.16)$$

Proof Formula (3.14) is obtained by comparing the coefficients of x^n in (3.13). Formula (3.15) follows by comparing the coefficients of x^{n-1} in (3.13).

Since $x=1$ corresponds to $z=1$ and $x=-1$ to $z=-1$ we easily find from (3.13) that

$$p_n(1) = \frac{\sqrt{2}\kappa_{2n}\Phi_{2n}(1)}{\sqrt{1+\Phi_{2n}(0)}}, \quad p_n(-1) = (-1)^n \frac{\sqrt{2}\kappa_{2n}\Phi_{2n}(-1)}{\sqrt{1+\Phi_{2n}(0)}}.$$

From the recurrence relation (3.6) we find

$$\Phi_{n+1}(1) = \left(\frac{\kappa_n}{\kappa_{n+1}}\right)^2 \Phi_n(1) + \Phi_{n+1}(0)\Phi_{n+1}(1).$$

Solving for $\Phi_{n+1}(1)$ and using (3.3) gives

$$\Phi_{n+1}(1) = [1+\Phi_{n+1}(0)]\Phi_n(1).$$

A similar reasoning for -1 gives

$$\Phi_{n+1}(-1) = -[1-(-1)^n\Phi_{n+1}(0)]\Phi_n(-1).$$

We thus find

$$R_n(1) = a_{n+1}\frac{p_{n+1}(1)}{p_n(1)} = a_{n+1}\frac{\kappa_{2n+2}}{\kappa_{2n}}\frac{\sqrt{1+\Phi_{2n}(0)}}{\sqrt{1+\Phi_{2n+2}(0)}}\frac{\Phi_{2n+2}(1)}{\Phi_{2n}(1)},$$

which readily gives

$$2R_n(1) = [1+\Phi_{2n}(0)][1+\Phi_{2n+1}(0)].$$

In a similar way we obtain

$$R_n(-1) = a_{n+1}\frac{p_{n+1}(-1)}{p_n(-1)} = -a_{n+1}\frac{\kappa_{2n+2}}{\kappa_{2n}}\frac{\sqrt{1+\Phi_{2n}(0)}}{\sqrt{1+\Phi_{2n+2}(0)}}\frac{\Phi_{2n+2}(-1)}{\Phi_{2n}(-1)}$$

and this gives

$$-2R_n(-1) = [1+\Phi_{2n}(0)][1-\Phi_{2n+1}(0)].$$

Subtracting $R_n(1)$ and $R_n(-1)$ then gives $\Phi_{2n}(0)$, whereas adding both quantities gives $\Phi_{2n+1}(0)$.

□

3.5 Examples

3.5.1 *Lebesgue measure* If we take Lebesgue measure on the unit circle, then $d\nu(\theta) = d\theta/(2\pi)$. The trigonometric moments are

$$\frac{1}{2\pi}\int_0^{2\pi} e^{ik\theta}\,d\theta = \delta_{k,0},$$

hence the Toeplitz matrices are all identity matrices and the orthonormal polynomials are $\varphi_n(z) = \Phi_n(z) = z^n$. Note that all the zeros coincide at the origin. Also these monic polynomials coincide with the monic orthogonal polynomials on any disk of radius $R > 0$ and center at the origin and also with the monic polynomials on any circle of radius $R > 0$ and center at the origin. We can only distinguish between these orthogonal polynomials by looking at the orthonormal polynomials, which explicitly contain the radius R.

3.5.2 *Bernstein-Szegő polynomials* Suppose that ν is absolutely continuous with respect to Lebesgue measure and that the Radon-Nikodym derivative $d\nu(\theta)/d\theta$ is the reciprocal of a positive trigonometric polynomial, i.e.,

$$d\nu(\theta) = \frac{d\theta}{2\pi q_k(\theta)}$$

where

$$q_k(\theta) = a_0 + a_1\cos\theta + b_1\sin\theta + \cdots + a_k\cos k\theta + b_k\sin k\theta, \qquad a_k, b_k \in \mathbb{R},$$

is positive for $\theta \in [0, 2\pi]$. Fejér and F. Riesz have shown that there exists a unique polynomial Q_k of degree k, which has no zeros in the open unit disk and for which $Q_k(0) > 0$, such that $q_k(\theta) = |Q_k(e^{ik\theta})|^2$. This can be shown by writing

$$2\cos n\theta = z^n + z^{-n}, \quad 2i\sin n\theta = z^n - z^{-n}, \qquad z = e^{i\theta},$$

so that

$$q_k(\theta) = z^{-k}G_{2k}(z),$$

where

$$G_{2k}(z) = a_0 z^k + \frac{1}{2}\sum_{j=1}^k a_j z^{k+j} + \frac{1}{2}\sum_{j=1}^k a_j z^{k-j} + \frac{1}{2i}\sum_{j=1}^k b_j z^{k+j} - \frac{1}{2i}\sum_{j=1}^k b_j z^{k-j}.$$

Clearly G_{2k} is a polynomial of degree $2k$ such that $G_{2k}^* = G_{2k}$. If z_l is a zero of G_{2k}, then $1/\overline{z_l}$ is also a zero of G_{2k}, so we can write

$$G_{2k}(z) = c\prod(z - z_l)(z - (\overline{z_l})^{-1})\prod(z - \xi_l)^2$$

where z_l are the zeros with $|z_l| > 1$ and ξ_l are the zeros on the unit circle, which have even multiplicity since q_k is a positive trigonometric polynomial. We thus have

$$q_k(\theta) = |q_k(\theta)| = |G_{2k}(z)| = |Q_k(z)|^2,$$

where

$$Q_k(z) = \sqrt{|c|/\prod|z_l|}\prod(z - z_l)\prod(z - \xi_l).$$

We now show that the orthonormal polynomials of degree greater than or equal to k are given by

$$\varphi_n(z) = z^{n-k}Q_k^*(z), \qquad n \geq k.$$

These polynomials are known as *Bernstein-Szegő polynomials*. Let r_m be an arbitrary polynomial of degree $m < n$, then

$$\int_0^{2\pi} \varphi_n(z)\overline{r_m(z)}\, d\nu(z) = \frac{1}{2\pi}\int_C \frac{z^{n-k}Q_k^*(z)\bar{r}(1/z)}{Q_k(z)z^{-k}Q_k^*(z)}\frac{dz}{iz} = \frac{1}{2\pi i}\int_C \frac{z^{n-1}\bar{r}_m(1/z)}{Q_k(z)}\, dz$$

and by Cauchy's theorem this integral is zero. The norm is computed easily by

$$\int_0^{2\pi} |\varphi_n(z)|^2\, d\nu(z) = \frac{1}{2\pi}\int_0^{2\pi} \frac{|Q_k^*(z)|^2}{|Q_k(z)|^2}d\theta = 1.$$

3.5.3 *Jacobi polynomials* If $d\mu(x) = w(x)\,dx$, with $w(x) = (1-x)^\alpha(1+x)^\beta$ $(\alpha,\beta > -1)$ then the orthogonal polynomials on the interval $[-1,1]$ are the Jacobi polynomials. The corresponding weight on the unit circle is $w(\cos\theta)|\sin\theta| = |\sin\theta/2|^{2\alpha+1}|\cos\theta/2|^{2\beta+1}$. From Szegő [1939] we find

$$\hat{p}_n(1) = 2^n\binom{n+\alpha}{n}\Big/\binom{2n+\alpha+\beta}{n}, \quad \hat{p}_n(-1) = (-1)^n 2^n\binom{n+\beta}{n}\Big/\binom{2n+\alpha+\beta}{n},$$

so that

$$\begin{aligned} R_n(1) &= \frac{4(n+\alpha+1)(n+\alpha+\beta+1)}{(2n+\alpha+\beta+2)(2n+\alpha+\beta+1)}, \\ R_n(-1) &= \frac{4(n+\beta+1)(n+\alpha+\beta+1)}{(2n+\alpha+\beta+2)(2n+\alpha+\beta+1)}. \end{aligned}$$

A simple calculation then gives the reflection coefficients for the *Jacobi polynomials on the unit circle*:

$$\Phi_{2n}(0) = \frac{\alpha+\beta+1}{2n+\alpha+\beta+1}, \quad \Phi_{2n+1}(0) = \frac{\alpha-\beta}{2n+\alpha+\beta+2}.$$

The special case $\beta = -1/2$, which is one way of getting Gegenbauer polynomials on the unit circle, gives

$$\Phi_n(0) = \frac{2\alpha+1}{2n+2\alpha+1}.$$

References

Andrews, G. E. and Askey, R. [1985] *Classical orthogonal polynomials*, in 'Polynômes Orthogonaux et Applications' (C. Brezinski et al, eds.), Lecture Notes in Mathematics **1171**, Springer-Verlag, New York, 36–62.

Askey, R. and Wilson, J. [1985] *Some basic hypergeometric orthogonal polynomials that generalize Jacobi polynomial*, Memoirs Amer. Math. Soc. **319**, Providence, RI.

Chihara, T. S. [1978] *An Introduction to Orthogonal Polynomials,* Gordon and Breach, New York.

Favard, J. A. [1935] *Sur les polynômes de Tchebicheff,* C.R. Acad. Sci. Paris **200**, 2052–2053.

Freud, G. [1971] *Orthogonal Polynomials,* Akadémiai Kiadó, Budapest and Pergamon Press, Oxford.

Geronimus, Ya. L. [1962] *Polynomials Orthogonal on a Circle and their Applications,* Amer. Math. Soc. Transl. **3**, 1–78.

Geronimus, Ya. L. [1960] *Polynomials Orthogonal on a Circle and Interval,* International Series of monographs in pure and applied mathematics, Pergamon Press, Oxford.

Grenander, U. and Szegő, G. [1984] *Toeplitz Forms and their Applications,* Chelsea Publishing Company, New York.

Hörmander, L. [1966] *An Introduction to Complex Analysis in Several Variables,* Van Nostrand, Princeton, NJ.

Horn, R. A. and Johnson, C. R. [1985] *Matrix Analysis,* Cambridge University Press.

Jones, W. B., Njåstad, O., and Thron, W. J. [1989] *Moment theory, orthogonal polynomials, quadrature, and continued fractions associated with the unit circle,* Bull. London Math. Soc. **21**, 113–152.

Koekoek, R. and Swarttouw, R. S. [1994] *The Askey-scheme of hypergeometric orthogonal polynomials and its q-analogue*, Reports of the Faculty of Technical Mathematics and Informatics **94-05**, Delft University of Technology, 120 pp.

Lubinsky, D. S. [1987] *A survey of general orthogonal polynomials for weights on finite and infinite intervals,* Acta Appl. Math. **10**, 237–296.

Lubinsky, D. S. [1993] *An update on orthogonal polynomials and weighted approximation on the real line*, Acta Appl. Math. **33**, 121–164.

Nevai, P. G. [1979] *Orthogonal Polynomials,* Memoirs Amer. Math. Soc. **213**, Providence, RI.

Nevai, P. [1986] *Géza Freud, orthogonal polynomials and Christoffel functions. A case study,* J. Approx. Theory **48**, 3–167.

Nevai, P. (ed.) [1990] *Orthogonal Polynomials: Theory and Practice,* NATO-ASI Series C **294**, Kluwer, Dordrecht.

Nikishin, E. M. and Sorokin, V. N. [1991] *Rational Approximations and Orthogonality,* Translations of Mathematical Monographs **92**, Amer. Math. Soc.

Rudin, W. [1978] *Real and Complex Analysis,* Tata McGraw-Hill, New Delhi.

Saff, E. B. [1990] *Orthogonal polynomials from a complex perspective,* in 'Orthogonal Polynomials: Theory and Practice' (P. Nevai, ed.), 363–393.

Stahl, H. and Totik, V. [1992] *General Orthogonal Polynomials,* Encyclopedia of Mathematics and its Applications **43**, Cambridge University Press.

Stieltjes, T. J. [1894] *Recherches sur les fractions continues,* Ann. Fac. Sci. Toulouse **8**, J1–122; **9** (1895), A1–47.

Szegő, G. [1939] *Orthogonal Polynomials,* Amer. Math. Soc. Colloq. Publ. **23**, Amer. Math. Soc., Providence, RI, (4th edition, 1975).

Van Assche, W. [1987] *Asymptotics for Orthogonal Polynomials,* Lecture Notes in Mathematics **1265**, Springer-Verlag, Berlin.

Fields Institute Communications
Volume **14**, 1997

On Orthogonal Polynomials in Several Variables

Yuan Xu
Department of Mathematics
University of Oregon
Eugene, Oregon 97403-1222 USA

Abstract. We report on the recent development on the general theory of orthogonal polynomials in several variables, in which results parallel to the theory of orthogonal polynomials in one variable are established using a vector-matrix notation.

1 Introduction

The main theme of this paper is the *general* theory of orthogonal polynomials in several variables, which deals with those properties shared by all systems of orthogonal polynomials. Thus, we shall not discuss the theory of particular systems and their applications.

Let Π^d be the set of polynomials in d variables on $\mathbb{R}^d$, and let Π_n^d be the subset of polynomials of total degree at most n. A linear functional $\mathcal{L}$ defined on Π^d is called square positive if $\mathcal{L}(p^2) > 0$ for all $p \in \Pi^d$, $p \neq 0$. Such a linear functional $\mathcal{L}$ induces an inner product on Π^d, defined by $\langle P, Q\rangle = \mathcal{L}(PQ)$. Two polynomials are said to be orthogonal with respect to $\mathcal{L}$ if their inner product is zero.

Given a square positive linear functional, we may apply the Gram-Schmidt orthogonalization process on the multiple sequence of monomials $x_1^{\alpha_1} \ldots x_d^{\alpha_d}$, $\mathbf{x} = (x_1, \ldots, x_d) \in \mathbb{R}^d$ and $\alpha_i \in \mathbb{N}_0$, to generate a sequence of orthogonal polynomials. In order to do so, however, it is necessary to order the multiple sequence in a simple one. There are many ways to do so; in general, different orderings will lead to different orthogonal systems. Thus, there is no unique system of orthogonal polynomials in several variables. Moreover, any system of orthogonal polynomials obtained by an ordering of the monomials is necessarily unsymmetric in the variables $x_1, \ldots, x_d$. These are essential difficulties in the study of orthogonal polynomials in several variables; they are mentioned in the introductory section of Chapter XII in the *Higher Transcendental Functions* (Erdélyi, et al. [1953, p. 265]). On the same page of the book it is also stated that "there does not seem to be an extensive general theory of

1991 *Mathematics Subject Classification.* Primary: 33C50, 42C05, 65D32; Secondary: 33D45, 41A10, 41A63.

Supported in part by NSF Grant #9302721 and #9500532.

orthogonal polynomials in several variables". A brief account of the general properties of orthogonal polynomials known up to then is summarized in Chapt. XII of Erdélyi, et al. [1953].

More recent accounts of general theory of orthogonal polynomials in several variables can be found in Krall and Sheffer [1967], Mysovskikh [1976] and Suetin [1988]. We will report on progress made in the past few years, from which a possible extensive general theory may emerge. In the center of this new development is a vector-matrix notation which helps us to overcome the difficulty caused by the non-uniqueness of orthogonal polynomials. For each $n > 0$ the set of polynomials of degree n that are orthogonal to all polynomials of lower degree froms a vector space V_n of dimension greater than one. The non-uniqueness of orthogonal polynomials means that there is no unique way of choosing a basis for V_n^d. In their work of extending the characterization of the classical orthogonal polynomials as eigenfunctions of second order differential operators from one to two variables, Krall and Sheffer [1967] suggested one way to overcome the difficulty; they remarked that if the results can be stated "in terms of $V_0^d, V_1^d, \ldots, V_n^d, \ldots$ rather than in terms of a particular basis in each V_n^d, a degree of uniqueness is restored". Kowalski [1982a] and [1982b] used a vector-matrix notation to prove an analog of Favard's theorem, but he seemed unaware of the paper of Krall and Sheffer. His theorem is simplified in Xu [1993a] using a modified vector-matrix notation. The latter notation is inspired by and very much reflects the principle of Krall and Sheffer; it has been adopted in further studies of the author Xu [1993b], [1994a-e] and [1994h], upon which the present report is based.

The paper is organized as follows. The next section is devoted to the notations and preliminaries. In Section 3, we discuss the three-term relation satisfied by vectors of orthogonal polynomials and the results around it. The three-term relation leads us naturally to define a family of block Jacobi matrices, which can be studied as a commuting family of self-adjoint operators; this connection is covered in Section 4. The zeros of orthogonal polynomials are defined as common zeros of a family of polynomials, they are related to numerical cubature formulae. In Section 5 we discuss the basic properties of common zeros and cubature formulae; deeper results are covered in Section 6. Finally, in Section 7 we discuss the preliminary results dealing with asymptotics and Fourier orthogonal expansion involving orthogonal polynomials in several variables.

2 Notations and preliminaries

We use the standard multiindex notation. Let $\mathbb{N}_0$ be the set of nonnegative integers. For $\alpha = (\alpha_1, \ldots, \alpha_d) \in \mathbb{N}_0^d$ and $\mathbf{x} = (x_1, \ldots, x_d) \in \mathbb{R}^d$ we write $\mathbf{x}^\alpha = x_1^{\alpha_1} \cdots x_d^{\alpha_d}$. The number $|\alpha| = \alpha_1 + \cdots + \alpha_d$ is called the total degree of $\mathbf{x}^\alpha$. For $n \in \mathbb{N}_0$, we denote by Π^d the set of all polynomials in d variables and by Π_n^d the subset of polynomials of total degree at most n; *i.e.*,

$$\Pi_n^d = \Big\{ \sum_{|\alpha| \le n} a_\alpha \mathbf{x}^\alpha : \quad a_\alpha \in \mathbb{R}, \quad \mathbf{x} \in \mathbb{R}^d \Big\}.$$

We denote by r_n^d the number of monomials of degree exactly n which is equal to the cardinality of the set $\{\alpha \in \mathbb{N}_0^d : |\alpha| = n\}$; it follows that

$$\dim \Pi_n^d = \binom{n+d}{d} \quad \text{and} \quad r_n^d = \binom{n+d-1}{n}.$$

For $n \in \mathbb{N}_0$ and $\mathbf{x} \in \mathbb{R}^d$, we denote by $\mathbf{x}^n = (\mathbf{x}^\alpha)_{|\alpha|=n}$ a vector of size r_n^d, where the monomials are arranged according to the lexicographical order of $\{\alpha \in \mathbb{N}_0^d : |\alpha| = n\}$.

A multisequence $s : \mathbb{N}_0^d \mapsto \mathbb{R}$ will be written in the form $s = \{s_\alpha\}_{\alpha \in \mathbb{N}_0^d}$. Such a sequence is called *positive definite* if for every tuple $(\beta^{(1)}, \dots, \beta^{(r)})$ of distinct multiindices $\beta^{(j)} \in \mathbb{N}_0^d$, $1 \le j \le r$, the matrix $(s_{\beta^{(i)}+\beta^{(j)}})_{i,j=1,\dots,r}$ has positive determinant. With each real d-sequence s one can associate a linear functional on Π^d defined by $\mathcal{L}(\mathbf{x}^\alpha) = s_\alpha$. If s is positive definite, we call the associated linear functional *square positive*. Since

$$\mathcal{L}(P^2) = \sum_{\alpha,\beta} s_{\alpha+\beta} a_\alpha a_\beta \qquad \text{where} \qquad P(\mathbf{x}) = \sum_\alpha a_\alpha \mathbf{x}^\alpha,$$

it follows that $\mathcal{L}$ is square positive if, and only if, $\mathcal{L}(P^2) > 0$ for all $P \in \Pi^d$ not identically zero. We remark that we do not deal with *semi square positive* which requires only $\mathcal{L}(P^2) \ge 0$; see Fuglede [1983] and Berg [1987] for discussion. Let $\mathcal{M} = \mathcal{M}(\mathbb{R}^d)$ denote the set of nonnegative Borel measures μ, defined on the σ-algebra of Borel sets, with an infinite support on $\mathbb{R}^d$ such that

$$\int_{\mathbb{R}^d} |\mathbf{x}^\alpha| \, d\mu(\mathbf{x}) < +\infty , \qquad \forall\, \alpha \in \mathbb{N}_0^d .$$

For any $\mu \in \mathcal{M}$ the moments of μ are defined as the real numbers $s_\alpha = \int_{\mathbb{R}^d} \mathbf{x}^\alpha d\mu$. A d-sequence $s = (s_\alpha)$ is called a moment sequence if there exists at least one measure μ whose moments are equal to s_α for all $\alpha \in \mathbb{N}_0^d$. If s is a moment sequence, the corresponding linear functional $\mathcal{L}$ is called a moment functional, which has an integral representation

$$\mathcal{L}(f) = \int_{\mathbb{R}^d} f(\mathbf{x}) \, d\mu(\mathbf{x}) , \qquad \mu \in \mathcal{M} . \tag{2.1}$$

Two measures in $\mathcal{M}$ are called equivalent if they have the same moments. If the equivalent class of measures having the same moments as μ consists of μ only, the measure μ is called determinate. If $\mathcal{L}$ is a moment functional of a determined measure, then the integral representation is unique. It is known that $\mathcal{L}$ is a moment functional if it is positive, which means that $\mathcal{L}(P) \ge 0$ whenever $P \ge 0$. For $d = 1$, $\mathcal{L}$ being positive is equivalent to $\mathcal{L}(P^2) \ge 0$. For $d > 1$, however, they are no longer equivalent, which is, in fact, the cause of many problems in the multidimensional moment problem (cf. Berg [1983], Berg and Thill [1991] and Fuglede [1983]).

A square positive linear functional $\mathcal{L}$ induces an inner product $\langle \cdot, \cdot \rangle$ on Π^d defined by

$$\langle P, Q \rangle = \mathcal{L}(PQ), \qquad P, Q \in \Pi^d.$$

For convenience, we shall always assume that $\mathcal{L}(1) = 1$. Two polynomials P and Q are said to be orthogonal with respect to $\mathcal{L}$, if $\langle P, Q \rangle = 0$. With respect to such an $\mathcal{L}$ we apply the Gram-Schmidt orthogonalization process on the monomials $\{\mathbf{x}^\alpha\}$ arranged as $\{\mathbf{x}^n\}_{n=0}^\infty$ to derive a sequence of orthonormal polynomials, denoted by $\{P_k^n\}_{k=1}^{r_n^d}{}_{n=0}^{\infty}$, where the superscript n means that $P_k^n \in \Pi_n^d$. We now introduce the vector notation that is fundamental in the development below:

$$\mathbb{P}_n(\mathbf{x}) = \left[P_1^n(\mathbf{x}), P_2^n(\mathbf{x}), \dots, P_{r_n^d}^n(\mathbf{x})\right]^T . \tag{2.2}$$

Using this notation, the orthonormality property of $\{P_k^n\}$ can be described as

$$\mathcal{L}(\mathbb{P}_n\mathbb{P}_m^T) = \delta_{m,n}I_{r_n^d}, \tag{2.3}$$

where I_n denotes the $n \times n$ identity matrix. For convenience, we sometimes call $\mathbb{P}_n$ orthonormal polynomials with respect to $\mathcal{L}$. When $\mathcal{L}$ has an integral representation (2.1) with $d\mu = W(\mathbf{x})d\mathbf{x}$, we call $\mathbb{P}_n$ orthonormal polynomials with respect to W instead of to $\mathcal{L}$. In terms of monomial vectors $\mathbf{x}^n$, we can write $\mathbb{P}_n$ as

$$\mathbb{P}_n = G_n\mathbf{x}^n + G_{n,n-1}\mathbf{x}^{n-1} + G_{n,n-2}\mathbf{x}^{n-2} + \dots \tag{2.4}$$

where $G_{n,i} : r_n^d \times r_{n-i}^d$ are matrices. We call $G_n = G_{n,n}$ the leading coefficient of $\mathbb{P}_n$, which can be seen to be invertible since $\mathcal{L}$ is square positive.

For each $k \geq 0$, let $V_k^d \subset \Pi_k^d$ be the set of polynomials spanned by $\mathbb{P}_k$, that is, spanned by the components of $\mathbb{P}_k$, together with zero. Then V_k^d is a vector space of dimension r_k^d which is orthogonal to all polynomials in Π_{k-1}^d. Clearly

$$\Pi_n^d = \bigoplus_{k=0}^{n} V_k^d \quad \text{and} \quad \Pi^d = \bigoplus_{k=0}^{\infty} V_k^d \ ,$$

and V_k^d are mutually orthogonal. As we mentioned before, the sequence of orthonormal polynomial is not unique. Actually, it is easy to see that each orthogonal matrix Q of order r_k^d gives rise to an orthonormal basis $Q\mathbb{P}_k$ of V_k^d and every orthonormal basis of V_k^d is of the form $Q\mathbb{P}_n$. One can also work with other bases of V_k that are not necessarily orthonormal. In particular, one basis consists of polynomials $\tilde{P}_\alpha^k$ of the form

$$\tilde{P}_\alpha^k = \mathbf{x}^\alpha + R_\alpha^{k-1}, \qquad |\alpha| = k, \qquad R_\alpha^{k-1} \in \Pi_{k-1}^d.$$

This basis is sometimes called monomial basis; in general, $\mathcal{L}(\tilde{P}_\alpha^k\tilde{P}_\beta^k) \neq 0$, although $\tilde{P}_\alpha^k$ are orthogonal to all polynomials of lower degrees. It is easy to see that the matrix $H_n = \mathcal{L}(\tilde{\mathbb{P}}_n\tilde{\mathbb{P}}_n^T)$ is positive definite and $\tilde{\mathbb{P}}_n = H_n^{-1/2}\mathbb{P}_n$. Because of the relation, most of the results below can be stated in terms of the monomial basis. We should mention that one can define orthogonal polynomials with respect to linear functionals that are not necessarily square positive. However, the square positiveness is necessary for obtaining orthonormal polynomials.

For a square positive linear functional expressible by (2.1), we can consider the orthogonal expansion of a function $f \in L_{d\mu}$. Using the vector notation $\mathbb{P}_n$, the Fourier orthogonal expansion of f with respect to a sequence of orthonormal polynomials $\{P_k^n\}$ is given by

$$f \sim \sum_{n=0}^{\infty} \mathbf{a}_n^T(f)\mathbb{P}_n, \qquad \mathbf{a}_n(f) = \int f(\mathbf{x})\mathbb{P}_n(\mathbf{x})d\mu \tag{2.5}$$

where $\mathbf{a}_n$ is a vector of $\mathbb{R}^{r_n^d}$. If we replace $\mathbf{a}_n^T\mathbb{P}_n$ by $\text{proj}_{V_n} f$, then the expansion can be viewed as in terms of the V_n^d and is independent of the particular basis of V_n^d. In fact, the n-th reproducing kernel of the orthonormal polynomials, defined by

$$\mathbf{K}_n(\mathbf{x},\mathbf{y}) = \sum_{k=0}^{n-1}\sum_{j=0}^{r_k^d} P_j^k(\mathbf{x})P_j^k(\mathbf{y}) = \sum_{k=0}^{n} \mathbb{P}_n^T(\mathbf{x})\mathbb{P}_n(\mathbf{y}), \tag{2.6}$$

is easily seen to depend on V_k^d rather than a particular basis of V_k^d. The n-th partial sum $S_n f$ of the expansion can be written in terms of $\mathbf{K}_n(\cdot,\cdot)$ as

$$S_n f = \sum_{k=0}^{n-1} \mathbf{a}_k^T(f)\mathbb{P}_k = \int \mathbf{K}_n(\cdot,\mathbf{y})f(\mathbf{y})d\mu. \tag{2.7}$$

From time to time, we may use results for orthogonal polynomials in one variable for motivation or comparison. We follow the standard notation in one variable (cf. Szegő [1975] and Chihara [1978]). The orthonormal polynomial of degree n on $\mathbb{R}$ is denoted by p_n. The three-term relation satisfied by p_n is denoted by

$$xp_n = a_n p_{n+1} + b_n p_n + a_{n-1}p_{n-1}, \qquad x \in \mathbb{R}, \tag{2.8}$$

where a_n and b_n are called the coefficients of the three-term relation.

3 Three-term relation

Our development of a general theory of orthogonal polynomials in several variables starts from a three-term relation in a vector-matrix notation very much like in the one variable theory.

Three-term relation. For $n \ge 0$, there exist matrices $A_{n,i} : r_n^d \times r_{n+1}^d$ and $B_{n,i} : r_n^d \times r_n^d$, such that

$$x_i\mathbb{P}_n = A_{n,i}\mathbb{P}_{n+1} + B_{n,i}\mathbb{P}_n + A_{n-1,i}^T\mathbb{P}_{n-1}, \qquad 1 \le i \le d, \tag{3.1}$$

where we define $\mathbb{P}_{-1} = 0$ and $A_{-1,i} = 0$.

In fact, since the components of $x_i\mathbb{P}_n$ are polynomials of degree $n+1$, they can be written in terms of linear combinations of $\mathbb{P}_{n+1}, \dots, \mathbb{P}_0$; the orthonormal property of $\mathbb{P}_n$ implies that only the coefficients of $\mathbb{P}_{n+1}$, $\mathbb{P}_n$ and $\mathbb{P}_{n-1}$ are nonzero. The result is the relation (3.1). Moreover, the matrices in the three-term relation are expressible as

$$A_{n,i} = \mathcal{L}(x_i\mathbb{P}_n\mathbb{P}_{n+1}^T) \quad \text{and} \quad B_{n,i} = \mathcal{L}(x_i\mathbb{P}_n\mathbb{P}_n^T). \tag{3.2}$$

As a consequence, the matrices $B_{n,i}$ are symmetric. If we are dealing with orthogonal polynomials, $\tilde{\mathbb{P}}_n$, which are not necessarily orthonormal, then the three-term relation takes the form

$$x_i\tilde{\mathbb{P}}_n = A_{n,i}\tilde{\mathbb{P}}_{n+1} + B_{n,i}\tilde{\mathbb{P}}_n + C_{n,i}^T\tilde{\mathbb{P}}_{n-1}, \qquad 1 \le i \le d, \tag{3.3}$$

where $C_{n,i} : r_n^d \times r_{n-1}^d$ is related to $A_{n,i}$ by

$$A_{n,i}H_{n+1} = H_n C_{n+1,i}, \qquad \text{where} \qquad H_n = \mathcal{L}(\tilde{\mathbb{P}}_n\tilde{\mathbb{P}}_n^T).$$

Moreover, comparing the highest coefficient matrices at both sides of (3.1), it follows that

$$A_{n,i}G_{n+1} = G_n L_{n,i}, \qquad 1 \le i \le d, \tag{3.4}$$

where $L_{n,i}$ are matrices of size $r_n^d \times r_{n+1}^d$ which are defined by

$$L_{n,i}\mathbf{x}^{n+1} = x_i\mathbf{x}^n, \qquad 1 \le i \le d.$$

Clearly, $\operatorname{rank} L_{n,i} = r_n^d$, and $\operatorname{rank} L_n = r_{n+1}^d$, where $L_n = (L_{n,1}^T|\dots|L_{n,d}^T)^T$. For example, for $d = 2$ we have

$$L_{n,1} = \begin{bmatrix} 1 & & \bigcirc & 0 \\ & \ddots & & \vdots \\ \bigcirc & & 1 & 0 \end{bmatrix} \quad \text{and} \quad L_{n,2} = \begin{bmatrix} 0 & 1 & & \bigcirc \\ \vdots & & \ddots & \\ 0 & \bigcirc & & 1 \end{bmatrix}.$$

From the relation (3.4) and the fact that G_n is invertible, it follows readily that the matrices $A_{n,i}$ satisfy *Rank conditions*. For $n \ge 0$, $\operatorname{rank} A_{n,i} = r_n^d$ for $1 \le i \le d$, and

$$\operatorname{rank} A_n = r_{n+1}^d, \qquad A_n = (A_{n,1}^T, \dots, A_{n,d}^T)^T. \tag{3.5}$$

The importance of the three-term relation is readily seen in the following analog of Favard's theorem of one variable. We extend the notation (2.2) to an arbitrary sequence of polynomials $\{P_k^n\}_{k=1}^{r_n^d}$.

Theorem 3.1 (Xu [1993a]). *Let $\{\mathbb{P}_n\}_{n=0}^\infty$, $\mathbb{P}_0 = 1$, be a sequence in Π^d. Then the following statements are equivalent:*

1. *There exists a linear functional which is square positive and which makes $\{\mathbb{P}_n\}_{n=0}^\infty$ an orthonormal basis in Π^d;*
2. *there exist matrices $A_{n,i} : r_n^d \times r_{n+1}^d$ and $B_{n,i} : r_n^d \times r_n^d$ such that*

   (1) *the polynomial vectors $\mathbb{P}_n$ satisfy the three-term relation* (3.1),

   (2) *the matrices in the relation satisfies the rank condition.*

An earlier version of this theorem is stated in Kowalski [1982b] with respect to the three-term relation (3.3) using the notation $\mathbf{x}\mathbb{P}_n = [x_1\mathbb{P}_n^T|\dots|x_d\mathbb{P}_n^T]^T$, where it is stated under an additional condition in part 2:

(3) *for an arbitrary sequence $D_0, D_1, \dots, D_n$ of matrices satisfying $D_k A_k = I$, the recursion*

$$J_0 = [1], J_{k+1} = D_k(C_{k+1,1}J_k^T|\dots|C_{k+1,1}J_k^T)^T, \qquad k = 0, 1, \dots,$$

produces nonsingular symmetric matrices J_k.

In Xu [1993a], the theorem is stated in the present form with respect to the three-term relation (3.3), where it is proved that the condition (3) is equivalent to that $\operatorname{rank} C_n = r_n^d$, here $C_n = (C_{k,1}|\dots|C_{k,d})$. The theorem in its present form for orthonormal polynomials is stated in Xu [1994a] for the first time.

The theorem is an analog of Favard's theorem in one variable, but it is not as strong as the classical Favard's theorem. It does not state, for example, when the linear functional $\mathcal{L}$ in the theorem will have an integral representation. We will address this question in the following section. For now, we concentrate on the three-term relation (3.1). It is an analog of the three-term relation in one variable; the fact that its coefficients are matrices reflect the complexity of the structure for $d \ge 2$.

In one variable, the sequence of orthogonal polynomials $\{p_n\}$ can also be viewed as a solution of the difference equation

$$y_{n+1} = a_n^{-1}(xy_n - b_ny_n - a_{n-1}y_{n-1}), \tag{3.6}$$

with initial values $p_0 = 1$ and $p_1 = a_1^{-1}(x - b_0)$. In particular, the three-term relation is also called *recurrence* relation, since p_n can be computed recursively through such an relation. It follows trivially that all solutions of (3.6) satisfy the three-term relation (2.8).

To derive an analog of (3.6) in several variables, we first note that the rank condition (3.5) implies that there exists a matrix $D_n^T : r_{n+1}^d \times d\, r_n^d$, which we write as $D_n^T = (D_{n,1}^T|\ldots|D_{n,d}^T)$ where $D_{n,i}$ are of size $r_n^d \times r_{n+1}^d$, such that

$$D_n^T A_n = \sum_{i=1}^{d} D_{n,i}^T A_{n,i} = I_{r_{n+1}^d}. \tag{3.7}$$

We note that such a D_n is not unique, as can be seen from the singular value decomposition of A_n. Multiplying the three-term relation (3.1) from the left by $D_{n,i}^T$ and summing up the d relations, we derived that $\mathbb{P}_n$ satisfies a recurrence relation

$$\begin{aligned}\mathbb{P}_{n+1}(\mathbf{x}) = \sum_{i=1}^{d} D_{n,i}^T x_i \mathbb{P}_n(\mathbf{x}) &- \left(\sum_{i=1}^{d} D_{n,i}^T B_{n,i}\right) \mathbb{P}_n(\mathbf{x}) \\ &- \left(\sum_{i=1}^{d} D_{n,i}^T A_{n-1,i}^T\right) \mathbb{P}_{n-1}(\mathbf{x})\end{aligned} \tag{3.8}$$

where $\mathbb{P}_{-1} = 0$, $\mathbb{P}_0 = 1$. For any given sequences $A_{n,i}$ and $B_{n,i}$, the relation (3.8) can be used to define a sequence of polynomials. However, in contrast to one variable, polynomials so defined may not satisfy the three-term relation (3.1), since (3.8) cannot be split into d relations (3.1) in general. The question is answered by the following theorem.

Theorem 3.2 (Xu [1994c]) *Let $\{\mathbb{P}_k\}_{k=0}^{\infty}$ be defined by* (3.8). *Then $\{\mathbb{P}_k\}_{k=0}^{\infty}$ satisfies the three-term relation* (3.1) *if, and only if, $B_{k,i}$ are symmetric, $A_{k,i}$ satisfy the rank condition* (3.5), *and together they satisfy the commuting conditions*

$$A_{k,i}A_{k+1,j} = A_{k,j}A_{k+1,i}\ , \tag{3.9a}$$

$$A_{k,i}B_{k+1,j} + B_{k,i}A_{k,j} = B_{k,j}A_{k,i} + A_{k,j}B_{k+1,i}\ , \tag{3.9b}$$

$$\begin{aligned}A_{k-1,i}^T A_{k-1,j} &+ B_{k,i}B_{k,j} + A_{k,i}A_{k,j}^T \\ &= A_{k-1,j}^T A_{k-1,i} + B_{k,j}B_{k,i} + A_{k,j}A_{k,i}^T\end{aligned} \tag{3.9c}$$

for $i \neq j$, $1 \leq i,j \leq d$, and $k \geq 0$, where $A_{-1,i} = 0$.

In one variable, one can start with a sequence of positive numbers a_n and an arbitrary sequence of real numbers b_n, use (3.6) to generate a sequence of polynomials, which is then automatically a sequence of orthogonal polynomials by Favard's theorem. For $d \geq 2$, such an approach is no longer feasible according to Theorem 3.2, since it is very difficult to tell when matrices will satisfy the commuting conditions (3.9). The reason why we call the equations in (3.9) commuting conditions will become clear in the next section. The necessity of these conditions follows easily from the fact that there are two ways to compute $\mathcal{L}(x_i x_j \mathbb{P}_n \mathbb{P}_n^T)$, $\mathcal{L}(x_i x_j \mathbb{P}_n \mathbb{P}_{n+1}^T)$ and $\mathcal{L}(x_i x_j \mathbb{P}_n \mathbb{P}_{n+2}^T)$ using the three-term relation. The sufficiency part is long and quite complicated, we refer to Xu [1994c].

In view of (3.6), we can also look at $\mathbb{P}_n$ as the solution of the multiparameter difference equations

$$x_i Y_k = A_{k,i} Y_{k+1} + B_{k,i} Y_k + A_{k-1,i}^T Y_{k-1}, \qquad 1 \le i \le d, \quad k \ge 1, \tag{3.10}$$

with the initial values

$$Y_0 = 1, \qquad Y_1 = A_0^{-1}(\mathbf{x} - B_0). \tag{3.11}$$

For $d = 1$, the difference equation (3.6) has two linearly independent solutions. One is $\{p_n\}$, which corresponds to the initial values $y_0 = 1$ and $y_{-1} = 0$, another is usually denoted by $\{q_n\}$, which corresponds to the initial values $y_0 = 0$ and $y_{-1} = 1$. The polynomials q_n are usually called the *associated polynomials*, or polynomials of the second kind. They play an important role in areas such as problem of moments, spectral theory of the Jacobi matrices, and continuous fractions (cf. Akheizer [1965]). For the multiparameter difference equations, one may easily guess that there will be $d+1$ linearly independent solutions. However, the following theorem says that $\mathbb{P}_n$ is essentially the only possible solution.

Theorem 3.3 (Xu [1994e]) *If the multiparameter difference equation* (3.10) *has a solution* $\mathbb{P} = \{\mathbb{P}_k\}_{k=0}^{\infty}$ *for the particular initial value* (3.11), *then all other solutions of* (3.10) *are multiples of* $\mathbb{P}$ *with the possible exception of the first component. More precisely, if* $Y = \{Y_n\}_{n=0}^{\infty}$ *is a solution of* (3.11), *then* $Y_n = h\mathbb{P}_n$ *for all* $n \ge 1$, *where* h *is a function independent of* n.

The theorem comes out somewhat surprising, especially in view of the important role played by the associated polynomials in one variable. It indicates that it is much harder to extract information from the three-term relation (3.1) for $d \ge 2$.

Among other consequences of the three-term relation, we mention an extension of the Christoffel-Darboux formula of one variable. Let $\mathbf{K}_n(\cdot,\cdot)$ be the reproducing kernel defined in (2.6). Then we have (Xu [1993a])

Christoffel-Darboux formula: For $n \ge 1$, $1 \le i \le d$,

$$\mathbf{K}_n(\mathbf{x},\mathbf{y}) = \frac{[A_{n-1,i}\mathbb{P}_n(\mathbf{x})]^T \mathbb{P}_{n-1}(\mathbf{y}) - \mathbb{P}_{n-1}^T(\mathbf{x})[A_{n-1,i}\mathbb{P}_n(\mathbf{y})]}{x_i - y_i}. \tag{3.12}$$

It is interesting to note that although the right hand side of the formula seems to depend on i, the left hand side shows that it does not. Because of (2.6), the Christoffel-Darboux formula is important in the study of Fourier orthogonal series. It also plays an important role in the study of the common zeros of orthogonal polynomials.

4 Block Jacobi Matrices

For $d = 1$, the linear functional in Favard's theorem is known to be given by an integral with respect to $d\phi$, where ϕ is a non-decreasing function with infinite support. There are several ways to establish this result, one of them uses the spectral theorem of self-adjoint operators. With the three-term relation (2.8) of one variable we associate a tridiagonal matrix J, customarily called the *Jacobi matrix*, with a_n on the main diagonal and b_n on the two subdiagonals, that acts as an operator on ℓ^2 (cf. Stone [1932]). This matrix has been studied extensively in operator theory and it plays a very important role in the study of orthogonal polynomials in one variable.

For $d \geq 1$, the coefficients of the three-term relation (3.1) can be used to define a family of block tridiagonal matrices (Xu [1994a]),

$$J_i = \begin{bmatrix} B_{0,i} & A_{0,i} & & \bigcirc \\ A_{0,i}^T & B_{1,i} & A_{1,i} & \\ & A_{1,i}^T & B_{2,i} & \ddots \\ \bigcirc & & \ddots & \ddots \end{bmatrix}, \qquad 1 \leq i \leq d. \tag{4.1}$$

We call them *block Jacobi matrices.* These matrices can be viewed as a family of linear operators which act via matrix multiplication on ℓ^2. The domain of J_i consists of all sequences in ℓ^2 for which matrix multiplication yields sequences in ℓ^2. Under proper conditions, the matrices J_i form a family of commuting self-adjoint operators, which will yield via spectral theorem that $\mathcal{L}$ in Theorem 3.1 has an integral representation. In order to proceed, we need some notion from the spectral theory of self-adjoint operators in a Hilbert space (cf. Riesz and Nagy [1955]).

Let $\mathcal{H}$ be a separable Hilbert space. Each self-adjoint operator T in $\mathcal{H}$ associates with a spectral measure E on $\mathbb{R}$ such that $T = \int x dE(x)$. E is a projection valued measure defined for the Borel sets of $\mathbb{R}$ such that $E(\mathbb{R})$ is the identity operator in $\mathcal{H}$ and $E(B \cap C) = E(B) \cap E(C)$ for Borel sets $B, C \subseteq \mathbb{R}$. For any $f \in \mathcal{H}$ the mapping $B \to \langle E(B)f, f\rangle$ is an ordinary measure defined for the Borel sets $B \subseteq \mathbb{R}$ and denoted $\langle Ef, f\rangle$. A family of operators $\{T_1, \ldots, T_d\}$ in $\mathcal{H}$ *commutes*, by definition, if their spectral measures commute, i.e. $E_i(B)E_j(C) = E_i(C)E_j(B)$ for any $i, j = 1, \ldots, d$ and any two Borel sets $B, C \subseteq \mathbb{R}$. If $T_1, \ldots, T_d$ commute, then $E = E_1 \otimes \cdots \otimes E_d$ is a spectral measure on $\mathbb{R}^d$ with values that are self-adjoint projections in $\mathcal{H}$. In particular, E is the unique measure such that

$$E(B_1 \times \cdots \times B_d) = E_1(B_1) \cdots E_d(B_d)$$

for any Borel sets $B_1, \ldots, B_d \subseteq \mathbb{R}$. The measure E is called the spectral measure of the commuting family $T_1, \ldots, T_d$. A vector $\Phi_0 \in \mathcal{H}$ is a *cyclic vector* in $\mathcal{H}$ with respect to the commuting family of self-adjoint operators $T_1, \ldots, T_d$ in $\mathcal{H}$ if the linear manifold $\{P(T_1, \ldots, T_d)\Phi_0,\ P \in \Pi^d\}$ is dense in $\mathcal{H}$. The spectral theorem for $T_1, \ldots, T_d$ states

If $T_1, \ldots, T_d$ are a commuting family of self-adjoint operators with a cyclic vector Φ_0, then $T_1, \ldots, T_d$ are unitarily equivalent to the multiplication operators $X_1, \ldots, X_d$,

$$(X_i f)(\mathbf{x}) = x_i f(\mathbf{x})\ , \qquad 1 \leq i \leq d\ ,$$

defined on $L^2(\mathbb{R}^d, \mu)$, where the measure μ is defined by $\mu(B) = \langle E(B)\Phi_0, \Phi_0\rangle$ for the Borel set $B \subset \mathbb{R}^d$.

The unitary equivalence means that there exists a unitary mapping $U : \mathcal{H} \to L^2(\mathbb{R}^d, \mu)$ such that $UT_iU^{-1} = X_i$, $1 \leq i \leq d$.

We apply the spectral theorem on the operators $J_1, \ldots, J_d$ on ℓ^2 defined by the block Jacobi matrices. The connection to orthogonal polynomials in several variables is as follows. Let $\{\psi_n\}_{n=0}^\infty$ be the canonical orthonormal basis for ℓ^2. We rewrite this basis as $\{\psi_n\}_{n=0}^\infty = \{\phi_j^k\}_{j=1}^{r_k^d}{}_{k=0}^{\infty}$ according to the lexicographical order, and introduce the formal vector notation $\Phi_k = [\phi_1^k, \ldots, \phi_{r_k^d}^k]^T$, $k \in \mathbb{N}_0$. Then, if the spectral theorem can be applied, the unitary mapping U maps Φ_n to $\mathbb{P}_n$ and the relation

$$\int \mathbb{P}_n(\mathbf{x})\mathbb{P}_m^T(\mathbf{x})\, d\mu(\mathbf{x}) = \langle \Phi_n, \Phi_m^T \rangle , \quad \text{where} \quad \mu(B) = \langle E(B)\Phi_0, \Phi_0 \rangle$$

establishes the integral representation of $\mathcal{L}$. The main task is therefore to show that $J_1, \ldots, J_d$ are commuting and self-adjoint. In Xu [1994a] this is done under the assumption that $J_1, \ldots, J_d$ are bounded operators, which also implies that the support set of the spectral measure is bounded. Moreover, it is shown that $J_1, \ldots, J_d$ are bounded if, and only if,

$$\sup_{k\geq 0} \|A_{k,i}\|_2 < +\infty \quad , \quad \sup_{k\geq 0} \|B_{k,i}\|_2 < +\infty \ , \qquad 1 \leq i \leq d \ , \tag{4.2}$$

where $\|\cdot\|_2$ is the matrix norm induced by the Euclidean norm for vectors. Thus, one can strengthen Favard's theorem as follows.

Theorem 4.1 (Xu [1994a]) *Let $\{\mathbb{P}_n\}_{n=0}^{\infty}$, $\mathbb{P}_0 = 1$, be a sequence in Π^d. Then the following statements are equivalent:*

1. *There exists a determinate measure $\mu \in \mathcal{M}$ with compact support in $\mathbb{R}^d$ such that $\{\mathbb{P}_n\}_{n=0}^{\infty}$ is orthonormal with respect to μ.*
2. *The statement 2) in Theorem* 3.1 *holds together with* (4.2).

For bounded self-adjoint operators $T_1, \ldots, T_d$, the commuting of the spectral measures is equivalent to the formal commuting $T_iT_j = T_jT_i$. For block Jacobi matrices, the formal commuting can be easily verified as equivalent to the conditions in (3.9), which is why we call (3.9) commuting conditions. However, there are examples (cf. Nelson [1959]) of unbounded self-adjoint operators with a common dense domain such that they formally commute but their spectral measures do not commute. Fortunately, a sufficient condition for the commuting of operators in Nelson [1959] can be applied to $J_1, \ldots, J_d$, which leads to

Theorem 4.2 (Xu [1993b]) *Let $\{\mathbb{P}_n\}_{n=0}^{\infty}$, $\mathbb{P}_0 = 1$, be a sequence in Π^d that satisfies the three-term relation* (3.1) *and the rank condition. If*

$$\sum_{k=0}^{\infty} \frac{1}{\|A_{n,i}\|_2} = \infty, \qquad 1 \leq i \leq d, \tag{4.3}$$

then there exists a determinate measure $\mu \in \mathcal{M}$ such that $\{\mathbb{P}_n\}$ is orthonormal with respect to μ.

It is worthwhile to point out that there is a close resemblance between the above set-up and the operator theoretic approach for moment problems, see Berg [1987] and Fuglede [1983]. For $d = 1$, the condition (4.3) is well-known; it implies the classical result of Carleman on the determinate of moment problem Akheizer [1965, p. 86]. For discussion of (4.3) in several variables we refer to Xu [1993b]. For further results concerning the block Jacobi matrices see Xu [1994b]; in particular, Dombrowski's formula for orthogonal polynomials of one variable (cf. Dombrowski [1990]) is extended to several variables in Xu [1994b]. The operator approach to the orthogonal polynomials in several variables also appears in Gekhtman and Kalyuzhny [1994].

For $d = 1$, the perturbation theory of self-adjoint operators can be applied to study the spectrum of Jacobi matrices, from which one can derive important results about the spectral measure. This is made possible by Weyl's theorem, which states that the essential spectrum of a self-adjoint operator does not change when a compact operator is added (cf. Riesz and Nagy [1955, p. 367]), and by the fact that the difference of two Jacobi matrices with respect to a large class of measures is a compact operator. The latter fact follows from the following limit relations on the coefficients of the three-term relation (2.8),

$$\lim_{n\to\infty} a_n = 1/2, \qquad \lim_{n\to\infty} b_n = 0, \tag{4.4}$$

which is satisfied by a wide class of measures, including all $\alpha \in \mathcal{M}$ such that $\alpha' > 0$ *a.e.* and $\operatorname{supp}\alpha = [-1,1]$. Moreover, the perturbation theorem shows that the relation (4.4) is a characteristic property of the class of measures in view of the spectrum. For the importance of the limit relation, we refer to Nevai [1979] and [1986] and the references given there. However, there seems to be no analog that plays the role of (4.4) in several variables. In fact, since the sizes of the coefficient matrices in the three-term relation (3.1) tend to infinity as n goes to infinity, one can only deal with the limit of $\|A_{n,i}\|$ or the limit of some other unitarily invariant characters, such as $\det G_n$ or $\|G_n\|$, which are related to $A_{n,i}$. It's highly unlikely that a limit relation as such will determine the spectrum of a family of block Jacobi matrices, which can be any geometric region in $\mathbb{R}^d$. In this respect, one interesting question is to understand the interrelation among the block Jacobi matrices of the same commuting family, since whether the essential spectrum is, say, a simplex or a ball has to be reflected as some interrelations.

So far, there is little work done in this direction. This is partly due to the problem within the operator theory. In fact, there are several definitions of joint spectrum for a family of commuting operators. Moreover, it is not clear whether there is a proper perturbation theory. From the discussion above, it is no surprise that the difference of two block Jacobi matrices from different measures is not a compact operator in general, which can be easily seen by examining the example of product measures. It is likely that the non-uniqueness of orthogonal polynomials plays a role here. For each given measure, the block Jacobi matrices defined above are unique up to a unitary transformation, which can be ignored when we deal with the spectral measure. However, when we deal with the difference of two families of block Jacobi matrices from different measures, the unitary transformation may play an important role.

Since the block Jacobi matrices can be viewed as a prototype of a family of commuting operators, the study of them may yield new results in operator theory.

5 Common zeros of orthogonal polynomials

Zeros of a polynomial in d-variables are algebraic varieties of dimension $d-1$ or less; they are difficult to deal with for $d \geq 2$. Moreover, in dealing with zeros of orthogonal polynomials, we are mainly interested in the zeros that are real. In one variable, zeros of orthogonal polynomial are all real and distinct. For orthogonal polynomials in several variables, however, the right notion seems to be the *common zeros* of a family of polynomials. The simplest case is to consider the common zeros of $\mathbb{P}_n$, or equivalently, zeros of V_n^d, which we will deal with in this section.

For $d = 1$, it is well-known that the zeros of orthogonal polynomials are the eigenvalues of a truncated Jacobi matrix (cf. Chihara [1978]). It is remarkable that the same is true in several variables. Let J_i be the block Jacobi matrices in the previous section. For $n \in \mathbb{N}_0$, the n-th truncated block Jacobi matrices, denoted by $J_{n,i}$, are given as follows

$$J_{n,i} = \begin{bmatrix} B_{0,i} & A_{0,i} & & & \bigcirc \\ A_{0,i}^T & B_{1,i} & A_{1,i} & & \\ & \ddots & \ddots & \ddots & \\ & & A_{n-3,i}^T & B_{n-2,i} & A_{n-2,i} \\ \bigcirc & & & A_{n-2,i}^T & B_{n-1,i} \end{bmatrix}, \qquad 1 \le i \le d\,.$$

We note that $J_{n,i}$ is a square matrix of order $N = \dim \Pi_{n-1}^d$. We say that $\Lambda = (\lambda_1, \dots, \lambda_d)^T \in \mathbb{R}^d$ is a *joint eigenvalue* of $J_{n,1}, \dots, J_{n,d}$, if there is an $\xi \neq 0$, $\xi \in \mathbb{R}^N$, such that $J_{n,i}\xi = \lambda_i \xi$ for $i = 1, \dots, d$; the vector ξ is called a *joint eigenvector* associated to Λ. By a common zero $\mathbf{x} \in \mathbb{R}^d$ of $\mathbb{P}_n$ we mean a zero for every component of $\mathbb{P}_n$, *i.e.* $P_j^n(\mathbf{x}) = 0$ for all $1 \le j \le r_n^d$. The common zeros of $\mathbb{P}_n$ are characterized in the following theorem, which is proved using the three-term relation (3.1) and the recurrence relation (3.8).

Theorem 5.1 (Xu [1994b]) *A point $\Lambda = (\lambda_1, \dots, \lambda_d)^T \in \mathbb{R}^d$ is a common zero of $\mathbb{P}_n$ if, and only if, it is a joint eigenvalue of $J_{n,1}, \dots, J_{n,d}$; moreover, a joint eigenvector of Λ is $(\mathbb{P}_0^T(\Lambda), \dots, \mathbb{P}_{n-1}^T(\Lambda))^T$.*

Since all $J_{n,i}$ are real symmetric matrices, the theorem immediately implies that common zeros of $\mathbb{P}_n$ are all real. Setting both variables in the Christoffel-Darboux formula as one common zero of $\mathbb{P}_n$, where a standard limiting process is necessary, it follows readily that one of the partial derivatives of $\mathbb{P}_n$ is not identically zero, which implies that all zeros of $\mathbb{P}_n$ are simple. Due to the size of $J_{n,i}$, there can be at most N common zeros of $\mathbb{P}_n$. Moreover, it follows that

Theorem 5.2 (Xu [1994b]) *The orthogonal polynomial $\mathbb{P}_n$ has $N = \dim \Pi_{n-1}^d$ distinct real common zeros if, and only if,*

$$A_{n-1,i}A_{n-1,j}^T = A_{n-1,j}A_{n-1,i}^T, \quad 1 \le i, j \le d. \tag{5.1}$$

For $d = 2$, the condition (5.1) is equivalent to a set of conditions derived much earlier by Mysovskikh [1976] using an entirely different method, where he used monic orthogonal polynomials and did not use the matrix notation. The theorem has important applications in numerical cubature formulae.

Common zeros of orthogonal polynomials in several variables are first studied in connection with cubature formulae, where the word *cubature* stands for the higher dimensional quadrature. For a square positive linear functional $\mathcal{L}$, a cubature formula of degree $2n-1$ is a linear functional

$$\mathcal{I}_n(f) = \sum_{k=1}^{N} \lambda_k f(\mathbf{x}_k)\,, \quad \lambda_k > 0, \quad \mathbf{x}_k \in \mathbb{R}^d, \tag{5.2}$$

defined on Π^d, such that $\mathcal{L}(f) = \mathcal{I}_n(f)$ whenever $f \in \Pi_{2n-1}^d$, and $\mathcal{L}(f^*) \neq \mathcal{I}_n(f^*)$ for at least one $f^* \in \Pi_{2n}$. The points $\mathbf{x}_1, \dots, \mathbf{x}_N$ are called *nodes* and the numbers $\lambda_1, \dots, \lambda_N$ are called *weights*. Such a formula is called minimal, if N, the number of nodes, is minimal among all cubature formulae of degree $2n-1$. It is easily seen (cf. Stroud [1971]) that

$$N \ge \dim \Pi_{n-1}^d \tag{5.3}$$

in general. Indeed, if the number of nodes N were less than $\dim \Pi^d_{n-1}$, then there would exist a polynomial $P \in \Pi^d_{n-1}$ vanishing on all nodes, which would then imply that $\mathcal{L}(P^2) = 0$. We call formulae attaining the lower bound (5.3) *Gaussian*, since for $d = 1$ these are the Gaussian quadrature formulae.

The problem of the existence of Gaussian cubature formulae was raised by Radon [1948]. The important contributions are made by, among others, Mysovskikh and his school (cf. Mysovskikh [1981]) and Möller ([1976] and [1979]). Let $\mathbb{P}_n$ be the vector of orthonormal polynomials with respect to $\mathcal{L}$. The following important theorem is due to Mysovskikh.

Theorem 5.3 (Mysovskikh [1976]) *A Gaussian cubature formula exists if and only if* $\mathbb{P}_n$ *has* $N = \dim \Pi^d_{n-1}$ *common zeros.*

Since matrix multiplication is not commutative, it is evident from Theorem 5.2 that $\mathbb{P}_n$ does not have $N = \dim \Pi^d_{n-1}$ common zeros in general, which means that Gaussian cubature formulae do not exist in general. In his ground-breaking work on minimal cubature formulae, Möller [1976] showed that there is an improved lower bound for the number of nodes. For $d = 2$ this bound states, in the formulation of Xu [1994h], that

$$N \geq \dim \Pi^2_{n-1} + \frac{1}{2} \operatorname{rank}(A_{n-1,1}A^T_{n-1,2} - A_{n-1,2}A^T_{n-1,1}), \tag{5.4}$$

which agrees with the lower bound (5.3) if and only if the Gaussian cubature formula exists. Moreover, Möller proved a surprising result which, for $d = 2$, states that

Theorem 5.4 (Möller [1976]) *For a centrally symmetric linear functional* $\mathcal{L}$ *the lower bound* (5.4) *holds with*

$$\operatorname{rank}(A_{n-1,1}A^T_{n-1,2} - A_{n-1,2}A^T_{n-1,1}) = 2[n/2]. \tag{5.5}$$

A linear functional $\mathcal{L}$ on Π^d is called centrally symmetric if it satisfies

$$\mathcal{L}(\mathbf{x}^\alpha) = 0, \quad \alpha \in \mathbb{N}^d, \quad |\alpha| = \text{odd integer}.$$

In the case that $\mathcal{L}$ is given by (2.1) with a weight function W, the central symmetry of $\mathcal{L}$ means that

$$\mathbf{x} \in \Omega \Rightarrow -\mathbf{x} \in \Omega \quad \text{and} \quad W(\mathbf{x}) = W(-\mathbf{x}),$$

where $\Omega \subset \mathbb{R}^d$ is the support set of W. Examples of centrally symmetric weight functions include the product weight functions $W(x, y) = w(x)w(y)$, where w is a symmetric function on a symmetric interval $[-a, a]$, and the radial weight functions $W(\mathbf{x}) = w(\|\mathbf{x}\|)$ on Euclidean balls centered at origin. In terms of the three-term relation (3.1), it is shown in Xu [1994h] that

a square positive linear functional $\mathcal{L}$ *is centrally symmetric if, and only if,* $B_{n,i} = 0$ *for all* $n \in \mathbb{N}_0$ *and* $1 \leq i \leq d$.

For $d > 2$, analogs of the lower bound (5.4) and the rank condition (5.5) are also given by Möller [1976] and [1979], see also Xu [1994h] where the bound is proved under the condition $B_{n,i}B_{n,j} = B_{n,j}B_{n,i}$. Moreover, the condition (5.5) has also been shown to be true for several families of non-centrally symmetric functions (Berens et al. [1995a] and Xu [1994g]). Whether the bound in (5.4) can be attained will be discussed in the following section. For now, we give a positive example in which the lower bound in (5.3) is attained.

From the result of Mysovskikh and Möller, it becomes clear that Gaussian cubature must be rare and there is no Gaussian cubature for most of the classical weight functions. In Mysovskikh and Chernitsina [1971], a Gaussian cubature formula of degree 5 is constructed numerically, which answers the original question of Radon. One natural question is whether there exists a weight function for which Gaussian cubature formulae exist for all n. The question is answered only recently in Berens et al. [1995b] and Schmid and Xu [1994], where two families of weight functions that admit Gaussian cubature formulae are given for $d = 2$ in Schmid and Xu [1994] and $d \geq 2$ in Berens et al. [1995b]. We describe the results below

We need to recall the definition of symmetric polynomials. A polynomial $P \in \Pi^d$ is called symmetric, if P is invariant under any permutation of its variables. In particular, the degree of P, considered as a polynomial of variable x_i, $1 \leq i \leq d$, remains unchanged, we denote it by $\tau(P)$. The elementary symmetric polynomials in Π^d are defined by

$$u_k := u_k(x_1, \dots x_d) = \sum_{1 \leq i_1 < \dots < i_k \leq d} x_{i_1} \cdots x_{i_k}, \qquad k = 1, 2, \dots, d,$$

and any symmetric polynomial P can be uniquely represented as

$$\sum_{\alpha_1 + \alpha_2 + \cdots + \alpha_d \leq \tau(P)} c_{\alpha_1, \dots, \alpha_n} u_1^{\alpha_1} \cdots u_d^{\alpha_d};$$

where $\tau(P)$ is the degree of P. Let $\mathbf{x} = (x_1, x_2, \dots, x_d)^T$ and $\mathbf{u} = (u_1, u_2, \dots, u_d)^T$, where u_k be the k-th symmetric polynomials in $x_1, \dots, x_d$. We consider the mapping $\mathbf{x} \mapsto \mathbf{u}$. The Jacobian of $\mathbf{u} = \mathbf{u}(\mathbf{x})$ can be expressed as

$$J(\mathbf{x}) := \det \frac{\partial \mathbf{u}}{\partial \mathbf{x}} = \prod_{1 \leq i < j \leq d} (x_i - x_j).$$

Since J^2 is a symmetric polynomial, we shall further use the notation $\Delta(\mathbf{u}) := J^2(\mathbf{x})$. Let $D = \{\mathbf{x} \in \mathbb{R}^d : x_1 < x_2 < \dots < x_d\}$. We define R to be the image of D under the mapping $\mathbf{x} \mapsto \mathbf{u}$; *i.e.*, $R = \mathbf{u}(D)$. Let μ be a nonnegative measure on $\mathbb{R}$ with finite moments and infinite support on $\mathbb{R}$. We define a measure $d\nu(\mathbf{u})$ on $R \subset \mathbb{R}^d$ by $d\nu(\mathbf{u}) = d\mu(\mathbf{x}) := d\mu(x_1) \cdots d\mu(x_d)$; *i.e.* ν is the image of the product measure under the mapping $\mathbf{x} \to \mathbf{u}$.

Theorem 5.5 (Berens et al. [1995b]) *Let ν be defined as above. The orthogonal polynomials $\mathbb{P}_n$ with respect to the measure $[\Delta(\mathbf{u})]^{\frac{1}{2}} d\nu(\mathbf{u})$ or $[\Delta(\mathbf{u})]^{-\frac{1}{2}} d\nu(\mathbf{u})$ have $N = \dim \Pi_{n-1}^d$ many common zeros. Consequently, there exist Gaussian cubature formulae for these measures.*

In fact, the above construction uses the structure of the tensor product of orthogonal polynomials and quadrature formulae. The tensor product is usually not suitable for generating cubature formulae, since the degree of the polynomials tends to be too high; however, the symmetric mapping $\mathbf{x} \mapsto \mathbf{u}$ maps the tensor product into a system of polynomials with proper degree. For example, let $\{p_n\}_{n=0}^{\infty}$ be the orthonormal polynomials with respect to $d\mu$. Then for $\alpha = (\alpha_1, \dots, \alpha_d)$, $0 \leq \alpha_1 \leq \dots \leq \alpha_d = n$, the polynomials P_α^n defined by,

$$P_\alpha^n(\mathbf{u}) = \sum_{\beta \in \{1,2,\dots,d\}} p_{\alpha_1}(x_{\beta_1}) \cdots p_{\alpha_d}(x_{\beta_d}),$$

where Σ is over all permutation β, are orthogonal with respect to $[\Delta(\mathbf{u})]^{-\frac{1}{2}} d\nu(\mathbf{u})$. For $\alpha = (\alpha_1, \ldots, \alpha_d)$, $0 \le \alpha_1 < \ldots < \alpha_d = n + d - 1$ and $n \in \mathbb{N}_0$,

$$P_\alpha^n(\mathbf{u}) = \frac{V_\alpha^n(\mathbf{x})}{J(\mathbf{x})}, \qquad V_\alpha^n(\mathbf{x}) = \det(p_{\alpha_i}(x_j))_{i,j=1}^d$$

are orthogonal with respect to $[\Delta(\mathbf{u})]^{\frac{1}{2}} d\nu(\mathbf{u})$. From this point of view, the symmetric orthogonal polynomials so defined are the most natural extensions of orthogonal polynomials in one variable.

Questions on cubature formulae of degree $2n-1$ can also be asked for cubature formulae of degree $2n-2$. Although there are interesting examples that distinguish the cubature formulae of even degree from those of odd degree, the theorems are of the simiiar nature. We will not go into details, but refer to the results in Schmid [1978], Xu [1994d] and [1994h].

6 Common zeros and cubature formulae

After the discussion in the previous section, it is natural to ask how many common zeros $\mathbb{P}_n$ can have if it does not have $\dim \Pi_{n-1}^d$ many. The answer, however, is a little surprising; the following result is proved in Xu [1994h]:

Let $\mathcal{L}$ be centrally symmetric. If n is an even integer, then $\mathbb{P}_n$ has no common zero; if n is an odd integer, then $\mathbb{P}_n$ has the origin as the only common zero.

This result indicates that it is often too much to ask for common zeros of *all* polynomials of $\mathbb{P}_n$, or *all* polynomials in V_n^d. Instead, one might look for common zeros of a part of V_n^d, such as common zeros of $\mathcal{U}_n^d$ spanned by

$$U^T\mathbb{P}_n, \qquad \text{where} \qquad U : r_n^d \times (r_n^d - \sigma)$$

where the matrix U has full rank $r_n^d - \sigma$. Such a consideration is important in the study of cubature formulae. Möller [1976] characterized the minimal cubature formulae that attain his lower bound (5.4). The nodes are shown to be common zeros of $U^T\mathbb{P}_n$ for some U with $\sigma = [n/2]$. His characterization is derived under the assumption that the span of $\{xU^T\mathbb{P}_n, yU^T\mathbb{P}_n\}$ contains r_{n+1}^2 many linearly independent polynomials of degree $n+1$; in other words, that $U^T\mathbb{P}_n$ generates the polynomial ideal that consists of all polynomials vanishing on the nodes of the cubature formula. Roughly speaking, such an assumption requires that $\mathcal{U}_n$ contains at least half of the polynomials in V_n^d; and to attain the lower bound (5.4), $\mathcal{U}_n$ has to have almost exactly half of the polynomials in V_n^d.

Clearly, such an assumption may not be satisfied in general. Indeed, the lower bound (5.4) is known to be attained only in a few cases (cf. Möller [1976], Morrow and Patterson [1978] and Cools and Schmid [1989]), and it has been shown to be not attainable for various classical weight functions (cf. Morrow and Patterson [1978], Verlinden and Cools [1992] and Cools and Schmid [1993]). When a cubature formula has more nodes than the lower bound (5.4), $\mathcal{U}_n$ will have fewer polynomials in V_n^d and the nodes of the cubature formula will be determined as common zeros of $U^T\mathbb{P}_n$ and some polynomials of degree greater than n; together they span the ideal determined by the nodes. Among these polynomials, those of degree $n+1$ are necessarily orthogonal to polynomials of degree $n-2$, since the cubature formula is of order $2n-1$. They are what we would call quasi-orthogonal polynomials of order 2. In general, polynomials that determine the nodes of a cubature formula may involve quasi-orthogonal polynomials of higher order and several consecutive degrees higher than n. The above consideration motivates the study of common zeros of a set of quasi-orthogonal polynomials in Xu [1994h].

For fixed n, we denote the vector of quasi-orthogonal polynomials of degree $n+s$, which are orthogonal to polynomials in Π^d_{n-1-s}, as

$$\mathbb{Q}_{n+s} = \mathbb{P}_{n+s} + \Gamma_{1,s}\mathbb{P}_{n+s-1} + \ldots + \Gamma_{2s,s}\mathbb{P}_{n-s}.$$

In Xu [1994h], the common zeros of the polynomials in the set

$$\mathcal{Q}_r = \{U_0^T\mathbb{P}_n, U_1^T\mathbb{Q}_{n+1}, \ldots, U_r^T\mathbb{Q}_{n+r}, \mathbb{Q}_{n+r+1}\}, \tag{6.1}$$

are studied, where U_k, $0 \le k \le r$, are matrices such that

$$U_k : r^d_{n+k} \times r^d_{n+k} - \sigma_k, \qquad \operatorname{rank} U_k = r^d_{n+k} - \sigma_k, \quad \sigma_k \ge 0.$$

We note that if U_k is the identity matrix, then the set of common zeros of $\mathcal{Q}_r$ will be the same as that of $\mathcal{Q}_{r-1}$. If a cubature formula is based on the zeros of $\mathcal{Q}_r$, then we say that the cubature formula is generated by the zeros of $\mathcal{Q}_r$. Moreover, we call $\mathcal{Q}_r$ *maximal*, if for $1 \le k \le r$ every polynomial in Π^d_{n+k} vanishing on all the common zeros of $\mathcal{Q}_r$ belongs to the linear space spanned by $U_0^T\mathbb{P}_n, U_1^T\mathbb{Q}_1, \ldots, U_{n+k}^T\mathbb{Q}_{n+k}$. The main result in Xu [1994h] characterizes the common zeros of $\mathcal{Q}_r$ that generates cubature formulas in terms of the nonlinear matrix equations satisfied by the matrices Γ_i and V_k, where V_k are orthogonal compliments of U_k, defined by

$$U_k^T V_k = 0, \qquad V_k : r^d_n \times \sigma_k.$$

The statement in the general case is rather complicated, we shall restrict ourselves to the case $r = 0$, where we deal with

$$\mathcal{Q}_0 = \{U^T\mathbb{P}_n, \mathbb{Q}_{n+1}\}, \qquad \mathbb{Q}_{n+1} = \mathbb{P}_{n+1} + \Gamma_1\mathbb{P}_n + \Gamma_2\mathbb{P}_{n-1}, \tag{6.2}$$

and we write U and V instead of U_0 and V_0 in this case. we have

Theorem 6.1 (Xu [1994h]) *The set $\mathcal{Q}_0 = \{U^T\mathbb{P}_n, \mathbb{Q}_{n+1}\}$ is maximal and it has $\dim \Pi^d_{n-1} + \sigma$ many pairwise distinct real zeros that generate a cubature formula of degree $2n-1$ if, and only if, there exists V such that $U^TV = 0$,*

$$\Gamma_2 = \sum_{i=1}^{d} D^T_{n,i}(I - VV^T)A^T_{n-1,i}, \tag{6.3}$$

and Γ_1 and V satisfy the following conditions:

$$A_{n-1,i}(VV^T - I)A^T_{n-1,j} = A_{n-1,j}(VV^T - I)A^T_{n-1,i}, \quad 1 \le i,j \le d, \tag{6.4}$$

$$(B_{n,i} - A_{n,i}\Gamma_1)VV^T = VV^T(B_{n,i} - \Gamma_1^T A^T_{n,i}), \qquad 1 \le i \le d, \tag{6.5}$$

and

$$\begin{aligned} &VV^TA^T_{n-1,i}A_{n-1,j}VV^T + (B_{n,i} - A_{n,i}\Gamma_1)VV^T(B_{n,j} - \Gamma_1^TA^T_{n,j}) \\ &= VV^TA^T_{n-1,j}A_{n-1,i}VV^T + (B_{n,j} - A_{n,j}\Gamma_1)VV^T(B_{n,i} - \Gamma_1^TA^T_{n,i}) \end{aligned} \tag{6.6}$$

for $1 \le i,j \le d$.

There is an analogous theorem for $\mathcal{Q}_r$ with many more matrix equations that involve $\Gamma_{i,k}$ and V_k. The lower bound (5.4) is attained if and only if $\Gamma_1 = 0$ and $\sigma = [n/2]$ in the above theorem; this special case has been treated by Möller. However, in the present general form, the theorem does not deal with just minimal cubature formulae; in fact, it characterizes *every* (positive) cubature formula generated by $\mathcal{Q}_0$. With the additional Γ_1 and the flexibility on the value of σ, it is likely that the equations (6.4)–(6.6) are solvable even if the special case that (5.4) is attained is not solvable. Still, there is no warrant that the system of equations is always solvable; the question depends on the coefficient matrices of the three-term relation. When the system is not solvable, one needs to consider $\mathcal{Q}_r$ for $r \geq 1$.

In Xu [1994h], the sufficiency part of (6.3)–(6.6) in the theorem is contained in two theorems, Theorem 4.1.4, which characterizes the common zeros of $\mathcal{Q}_0$, and Theorem 4.4.3, which established the existence of the cubature formula, necessarily positive, generated by the common zeros. The first theorem also contains a statement of the necessity part which, however, has an error; namely, the clause *that generates a cubature formula of degree* $2n-1$ is missing. The same clause should also be added to the statement of Theorem 7.1.4 in Xu [1994h] which deals with $\mathcal{Q}_r$ in general. The necessity part is the easier half; we give a complete proof below.

Proof of the necessity part of Theorem 6.1. Since $\mathbf{x}_k$ generates a cubature formula, the linear functional $\mathcal{I}_n(f)$ in (5.2) is square positive. Moreover, that $\mathcal{Q}_0$ is maximal implies that the matrix $\mathcal{I}_n(\mathbb{P}_n\mathbb{P}_n^T)$ is of rank σ. Therefore, there exists a matrix $V : r_n^d \times \sigma$ of rank σ such that $\mathcal{I}_n(\mathbb{P}_n\mathbb{P}_n^T) = VV^T$. Since $U^T\mathbb{P}_n(\mathbf{x}_k) = 0$ and V is of full rank, it follows easily that $U^TV = 0$. For this V we verify the validity of (6.3)–(6.6) by computing the matrices $\mathcal{I}_n(x_i\mathbb{P}_{n-1}\mathbb{P}_n^T)$ and $\mathcal{I}_n(x_ix_j\mathbb{P}_{n-1}\mathbb{P}_{n-1}^T)$, $\mathcal{I}_n(x_i\mathbb{P}_n\mathbb{P}_n^T)$ and $\mathcal{I}_n(x_ix_j\mathbb{P}_n\mathbb{P}_n^T)$, respectively. Since the cubature formula is of degree $2n-1$, it follows readily that $\mathcal{I}_n(\mathbb{P}_{n-1}\mathbb{P}_n^T) = \mathcal{L}(\mathbb{P}_{n-1}\mathbb{P}_n^T) = 0$. Moreover, since $\mathbb{Q}_{n+1}$ vanishes on $\mathbf{x}_k$, it follows that $\mathcal{I}_n(\mathbb{Q}_{n+1}\mathbb{P}_n^T) = 0$. Therefore, using the three-term relation and the definition of $\mathbb{Q}_{n+1}$, we have, for example,

$$\begin{aligned}\mathcal{I}_n(x_i\mathbb{P}_n\mathbb{P}_n^T) =& A_{n,i}\mathcal{I}_n(\mathbb{P}_{n+1}\mathbb{P}_n^T) + B_{n,i}\mathcal{I}_n(\mathbb{P}_n\mathbb{P}_n^T) \\ =& A_{n,i}\mathcal{I}_n[(\mathbb{Q}_{n+1} - \Gamma_1\mathbb{P}_n - \Gamma_2\mathbb{P}_{n-1})\mathbb{P}_n^T] + B_{n,i}\mathcal{I}_n(\mathbb{P}_n\mathbb{P}_n^T) \\ =& (B_{n,i} - A_{n,i}\Gamma_1)VV^T\end{aligned}$$

from which (6.5) follows since the matrix on the left hand side is clearly symmetric. Other matrices are computed similarly, but the computation involves two ways of using the three-term relation; the desired equations are simply the consequence that the two ways should yield the same result. □

In the proof of this direction in Xu [1994h], we assumed that the matrices $S_{n,i}$ defined below are symmetric, which can be justified, when the additional clause is assumed, by computing $\mathcal{I}_n(x_i\mathbb{P}_{n-1}\mathbb{P}_n^T)$ and $\mathcal{I}_n(x_i\mathbb{P}_n\mathbb{P}_n^T)$ as in the above proof.

The proof of the other direction of the theorem is long. In Xu [1994h] it is proved following a approach which is very different from the method introduced for the special case in Möller [1976] and followed by others (cf. Morrow and Patterson [1978] and Schmid [1978]). It is interesting to mention that the approach in Xu [1994h] resembles the approach in one variable very closely (cf. Davis and Rabinowitz [1975] and Xu [1994f]). Assume that (6.4)–(6.6) are satisfied by V_0 and Γ_1, then the approach consists of three steps. As the first step, the common zeros are shown to be exist as the joint eigenvalues of the matrices $S_{n,i}$, defined by

$$S_{n,i} = \begin{bmatrix} B_{0,i} & A_{0,i} & & & \bigcirc \\ A_{0,i}^T & B_{1,i} & A_{1,i} & & \\ & \ddots & \ddots & \ddots & \\ & & & B_{n-1,i} & A_{n-1,i}V \\ \bigcirc & & & V^+A_{n-1,i}^{*T} & V^+B_{n,i}^*V \end{bmatrix}, \qquad 1 \le i \le d,$$

where $B_{n,i}^* = B_{n,i} - A_{n,i}\Gamma_1$ and $A_{n-1,i}^* = A_{n-1,i}^T - A_{n,i}\Gamma_1$. All common zeros are real, since $S_{n,i}$ are symmetric according to (6.4) and (6.5), and they are pairwise distinct, by a modified Christoffel-Darboux formula for

$$\mathbf{K}_n^{(0)}(\mathbf{x}, \mathbf{y}) = \mathbf{K}_n(\mathbf{x}, \mathbf{y}) + [V^+\mathbb{P}_n(\mathbf{x})]^T V^+\mathbb{P}_n(\mathbf{y}),$$

where $V^+ = (V^TV)^{-1}V^T$ is the generalized inverse of V. It is worthwhile to point out that the theorem for the special case in Möller [1976] is stated under the restriction that the zeros are all real, which is removed later in Schmid [1978] as a consequence of a theorem from real algebraic ideal theory; it is not clear whether the theorem from the ideal theory can possibly be applied to the present general setting. Back to the proof of the theorem, the second step shows that the Lagrange interpolation based on the common zeros of $\mathcal{Q}_0$ exists and is unique in the polynomial subspace $\Pi_{n-1}^d \cup \operatorname{span}\{V^+\mathbb{P}_n\}$; moreover, if the common zeros are denoted by $\mathbf{x}_1, \ldots, \mathbf{x}_N$, then the interpolation polynomial, denoted by L_nf, can be written down explicitly as

$$L_n(f, \mathbf{x}) = \sum_{k=1}^{N} f(\mathbf{x}_k) \frac{\mathbf{K}_n^{(0)}(\mathbf{x}, \mathbf{x}_k)}{\mathbf{K}_n^{(0)}(\mathbf{x}_k, \mathbf{x}_k)}, \tag{6.7}$$

where the interpolation $L_n(f, \mathbf{x}_k) = f(\mathbf{x}_k)$, $1 \le k \le N$, is verified via the modified Christoffel-Darboux formula. In the third step, the cubature formula generated by the common zeros is shown to exist by applying the linear functional $\mathcal{L}$ on L_nf. As a by product, it follows readily that the cubature weights are given by the formula

$$\lambda_k = [\mathbf{K}_n^{(0)}(\mathbf{x}_k, \mathbf{x}_k)]^{-1},$$

which are clearly positive and the formula is shown for the first time in Xu [1994h]. It is worthwhile to mention that the Lagrange interpolation polynomials defined in (6.7) are of interests, since interpolation in several variables is very difficult in general and compact formulae are rare. For the product Chebyshev polynomials of the first kind, there exists a well-defined sequence of polynomials L_nf which is proved in Xu [199xa] to converge to f in weighted L^p norm, $0 < p < \infty$, for every continuous function f.

The theorem gives a complete characterization for the common zeros of $\mathcal{Q}_0$ that generates a cubature formula. A central question is whether $\mathcal{Q}_0$ always has enough common zeros for some U and Γ_1, which is equivalent to ask whether (6.4)–(6.6) is always solvable. So far, they are solved only for the product Chebyshev weight functions, $(1-x^2)^{1/2}(1-y^2)^{1/2}$ and $(1-x^2)^{-1/2}(1-y^2)^{-1/2}$, and some simple weight functions for small n; moreover, the previous efforts are mostly on the special case that (5.4) is attained, which implies that $\Gamma_1 = 0$, and there are a number of negative results for the case. There is reason to be optimistic about solving (6.4)–(6.6) with some nonzero Γ_1, although it is likely that the resulting cubature formulae may not be the minimal ones. It should be mentioned that the system (6.4)–(6.6) in its full generality has not yet been explored.

7 Fourier orthogonal series and asymptotics

As we mentioned in Section 5, there is no analog of the limit relations (4.4) in several variable. These relations play an important role in proving the asymptotic relations related to orthogonal polynomials of one variable. Let $\mathbf{K}_n(\cdot,\cdot)$ be the reproducing kernel function. The function

$$\Lambda_n(\mathbf{x}) = [\mathbf{K}_n(\mathbf{x},\mathbf{x})]^{-1},$$

is called the *Christoffel function*; it has the following fundamental property.

If $\mathcal{L}$ is square positive, then

$$[\mathbf{K}_n(\mathbf{x},\mathbf{x})]^{-1} = \min\{\mathcal{L}(P^2) : P \in \Pi_{n-1}^d, \quad P(\mathbf{x}) = 1\}.$$

For $d = 1$, this function plays a significant role in the study of orthogonal polynomials of one variable, which we refer to the extensive survey Nevai [1986]. The limit relation (4.4) is used in proving the following important property of λ_n, the usual notation for Λ_n in $d = 1$,

$$\lim_{n\to\infty} n\lambda_n(x) = \pi\alpha'(x)(1-x^2)^{1/2}, \qquad a.e.\ x \in [-1,1]$$

for measures α belonging to Szegő's class, i.e., $\log\alpha'(\cos t) \in L^1[0,2\pi]$ (Máté et al. [1991]). Such a relation, like many other limit relations involving orthogonal polynomials on the real line, is proved by first establishing an analogous result for orthogonal polynomials on the unit circle, then coming back to the real line by using the so-called $*$-transform in Szegő's theory. Although part of Szegő's theory undoubtly can be extended to the setting in several complex variables, a moment reflection shows that $*$-transform as a bridge between the real and the complex does not extend to several variables. So far, only very limited effort has been put into proving the limit relation in several variables; namely, certain product weight functions in Xu [1995] and the radial weight function $(1-|\mathbf{x}|^2)^\alpha$ on the unit ball in Bos [1994]. In general, for $\Lambda_n(\cdot)$ associated with orthogonal polynomials with respect to W, the limit relation should take the form

$$\lim_{n\to\infty} \binom{n+d}{d} \Lambda_n(\mathbf{x}) = C_d(\mathbf{x})W(\mathbf{x})$$

where C_d is independent of W. For the product weight functions in Xu [1995], we have $C_d(\mathbf{x}) = 1/W_0(\mathbf{x})$ where W_0 is the product Chebyshev weight function of the first kind,

$$W_0(\mathbf{x}) = \frac{1}{\pi^d}\frac{1}{\sqrt{1-x_1^2}}\cdots\frac{1}{\sqrt{1-x_d^2}}, \qquad \mathbf{x} = (x_1,\ldots,x_d) \in [-1,1]^d;$$

for the radial weight function $(1-|\mathbf{x}|^2)^\alpha$ on the unit ball in $\mathbb{R}^d$, we have that $C_d(\mathbf{x}) = c_d/(1-|\mathbf{x}|^2)^{-1/2}$ where c_d is a constant chosen so that the integral of C_d is 1. One interesting remark is that for the weight function $(1-|\mathbf{x}|^2)^\alpha$ the limit

$$\lim_{n\to\infty} \frac{1}{\binom{n+d-1}{d-1}} \mathbb{P}_n^T(\mathbf{x})\mathbb{P}_n(\mathbf{x}) = \frac{1}{C_d(\mathbf{x})W(\mathbf{x})}$$

exists, which is not true for $d = 1$, where the asymptotic of p_n is much more complicated. Of course, with $\mathbb{P}_n^T\mathbb{P}_n$ we are dealing with a sum instead of an individual term.

On the other hand, the order of Λ_n as n goes to infinity is enough to give a sufficient condition for the convergence of orthogonal series. For many weight functions with compact support, it is shown in Xu [1995] that $\Lambda_n(\mathbf{x}) = \mathcal{O}(n^{-d})$. The sufficient condition for the convergence of $S_n(f)$ is derived using the L^2 theory as in the case of $d = 1$ (cf. Nevai [1986]). As an example, we give one of the consequences as follows.

Theorem 7.1 (Xu [1995]) *Let $\mu \in \mathcal{M}(\mathbb{R}^d)$,* $\operatorname{supp}\mu = [-1,1]^d$, *and suppose that $[n^d\Lambda_n(\mathbf{x})]^{-1}$ is bounded uniformly on a subset Δ of $[-1,1]^d$. Suppose $f \in C^{[d/2]}([-1,1]^d)$ and each of its $[d/2]$-th derivatives satisfies*

$$|D^{[d/2]}f(\mathbf{x}) - D^{[d/2]}f(\mathbf{y})| \le ch^\beta, \quad |\mathbf{x}-\mathbf{y}| \le h,$$

where for odd d, $\beta > 1/2$, and for even d, $\beta > 0$. Then $S_n(f)$ converges uniformly and absolutely to f on Δ.

Because of the dependence of the asymptotic order of Λ_n on the dimension, the requirement on the class of functions in the theorem becomes stronger for higher dimension. The use of L^2 theory also implies similar results for the first Cesáro means of the Fourier orthogonal series. These results provide sufficient conditions for the convergence with little restriction on the class of measures; being so, they are usually not sharp. To fully understand the expansion of Fourier orthogonal series, one needs to go back to the reproducing kernel function $\mathbf{K}_n(\cdot,\cdot)$.

For $d = 1$, the Christoffel-Darboux formula (3.12) provides a compact formula of $\mathbf{K}_n(\cdot,\cdot)$ that contains only one term. For $d \ge 2$, however, the nominator in the formula (3.12) is a sum of $r_{n-1}^d = \mathcal{O}(n^{d-1})$ many terms, while the number of terms in $\mathbf{K}_n(\cdot,\cdot)$ is $\dim \Pi_{n-1}^d = \mathcal{O}(n^d)$. In order to understand the behavior of $S_n(f)$, it is necessary to have a compact formula for $\mathbf{K}_n(\cdot,\cdot)$. The first such compact formula is given in Xu [1995] for the product Chebyshev weight function of the first kind. Let $\mathbf{x} = (\cos\theta_1, \ldots, \cos\theta_d)$ and $\mathbf{y} = (\cos\phi_1, \ldots, \cos\phi_d)$. Then we have

for the product Chebyshev weight function W_0,

$$\mathbf{K}_n(\mathbf{x},\mathbf{y}) = \sum_{(\varepsilon_1,\ldots,\varepsilon_d)\in\{-1,1\}^d} D_{n,d}(\theta_1+\varepsilon_1\phi_1, \ldots, \theta_d+\varepsilon_d\phi_d), \tag{7.1}$$

where the function $D_{n,d}$ is a divided difference

$$D_{n,d}(\theta_1,\ldots,\theta_d) = [\cos\theta_1,\ldots,\cos\theta_d]\, G_{n,d},$$

and

$$G_{n,d}(\cos t) = (-1)^{[\frac{d-1}{2}]} 2\cos\frac{t}{2}(\sin t)^{d-2} \begin{cases} \cos(n-\frac{1}{2})t, & \text{for } d \text{ even}, \\ \sin(n-\frac{1}{2})t, & \text{for } d \text{ odd}. \end{cases}$$

In fact, the function $D_{n,d}$ is the Dirichlet kernel of the ℓ-1 summability of the multiple Fourier series on $\mathbb{T}^d$;

$$D_{n,d}(\Theta) = \sum_{|\alpha|_1 \le n-1} e^{i\alpha\cdot\Theta}, \qquad \Theta \in \mathbb{T}^d,$$

where $|\alpha|_1 = |\alpha_1| + \ldots + |\alpha_d|$ for $\alpha \in \mathbb{Z}^d$ and we use $|\cdot|$ to denote Euclidean norm in this section. It turns out that the ℓ-1 summability is very different from the usual summability of the spherical means of multiple Fourier series. For example, there is no critical index for the Riesz means in the ℓ-1 summability. We refer to Berens and Xu [1995], [1996] and the references there for the results. Here we just point out that overall the ℓ-1 summability is very much similar to the summability of Fourier series in one variable, which seems to indicate that we can expect a summability theory for orthogonal series in several variable similar to that of one variable. We quote one result here.

Theorem 7.2 (Berens and Xu [1996]) *For W_0, the Cesàro $(C, 2d-1)$ means of $S_n(f)$ define a positive operator; the order of summability is best possible in the sense that the (C, δ) means are not positive for $0 < \delta < 2d-1$.*

For $d = 1$ this is Fejér's theorem which states that the arithmetic means of the Fourier partial sum is positive. The proof of the theorem uses an inequality of Askey-Gasper on Jacobi polynomials (cf. Askey [1975]).

For the product Gegenbauer weight functions $(1-x_1^2)^\alpha \ldots (1-x_d^2)^\alpha$ on $[-1,1]^d$, the compact formula similar to that of (7.1) holds for $2\alpha \in \mathbb{N}_0$; the formula for other values of α has yet to be found. In fact, there is little results known for the summability of Fourier orthogonal expansion with respect to these weight functions. For the weight function

$$W_\mu(\mathbf{x}) = w_\mu(1-|\mathbf{x}|^2)^{\mu-1/2}, \qquad \mathbf{x} \in B^d = \{\mathbf{x} \in \mathbb{R}^d : |\mathbf{x}|^2 = x_1^2 + \ldots + x_d^2 \le 1\},$$

where w_μ is the normalization constant so that the integral of W_μ over B^d is 1, there is an interesting compact formula of $\mathbf{K}_n(\cdot,\cdot)$ which has been discovered by the author recently in [199xb]. In fact, a compact formula holds already for $\mathbb{P}_n^T(\mathbf{x})\mathbb{P}_n(\mathbf{y})$.

Let C_n^λ denotes the usual Gegenbauer polynomials. Then the compact formula for $\mathbb{P}_n^T(\mathbf{x})\mathbb{P}_n(\mathbf{y})$ is as follows.

$$\begin{aligned}\mathbb{P}_n^T(\mathbf{x})\mathbb{P}_n(\mathbf{y}) = & \frac{n+\mu+\frac{d-1}{2}}{\mu+\frac{d-1}{2}} \int_{-1}^{1} C_n^{\mu+\frac{d+1}{2}}(\mathbf{x}\cdot\mathbf{y} + \sqrt{1-|\mathbf{x}|^2}\sqrt{1-|\mathbf{y}|^2}\,t) \\ & \times (1-t^2)^{\mu-1}dt \Big/ \int_{-1}^{1} (1-t^2)^{\mu-1}dt, \qquad \mu > 0 \qquad \mathbf{x},\mathbf{y} \in B^d.\end{aligned}$$

A simpler formula holds for the limiting case $\mu = 0$. It's worthwhile to mention that for $d = 1$ this formula coincides with the product formula of the ultraspherical polynomials. Moreover, by setting $|\mathbf{y}| = 1$, the integral in the formula can be removed; we see the formula resembles the addition formula for the spherical harmonics (cf. Erdélyi et al. [1953, Vol. II, p. 244, (2)]). From the formula, a theorem similar to that of Theorem 7.2 follows. More importantly, it allows us to give a complete answer for the Cesáro summability of the Fourier orthogonal series.

Theorem 7.3 (Xu [199xb]) *Let f be continuous on the closed ball B^d. The expansion of f in the Fourier orthogonal series with respect to W_μ is uniformly (C, δ) summable on B^d if, and only if, $\delta > \mu + \frac{d-1}{2}$.*

Other properties and results have been derived from the compact formula in an ongoing work of the author.

Comparing to our understanding of the classical orthogonal polynomials in one variable, we know little about their extensions in several variables. The orthogonal polynomials with respect to the classical weight functions, starting from those in Koornwinder [1974], should be studied in detail. The knowledge we obtain from studying them not only will enhance our understanding of the general theory, it may also lead to new directions in developing a general theory of orthogonal polynomials in several variables. In this respect, we should at least mention the theory of orthogonal polynomials on spheres with respect to measures invariant under a reflection group. This important theory has been developed by Dunkl recently (cf. Dunkl [1989] and [1991]). Among other things, it opens a way to study orthogonal polynomials on spheres and balls with respect to a large class of weight functions.

References

Akheizer, N. I. [1965] *The classical moment problem and some related questions in Analysis*, Hafner, New York.

Askey, R. [1975] *Orthogonal polynomials and special functions*, SIAM, Philadelphia.

Berg, C. [1987] *The multidimensional moment problem and semigroups*, Moments in Mathematics, Proc. Sympos. Appl. Math., vol. 37, Amer. Math. Soc., Providence, RI, pp. 110–124.

Berg, C. and Thill, M. [1991] *Rotation invariant moment problems,* Acta Math **167**, 207–227.

Berens, H. and Xu, Y. [1995] *ℓ-1 summability for multivariate Fourier integral*, Approximation Theory VIII (C. K. Chui and L. Schumaker, eds.), World Sci. Publ. Co., pp. 55–62.

Berens, H. and Xu, Y. [1996] *Fejér means for multivariate Fourier series,* Math. Z. **221**, 449-465.

Berens, H., Schmid, H. and Xu, Y. [1995a] *On twodimensional definite orthogonal systems and on a lower bound for the number of nodes of associated cubature formulae,* SIAM J. Math. Anal. **26**, 468–487.

Berens, H., Schmid, H. and Xu, Y. [1995b] *Multivariate Gaussian cubature formulae,* Arch. Math. **64**, 26–32.

Bos, L. [1994] *Asymptotics for the Christoffel function for Jacobi like weights on a ball in $\mathbb{R}^m$,* New Zealand J. Math. **23**, 99–109.

Chihara, T. S. [1978] *An introduction to orthogonal polynomials*, Gordon and Breach, New York.

Cools, R. and Schmid, H. J. [1989] *Minimal cubature formulae of degree $2k-1$ for two classical functions,* Computing **43**, 141–157.

Cools, R. and Schmid, H. J. [1993] *A new bound for the number of nodes in cubature formulae of degree $4n+1$ for some circularly symmetric integrals*, Numerical Integration IV (H. Braßand G. Hämmerlin, eds.), Birkhüser, Basel, pp. 57–66.

Davis, P. J. and Rabinowitz, P. [1975] *Methods of numerical integration*, Academic Press, New York.

Dombrowski, J. [1990] *Orthogonal polynomials and functional analysis*, Orthogonal Polynomials (P. Nevai, ed.), Kluwer Acad. Pub., Netherlands, pp. 147–161.

Dunkl, C. [1989] *Differential-difference operators associated to reflection groups,* Trans. Amer. Math. Soc. **311**, 167-183.

Dunkl, C. [1991] *Integral kernels with reflection group invariance,* Can. J. Math. **43**, 1213-1227.

Erdélyi, A., Magnus, W., Oberhettinger, F. and Tricomi, F. G. [1953] *Higher transcendental functions*, vol. 2, McGraw-Hill, New York.

Engles, H. [1980] *Numerical quadrature and cubature*, Academic Press, New York.

Fuglede, B. [1983] *The multidimensional moment problem,* Expo. Math. **1**, 47–65.

Gekhtman, M. I. and Kalyuzhny, A. A. [1994] *On the orthogonal polynomials in several variables,* Integr. Equat. Oper. Theory **19**, 404–418.

Koornwinder, T. [1975] *Two-variable analogues of the classical orthogonal polynomials*, Theory and Application of Special Functions (R. A. Askey, ed.), Academic Press, New York, pp. 435–495.

Kowalski, M. A. [1982a] *The recursion formulas for orthogonal polynomials in n variables,* SIAM J. Math. Anal. **13**, 309–315.

Kowalski, M. A. [1982b] *Orthogonality and recursion formulas for polynomials in n variables,* SIAM J. Math. Anal. **13**, 316–323.

Krall, H. L. and Sheffer, I. M. [1967] *Orthogonal polynomials in two variables,* Ann. Mat. Pura. Appl. **76**, no. 4, 325–376.

Máté, A, Nevai, P. and Totik, V. [1991] *Szegő's extremum problem on the unit circle,* Ann. of Math. **134**, 433–453.

Möller, H. M. [1976] *Kubaturformeln mit minimaler Knotenzahl,* Numer. Math. **35**, 185–200.

Möller, H. M. [1979] *Lower bounds for the number of nodes in cubature formulae,* Numerical Integration, ISNM vol. 45 (G. Hämmerlin, ed.), Birkhäuser, Basel.

Morrow, C. R. and Patterson, T. N. L. [1978] *Construction of algebraic cubature rules using polynomial ideal theory,* SIAM J. Numer. Anal. **15**, 953–976.

Mysovskikh, I. P. [1976] *Numerical characteristics of orthogonal polynomials in two variables,* Vestnik Leningrad Univ. Math. **3**, 323–332.

Mysovskikh, I. P. [1981] *Interpolatory cubature formulas*, "Nauka", Moscow. (Russian)

Mysovskikh, I. P. and Chernitsina, V. Ya. [1971] *Answer to a question of Radon,* Dokl. Akad. Nauk SSSR **198**, 537–539.

Nevai, P. [1979] *Orthogonal Polynomials,* Memoirs of Amer. Math. Soc. **213**, Amer. Math. Soc., Providence, RI.

Nevai, P. [1986] *Géza Freud, orthogonal polynomials and Christoffel functions. A case study,* J. Approx. Theory **48**, 3–167.

Nelson, E. [1959] *Analytic vectors,* Ann. Math. **70**, 572–615.

Riesz, F. and Sz.-Nagy, B. [1955] *Functional analysis*, Ungar, New York.

Radon, J. [1948] *Zur mechanischen Kubatur,* Monatsh. Math. **52**, 286–300.

Schmid, H. J. [1978] *On cubature formula with a minimal number of knots,* Numer. Math. **31**, 282–297.

Schmid, H. J. [1983] *Interpolatorische Kubaturformeln,* Diss. Math. **CCXX**, 1-122.

Schmid, H. J. and Xu, Y. [1994] *On Bivariate Gaussian cubature formula,* Proc. Amer. Math. Soc. **122**, 833–842.

Stone, M. [1932] *Linear transformations in Hilbert space and their applications to analysis*, reprinted 1983, Amer. Math. Soc., Colloq. Publ. vol.15, Providence, RI.

Stroud, A. [1971] *Approximate calculation of multiple integrals*, Prentice Hall, Englewood Cliffs, NJ.

Suetin, P. K. [1988] *Orthogonal polynomials in two variables*, "Nauka", Moscow. (Russian)

Szegő, H. [1975] *Orthogonal polynomials*, 4th ed., Amer. Math. Soc. Colloq. Publ. vol.23, Providence, RI.

Verlinden, P. and Cools, R. [1992] *On cubature formulae of degree $4k+1$ attaining Möller's lower bound for integrals with circular symmetry,* Numer. Math. **61**, 395–407.

Xu, Y. [1992] *Gaussian cubature and bivariate polynomial interpolation,* Math. Comp. **59**, 547–555.

Xu, Y. [1993a] *On multivariate orthogonal polynomials,* SIAM J. Math. Anal. **24**, 783–794.

Xu, Y. [1993b] *Unbounded commuting operators and multivariate orthogonal polynomials,* Proc. Amer. Math. Soc. **119**, 1223-1231.

Xu, Y. [1994a] *Multivariate orthogonal polynomials and operator theory,* Trans. Amer. Math. Soc. **343**, 193-202.

Xu, Y. [1994b] *Block Jacobi matrices and zeros of multivariate orthogonal polynomials,* Trans. Amer. Math. Soc. **342**, 855–866.

Xu, Y. [1994c] *Recurrence formulas for multivariate orthogonal polynomials,* Math. Comp. **62**, 687–702.

Xu, Y. [1994d] *On zeros of multivariate quasi-orthogonal polynomials and Gaussian cubature formulae,* SIAM J. Math. Anal. **25**, 991–1001.

Xu, Y. [1994e] *Solutions of three-term relations in several variables,* Proc. Amer. Math. Soc. **122**, 151–155.

Xu, Y. [1994f] *A characterization of positive quadrature formulae,* Math. Comp. **62**, 703–718.

Xu, Y. [1994g] *On a class of bivariate orthogonal polynomials and cubature formula,* Numer. Math. **69**, 233–241.

Xu, Y. [1994h] *Common zeros of polynomials in several variables and higher dimensional quadrature*, Pitman Research Notes in Mathematics Series, Longman, Essex.

Xu, Y. [1995] *Christoffel functions and Fourier series for multivariate orthogonal polynomials,* J. Approx. Theory **82**, 205-239.

Xu, Y. [199xa] *Lagrange interpolation on Chebyshev points of two variables,* J. Approx. Theory (to appear).

Xu, Y. [199xb] *Summability of Fourier orthogonal series for Jacobi weight on a ball in* $\mathbb{R}^d$, (to appear).

Appendix I

Program List of Speakers and Topics

Arnold Adelberg	p-adic integer order Bernoulli polynomials
Krishnaswami Alladi	Refinements of Rogers-Ramanujan type identitities
Natig Atakishiyev	Ramanujan-type continuous measures for classical q-polynomials
Herman Bavinck	Difference and differential equations for polynomials orthogonal with respect to a classical weight function at which a point mass at 0 is added (a comparison of the Meixner case and the Laguerre case)
Grahame Bennett	From coin tossing to the Jacobi polynomials and beyond
Bruce Berndt	Ramanujan was a radical
Gaurav Bhatnagar	Matrix inversion in $U(n)$
Amilcar Branquinho	A non-homogenous linear differential equation that has orthogonal polynomial solutions
William Connett	Radial prolate spheriodal wave functions: product formulas and signed hypergroups
Shaun Cooper	The Macdonald identities for G_2 and some extensions
Mathijs Dijkhuizen	A family of quantum projective spaces and related q-hypergeometric polynomials
Charles Dunkl	Pairings and integrals
Dennis Eichhorn	Finding new congruences of partition functions
Pavel Etingof	1. Representation-theoretic proof of identitites with Macdonald's polynomials 2. Affine Jack and Macdonald polynomials 3. Galois groups and connection matrices of q-difference equations 4. Quantization of Lie bialgebras
Jörg Feldvoss	Representation theory of quantum groups at a root of unity
Roberto Floreanini	Lie symmetries of difference equations
Paul Floris	Addition formula for q-disk polynomials
Ana P. Foulquie	Strong asymptotics for polynomials orthogonal with respect to a discrete Sobolev inner product on the

	support of the measure of orthogonality
Frank Garvan	Approximation to π via modular equations for the Dedekind η-functions
George Gasper	q-Series
R. Wm. Bill Gosper	Miscellaneous mail to math fun
Arcadii Grinshpan	Exponential convolutions and q-functions
F. Alberto Grünbaum	The bispectral problem for tridiagonal matrices
Dharma Gupta	Some new q-series identities
Robert A. Gustafson	Multivariable orthogonal polynomials generalizing Askey-Wilson polynomials and symmetric functions
Brian Hagler	Orthogonal Laurent polynomials and Apt-fractions
James Haglund	Rook theory and hypergeometric series
David Jackson	Transforming the genus series for maps
Mendel Kapilevich	Operator approach to the theory of some multivariable special functions
Roelof Koekoek	Differential equations for Sobolev-type Laguerre polynomials
H. T. Erik Koelink	Addition formulas
Tom Koornwinder	Quantum groups and q-special functions
Christian Krattenthaler	A_r and C_r basic hypergeometric series and combinatorics
Kil H. Kwon	Orthogonal polynomials and differential equations of spectral type
Boris Levitan	On the spectral theory of vector-valued Sturm-Liouville equation
Xin Li	A separation theorem of Markov-Stieltjes type for orthogonal Laurent polynomials
Carl Libis	Minimal interpolation and pre-orthogonal polynomials
Gregory Matviyenko	On the Jacobi polynomials $P_n^{(\alpha,-1)}$
Richard McIntosh	An asymptotic involution for q-series
Katsuhisa Mimachi	Singular vectors of the Virasoro algebra in terms of Jack symmetric polynomials
Paul Nevai	1. Poincare's theorem with applications to orthogonal polynomials 2. Rakhmanov's theorem 3. Christoffel functions
Masatoshi Noumi	1. Quantum symmetric spaces and their spherical functions 2. q-hypergeometric series associated with quantum Grassmannians 3. Dunkl operators for (multivariable) Askey-Wilson polynomials
Eric Opdam	Special functions and graded Hecke algebras
Peter Paule	Computer algebra and q-hypergeometric summation
Zoltan Reti	Deblurring images blurred by the discrete Gaussian
Frode Rønning	Geometric properties of certain integral transforms with applications to special functions
André Ronveaux	Recurrence relations for connection coefficient between

	q-orthogonal polynomials of discrete variables in the non-uniform exponential lattice
Alan Schwartz	Product formulas and hypergroups for polynomials orthogonal on a simplex and on a parabolic biangle
Vyacheslav Spiridonov	Schrödinger equation and q-special functions
Dennis Stanton	Positivity conjectures and theorems
Jasper Stokman	Multivariable big and little q-Jacobi polynomials
Sergei Suslov	Askey-Wilson type operators and their inverses
Walter Van Assche	Orthogonal polynomials
Luc Vinet	Algebraic interpretation of certain continuous q-polynomials
Herbert Wilf	A high-tech proof of the Mills-Robbins-Rumsey determinant evaluation
K. Bernardo Wolf	Approximations on a finite set of points through Kravchuk functions
Yuan Xu	Orthogonal polynomials in several variables
Lily Yen	Algorithm for proving q-hypergeometric series and combinatorics
Doron Zeilberger	Computer algebra and special functions

Appendix II

List of Participants

Adelberg, Arnold, Grinnell College, Grinnell, Iowa
Al-Salam, Waleed A., University of Alberta, Edmonton, Alberta
Alladi, Krishnaswami, University of Florida, Gainesville, Florida
Allaway, William R., Lakehead University, Thunder Bay, Ontario
Andrews, George E., Pennsylvania State University, University Park, Pennsylvania
Askey, Richard, University of Wisconsin, Madison, Wisconsin
Atakishiyev, Natig M., Universidad Nacional Autonoma de Mexico, Cuernavaca Morelos, Mexico
Bavinck, Herman, Delft University of Technology, Delft, The Netherlands
Belyi, Sergey, University of South Florida, Tampa, Florida
Bennett, Grahame, Indiana University, Bloomington, Indiana
Berenhaut, Kenneth S., University of Georgia, Athens, Georgia
Berndt, Bruce C., University of Illinois, Urbana, Illinois
Bhatnagar, Gaurav, The Ohio State University, Columbus, Ohio
Bowman, Douglas, University of Illinois, Urbana-Champaign, Urbana, Illinois
Branquinho, Amilcar J. P. L., Universidade de Coimbra, Coimbra, Portugal
Brown, Daniel, University of Waterloo, Waterloo, Ontario
Bustoz, Joaquin, Arizona State University, Chandler, Arizona
Carlson, Keith A., University of Central Florida, Orlando, Florida
Chen, Li-chen, Soochow University, Taipei, Taiwan
Chen, Yang, Imperial College, London, England
Choi, Youn-Seo, University of Illinois at Urbana-Champaign, Urbana, Illinois
Coleman, James, University of Toronto, Toronto, Ontario
Connett, William C., University of Missoui, St. Louis, Missouri
Cooper, Shaun, University of Wisconsin, Madison, Wisconsin
de Jeu, Marcel F. E., Massachusetts Institute of Technology, Cambridge, Massachusetts
DeFaxio, Mark V., York University, North York, Ontario
Dijkhuizen, Mathijs S., Kobe University, Kobe, Japan
Dombrowski, Joanne, Wright State University, Dayton, Ohio

Dragnev, Peter, University of South Florida, Tampa, Florida
Dunkl, Charles F., University of Virginia, Charlottesville, Virginia
Eichhorn, Dennis, University of Illinois at Urbana-Champaign, Urbana, Illinois
Elbert, Arpad, Mathematical Institute of HAS, Budapest, Hungary
Etingof, Pavel I. Harvard University, Cambridge, Massachusetts
Ezio, Venturino, University of Iowa, Iowa City, Iowa
Feldvoss, Jörg, University of Hamburg, Hamburg, Germany
Floreanini, Roberto, INFN, Trieste, Italy
Floris, Paul G. A., Leiden University, Leiden, The Netherlands
Foulquie Moreno, Ana Pilar, Carlos III De Madrid, Leganese, Madrid, Spain
Garvan, Frank, University of Florida, Gainesville, Florida
Gasper, George, Northwestern University, Evanston, Illinois
Godoy, Eduardo, Universidad de Vigo, Vigo, Spain
Gosper, R. Wm., Macsyma, Inc., Los Altos Hills, California
Grinshpan, Arcadii, University of South Florida, Tampa, Florida
Grünbaum, F. Alberto, University of California, Berkeley, California
Gupta, Dharma P., University of Toronto, Toronto, Ontario
Gustafson, Robert A., Texas A&M University, College Station, Texas
Hagler, Brian Allan, University of Colorado at Boulder, Boulder, Colorado
Haglund, James, University of Illinois, Urbana, Illinois
Ismail, Mourad, University of South Florida, Tampa, Florida
Ivanov, Ivan, University of South Florida, Tampa, Florida
Jackson, David M., University of Waterloo, Waterloo, Ontario
Johnson, Warren, Penn State University, University Park, Pennsylvania
Kapilevich, Mendel B., Brighton, Massachusetts
Koekoek, Roelof, Delft University of Technology, Delft, The Netherlands
Koelink, Hendrik T., Universiteit van Amsterdam, Amsterdam, The Netherlands
Koornwinder, Tom H., University of Amsterdam, Amsterdam, The Netherlands
Krattenthaler, Christian, Universität Wien, Vienna, Austria
Kulkarni, Devadatta M., Oakland University, Rochester, Michigan
Kwon, Kil H., KAIST, Taejon, Korea
Lapointe, Luc, University of Montreal, Montreal, Quebec
Le Tourneux, Jean, University of Montreal, Montreal, Quebec
Levitan, Boris M., University of Minnesota, Minneapolis, Minnesota
Li, Xin, University of Central Florida, Orlando, Florida
Libis, Carl, Idaho State Univewrsity, Pocatello, Idaho
Lorentzen, Lisa, University of Trondheim, Trondheim, Norway
Maslen, David Keith, Max-Planck-Institut für Mathematik, Bonn, Germany
Masson, David R., University of Toronto, Toronto, Ontario
Matviyenko, Gregory, Yale University, New Haven, Connecticut
McIntosh, Richard, University of Regina, Regina, Saskatchewan
Mimachi, Katsuhisa, Kyushu University, Fukuoka, Japan
Muldoon, Martin E., York University, North York, Ontario

Nemes, Istvan, Johannes Kepler University, Linz, Austria
Nevai, Paul, Ohio State University, Columbus, Ohio
Nikolova, Inna, University of South Florida, Tampa, Florida
Njastad, Olav, University of Trondheim, Trondheim, Norway
Noumi, Masatoshi, Kobe University, Kobe, Japan
Opdam, Eric, University of Leiden, Leiden, The Netherlands
Paule, Peter, Johannes Kepler University, Linz, Austria
Petersen, Vigdis, University of Trondheim, Dragvoll, Norway
Rahman, Mizan, Carleton University, Ottawa, Ontario
Reti, Zoltan, University of Florida, Gainesville, Florida
Riser, Stephanie, Ohio State Univewrsity, Columbus, Ohio
Rønnig, Frode, Sor Trøndelag College, Trondheim, Norway
Ronveaux, André, Facultes Universitaires Notre-Dame de la Paix, Namur, Belgium
Rooney, Paul G., University of Toronto, Toronto, Ontario
Roy, Ranjan, Beloit College, Beloit, Wisconsin
Schwartz, Alan L. University of Missouri, St. Louis, Missouri
Simeonov, Plamen, University of South Florida, Tampa, Florida
Spiridonov, Vyacheslav, University of Montreal, Montreal, Quebec
Stanton, Dennis, University of Minnesota, Minneapolis, Minnesota
Stokman, Jasper V., University of Amsterdam, Amsterdam, The Netherlands
Sugitani, Tetsuya, University of Tokyo, Tokyo, Japan
Suslov, Sergei K., University of Toronto, Toronto, Ontario
Swarttouw, René F., Free University, Amsterdam, The Netherlands
Tariq, Qazi M., Carleton University, Ottawa, Ontario
Van Assche, Walter, Katholieke Universiteit, Leuven, Belgium
Vinet, Luc, University of Montreal, Montreal, Quebec
Waadeland, Haakon, University of Trondheim, Dragvoll, Norway
Wilf, Herbert S., University of Pennsylvania, Philadelphia, Pennsylvania
Witherspoon, Sarah, University of Toronto, Toronto, Ontario
Wolf, Kurt B., Universidad Nacional Autonoma de Mexico, Cuernavaca, Mexico
Xu, Yuan, University of Oregon, Eugene, Oregon
Yang, Kung-Wei, Western Michigan University, Kalamazoo, Michigan
Yen, Lily, University of Waterloo, Waterloo, Ontario
Zeilberger, Doron, Temple University, Philadelphia, Pennsylvania
Zeng, Jean Jiang, Universite Louis Pasteur, Strasbourg, France
Zhang, Ruiming, York University, North York, Ontario